GOLD, SILVER, URANIUM, AND COAL

GEOLOGY, MINING, EXTRACTION AND THE ENVIRONMENT

Editors
Maurice C. Fuerstenau
Professor and Chairman
Department of Metallurgical Engineering
South Dakota School of Mines and Technology
Rapid City, South Dakota

Bruce R. Palmer
Professor of Metallurgy
South Dakota School of Mines and Technology
Rapid City, South Dakota

Published by
The American Institute of Mining, Metallurgical and Petroleum Engineers, Inc.
New York, New York • 1983

Printed in the United States of America
by BookCrafters, Chelsea, Michigan

Library of Congress Catalog Card Number 82-073914
ISBN 0-89520-406-1

Conference Committee

Maurice C. Fuerstenau
Chairman

Bruce R. Palmer
Vice Chairman

Wendell E. Dunn
John D. Erickson
Albert P. Gilles
Jack C. Haptonstall
Charles A. Kliche
Robert P. Otto
Charles L. Tesch

Session Chairmen

C. Dudley Blancke
George E. Calahan
Harlan E. Gellhaus
Charles J. Gessner
Kenneth B. Hall
Peter R. Oslund
Bruce R. Palmer
Allen S. Winters

Foreword

"The First Western Regional Conference on Gold, Silver, Uranium, and Coal" was convened in Rapid City, South Dakota on September 18-20, 1980. The Black Hills Section of AIME organized this conference to focus on significant advances that have occurred in the geology, mining, and extraction of gold, silver, uranium, and coal.

Among the substantial technical advances in the production of gold and silver during the past two decades is the carbon-in-pulp process which has impacted significantly on production cost and has made processing of fine clay-bearing ores feasible. Additionally, vertical crater retreat stoping has had an important influence on the cost of extracting ores from underground gold mines. In the case of uranium, roll front models have been developed and are now used to delineate complex ore bodies. Both in-situ leaching as well as heap leaching are now in use and have extended reserves of uranium energy sources by allowing recovery from large quantities of low-grade ore. Development of western coal resources has resulted in the opening of vast open pit mines. New advances have focused on production planning from these mines as well as on methods for transporting this material to distant markets. These and many other important and new technological developments in the gold, silver, uranium, and coal industries are contained in this publication of the conference proceedings.

This book is the result of the efforts of many authors and employers. Their contributions are greatly appreciated, especially those companies that released technical data for publication. The authors also gratefully acknowledge the assistance of Mrs. Gloria Pederson and Mrs. Doris Lushbough in typing the manuscripts, Mr. Larry Fairbank in preparation of the figures, and Mr. Keith Ikerd and Mr. Daniel Goodart for their assistance with the photographic work.

Rapid City, South Dakota
October 1, 1982

Maurice C. Fuerstenau
Bruce R. Palmer

Table of Contents

Keynote Address

Session Chairman
M.C. Fuerstenau
South Dakota School of Mines
and Technology
Rapid City, South Dakota

Chapter 1

WHAT LIES AHEAD FOR THE GOLD, SILVER, URANIUM AND COAL INDUSTRIES?

Harry M. Conger
President and Chief Executive Officer
Homestake Mining Company
San Francisco, California

It is a pleasure for me to keynote this first Western Regional Minerals and Coal Conference here in the Black Hills of South Dakota. It is pleasing for two main reasons. First, Homestake Mining Company has been associated with the Black Hills since 1875 when gold was discovered following Custer's trek through the Hills during 1874. Second, the future mineral development of the Black Hills will depend, to a great extent, on minerals and metals covered in this conference.

The metals and minerals that are featured in this conference have unique similarities--and important differences--and therefore, will be of great interest to all who are involved in our industry in this area of the country.

Precious metals have always been, and likely will always be, subject to political influence; they are only partially affected by the demands of our industrial economy. Energy minerals, while more directly affected by industrial stimulus, are also politically sensitive. In this regard, the need for a minerals policy affecting the energy sphere has been made very clear to us in the last five years. More importantly, the lack of a national policy and purpose relating to our total future energy needs has severely affected the overall economy and security of this country.

Very briefly today, I would like to outline for you my thoughts on the future of these most important minerals. I will start by a discussion of precious metals, primarily my favorite precious metal, gold. A brief history of gold will be helpful in setting a perspective for forecasting its future although it is difficult to give a brief description of a metal that has been prized by man for over 6,000 years. Until recently, as you know, gold was held by the Federal government as a metal to be owned by the government only and not by private ownership. This started in 1934 when it became illegal for citizens to own more than a small quantity of gold. Gold could not be traded freely, and its price was set at $35/oz. The restriction on gold ownership remained in effect until 1974 when it finally became legal for Americans to own and trade gold again.

Immediately thereafter, the price of gold rapidly escalated to nearly $200/oz., then de-escalated rapidly to around $103/oz. in September of 1976. Since that time, gold has continued to escalate. The U.S. dollar equivalent of an ounce of gold as of this time is $650, after having reached $850 in January of this year.

As I mentioned earlier, gold is a metal that has value both as an industrial metal and as a monetary metal. This makes it unique among nearly all the metals and minerals that man mines today. With this multiple use, it is more difficult to forecast what its supply/demand relationship truly is.

In the last two or three years, gold has been a very accurate indicator of political stability (or instability) and is now widely recognized as a store of value. Indeed, for the first time in years, there is much talk today that the United States government should follow the lead of several European governments, and the European community in general, and re-institute gold as a backing for government currency. I think gold is once again becoming a de facto backing of the major currencies of the world.

In addition to gold's use in a monetary sense, it has always been valued in an industrial sense, particularly in the jewelry sector. Recent wide price fluctuations in the metal have had an adverse effect on the jewelry industry, but I feel that as the public perceives why gold has fluctuated when measured in various currencies, they will once again be willing to divert money into gold in the form of jewelry, seeing it not only as adornment, which has always been enjoyed, but also as an investment that has a high degree of security. So, I feel that within the next 12 to 18 months, jewelry manufacture will start to increase.

Because of gold's unique physical properties, its industrial demand should continue to increase in all areas of high technology manufacturing.

So much for the demand and price side of gold. Now let's look at the supply side. As the price of gold has increased in the last few years, total gold production has fallen. This is a phenomenon that is very easy to explain. As the price of any metal increases, the miner usually has the opportunity to mine lower grade ore at a profit. Since most of the gold mining operations in the world operate at full capacity all the time because there is always total offtake, any reduction in the grade of the ore results in a lesser quantity of ounces being produced. A review of production records and corresponding gold price substantiate this.

Higher prices for gold also stimulate and increase aggressive exploration programs to both reactivate old mines and districts, and search new areas for discovery. This higher level of exploration will result in new properties being brought into production in the next five to ten years. In my judgment, however, the net change in gold production will be very small since these two conditions--lower grade and new discoveries--will be offsetting. In other words, I think we will continue to see a reduction in annual ounces produced in the free world over the near term, but as new properties come on, annual world production will start to increase. But it will not be likely to exceed the levels that we have seen in the last four or five years.

The effects of this production on the supply and demand of gold will be negligible. Gold is not a supply/demand metal in the true sense because of the huge overhang of gold stocks above ground in the world. Probably over 80% of all the gold mined in the last 6,000 years is still held by individuals or governments and, given a sufficient price incentive, would probably come back into the market again. So the annual increases that we see now, and will likely see in the future, will be but a small percentage, perhaps no more than 1% of the total gold held in the world today. I believe that the combination of all these factors will allow gold to maintain its historical position as a true measure of value between governments, between their currencies, and between individuals who wish to exchange things of value. Therefore, gold's relative worth as measured in various currencies will continue to match inflationary changes and, therefore, will continue to reward miners who search for, develop and produce it.

Silver is sometimes referred to as "poor man's gold" and, as such, plays a role similar to gold. However, silver has a much higher industrial use than does gold and, therefore, it follows a truer supply/demand path. In looking backward for the purpose of forecasting silver's performance, the most significant factor is clearly that new mine production has been substantially less than the annual quantity consumed. To illustrate, although consumption dropped from 451 million ounces in 1969 to 433 million ounces in 1979, mine production increased from 241 million ounces in 1969 to 271 million ounces in 1979. Supply from other sources decreased 46% of total in 1969 to 37% in 1979.

Even more spectacular during this period of time was the increase in silver prices. In 1972, silver was selling for $1.55 an ounce, and in 1979 silver averaged $11.09 per ounce. This increase in price has had a dramatic effect on silver in many respects. First, it has made the mining of silver extremely profitable during these periods of high prices and particularly during very short periods of rapidly escalating prices. Second, it has caused a disintermediation in the customary and historic consumption patterns. Third, it has caused an increase in exploration for new silver production sources.

On the supply side of silver, we have a picture that differs significantly from gold. While new-mined gold comes of the result of operating mines that produce solely gold, (i.e. Homestake Mine, South African mines, placer mines, Canadian mines) new-mined silver comes from the following sources:

38% of production is from mines operated as silver mines
34% of silver is a by-product of copper mining
13% of silver is a by-product of lead mining
4% of silver is a by-product of gold mining
9% of silver is a by-product of copper, lead, zinc mining

Because of this distribution, silver production is not necessarily at full capacity at all times as gold is. For instance, during a depressed copper market, by-product silver production decreases even during periods of high silver prices.

U.S. silver production has remained very constant at about 45 million ounces per year during the last 10 years in spite of its dramatic rise in price over the last several years. The new higher prices are stimulating exploration and development of new properties, but this production will be offset somewhat by that portion of the industry that mines silver as their primary ore (about 40% of total production), which will take advantage of mining lower grade ore profitably.

For the forseeable future, it is likely that silver prices will continue to be somewhat volatile. In the judgment of most silver analyists, silver prices will continue to escalate at least as fast as the overall economy is inflated and, therefore, silver represents a metal worth exploring for and developing. Perhaps one of the greatest threats to future prices of silver would be a substitute for silver in the photographic process, since nearly 40% of silver consumption in the free world goes into this industry alone.

Now, let's turn our attention to the energy minerals. While gold is probably the oldest metal sought and treasured by man, uranium is one of the newest metals to come into use. The uranium industry as we know it today started in the early 1950's when the Atomic Energy Commission was seeking secure sources of uranium for weapons. During this period the budding uranium industry enjoyed a period of prosperity when government subsidized prices attracted a large number of individuals and companies into the business of exploring, developing and producing uranium in the Western United States, particularly the Colorado Plateau and the Ambrosia Lake area of New Mexico.

Also during this period, the government started to encourage utilities to look to nuclear power as a replacement for fossil fuel power generation, and stimulated this activity to the extent that many utilities did conduct studies and eventually started building nuclear reactors. However, during the mid-60's, the AEC started to phase out its contracts with the uranium suppliers, and the industry went into a recession. Very little exploration work went on, and few new mines were developed. Then, towards the end of the '60's, it became evident that the U.S. was going to face an energy shortage and that nuclear power could and should play a significant role in supplying base power loads for most of the utilities in the country, particularly in the areas where coal-fired and natural gas-fired units were not practical.

By 1971-72, this activity by the government had stimulated industry to the point where exploration had increased to a very high level, and many new deposits of uranium were located and plans for development were proceeding. It was at this point that the price of uranium, which had been held at artificially low levels all during the 1960's, '70, '71, started to escalate very rapidly. Factors influencing this rapid increase were the announced plans of utilities to more than double the number of reactors, estimates of increased future energy demands by the utilities, and an apparent shortage of supply of yellowcake to fuel new reactors, as well as to sustain reactors newly constructed or being constructed.

All of these events came into focus very clearly in 1973-74 when the price of uranium moved from around $8/lb. to +$40/lb. This dramatic price movement encouraged the mining industry to explore for and start to develop new projects which heretofore had been either marginal or sub-marginal in the earlier economics but which now looked very attractive.

In 1975 and 1976, a new factor entered the uranium picture. This might best be described as an environmentally-inspired desire to eventually take the utility industry entirely off nuclear power by reducing consumption and substituting other power sources, generically called renewable resources. Early on, the forces at work here were not easily identified, but in the last 18 to 20 months, they have coalesced into an effective group that has been very successful in causing a large segment of the American public to question the viability of nuclear power for the United States. They base their argument on these factors:

1) Health of the population living in the vicinity of any part of the fuel cycle.

2) Waste disposal.

3) Nuclear proliferation.

4) Future zero growth so no need to develop further energy sources.

Their theme has been that we must have zero risk for thousands of years. This activity has resulted in massive government regulation on all parts of the fuel cycle, adding enormous costs to the consumer.

The anti-nuclear group has been extremely effective with their arguments in spite of the fact that the nuclear industry has enjoyed the best safety record of any new major technology. Even Three Mile Island was an outstanding success for the industry, since the systems designed for total containment worked so well (in spite of the operators) that there was not a single reported injury. But what has been reported is the possibility of harm and not the reality of safety. So where do we go from here?

I believe that the following points can be proven to be fact and will ultimately prevail in the nuclear debate:

1) That the nuclear industry is safe and economically viable.

2) That the United States must be self-sufficient in energy and, therefore, must use its abundant uranium and coal reserves.

3) That the proponents of renewable energy resources as the energy panacea are playing a cruel hoax on the lay public.

4) That a zero growth philosophy (and therefore a reduction in energy needs) smacks of "I got mine so to heck with the rest of you," and I say the vast majority of the United States public is not in sympathy with that attitude.

Based on these arguments, I think the nuclear industry will survive and prosper.

Turning to coal--which is far and away our most abundant domestic source of energy--we find that the United States now produces one fifth of the world's annual coal supply. Our reserves are equal to about one-third of the world total. In many respects, we are the Middle East of coal. But this energy mineral is not without its challenges, environmental problems in both the mining and burning of coal paramounting them.

Nevertheless, when compared to oil and gas, coal is by far the more abundant and less expensive fuel available to electric utilities today. As of April, relative fossil fuel costs at utility plants in cents per million BTUs delivered were 133 cents for coal, 402 cents for oil, and 203 cents for gas. Cost and availability have been the major factors in the growth of coal in the electric utilities industry since 1972. Coal use is expected to double from present levels to more than 1.1 billion tons by 1995, according to EPA projections.

Industrial coal consumption is also expected to rise from the greatly reduced levels that prevailed in the post World War II period as a result of abundant cheap gas and a need to improve air quality, especially in urban areas. A reconversion to coal due to price and supply is in progress but will be slow as a result of the very high capital costs for boilers. Nevertheless, current industrial coal use will double or even triple by 1995.

The greater use of coal as an energy source is absolutely necessary for our continued economic progress--and, in my view, the objections that have been raised on environmental grounds can and must be coped with. We know how to mine coal safely and efficiently with due regard for environmental considerations. We have the technological know-how to burn coal with no untoward effect on air quality. What remains is to develop the commitment and the programs that are necessary to get this job done. Given our country's energy needs, and especially our need for a higher degree of independence from foreign imports, we must do no less.

Gold and Silver I

Session Chairman
Allen S. Winters
Homestake Mining Company
Lead, South Dakota

Chapter 2

REVIEW OF UNCONFORMITY CONTROL ON GOLD-BEARING CONGLOMERATES

Andrew Button

Department of Geology and Geological Engineering
South Dakota School of Mines and Technology
Rapid City, South Dakota

INTRODUCTION

This review deals with the controls exerted by unconformities on the stratigraphic and areal distribution of gold (and uranium) in Precambrian conglomerates. The relations described are drawn mainly from Precambrian basins in southern Africa, particularly the Witwatersrand Basin.

The mining industry of South Africa started in earnest with the discovery of diamonds in 1866. Gold was first discovered in 1870, and in March of 1886, two out-of-luck prospectors stumbled across what was to become the world's greatest goldfield, in the vicinity of Johannesburg. George Harrison and George Walker sold their claims for a pittance and were heard of no more.

The Witwatersrand goldfield has been in continuous operation since that time, some 150 mines having operated at one time or another. Deep mining technology developed here has allowed operations to proceed to depths approaching 13,000 feet.

The Witwatersrand has produced some 55% of all the gold ever mined, and currently produces 60% of the world's supply, some 650,000 kg Au per year. To 1971, the Witwatersrand had produced some $32 billion of gold and $2 billion of uranium (when average prices were less than $35/oz. for Au and $10/lb. for U_3O_8). At present prices, the annual Witwatersrand production of close to 700,000 kg would be worth about $10 billion.

Uranium was discovered in the Witwatersrand in 1923. In the $15-30/lb. bracket, the basin contains a reasonably assured plus estimated reserve of 350,000 metric tons U_3O_8. Current U production is about 8 million pounds per year, about 11.5% of the western world's production.

Overall, the basin has yielded well over 3 X 10^9 metric tons of ore, grading 7-21 ppm Au (average 10 ppm).

GEOLOGIC SETTING

The Witwatersrand Basin is situated in central South Africa. It is developed on the Kaapvaal Block, a crustal unit of great stability and antiquity (greater than 3,000 million years). This craton is composed of a basement of granitic and gneissic rocks with greenstone belts. It is surrounded by metamorphic mobile belts, such as the Limopop Belt (3,700 million years) and the Mozambique Belt (1,200 million years). The sources of gold and of uranium are thought to be the Archean greenstone belts and younger-Archean granitic rocks, respectively.

STRATIGRAPHIC PERSPECTIVE AND AGE

The Witwatersrand is just one of several Precambrian basins, the oldest of which is the 3,000 million year old Pongola Basin. This is followed by the Witwatersrand, Ventersdorp, Transvaal and Waterberg Supergroups (1,800 million years).

The Witwatersrand Basin rests on 2,800 million year old lavas of the Dominion Reef Group and is overlain by the 2,640 million year old Ventersdorp Supergroup. Its age is presently bracketed between these limits.

SIZE

Only a very small fraction of the perimeter of the basin is seen on outcrop. The remainder has been delimited by many thousand cored boreholes, some of them extending to depths of 15,000 feet.

The basin has a long axis dimension of about 400 km and a short axis dimension of about 200 km. The total extent of the basin has not yet been fully determined by drilling. It could extend for over 200 km further on either end, giving a total length of some 800 km (500 miles).

STRUCTURE

The dominant structural fabric of the basin trends east and northeast and is manifested by both "longitudinal" folds and faults.

This fabric is overprinted by northwest-trending folds. The resulting fold interference pattern has produced a series of domes and basins. The major goldfields are located in structural basins, nestled between domes. Both domes and basins were active during and after sedimentation.

The basin was formed in part in response to faulting. The most active fault movement was on the northwest side of the basin, which was continually uplifted along a set of basin-bounding faults. This accounts for the fact that the thickest, coarsest accumulations (with the greatest concentrations of gold and uranium) are located along the northwest side of the basin.

Faulting continued after basin formation. Some normal faults have displacements of 5,000 m, while many have displacements of 1,000-2,000 m. Faulting has had marked effects on mining, some positive, some negative. Large horst blocks were erosively stripped of payable conglomerate over extensive areas. Minor faulting is a continuous complicating factor in mine planning. Conversely, faulting has brought "reef" to mineable depths in some areas. Younger Witwatersrand and Ventersdorp fans containing gold are being mined in graben valleys adjacent to some of the ancient horst blocks.

STRATIGRAPHIC SETTING

The Witwatersrand Supergroup rests nonconformably on an Archean basement composed of gneisses, granitic rocks and greenstone belts.

The Witwatersrand Supergroup is from 7,000 to 8,000 m thick where best developed. It comprises an upper and a lower division, separated by a profound angular unconformity. The lower division (up to 4,600 m) comprises shale, quartzite and subgreywacke with minor diamictite, iron-formation and basalt. Only a few conglomerates have been mined from this division. The upper division (3,000 m) comprises quartz wacke, some conglomerate, basalt and a single shale formation. All but a fraction of the total reserves of gold and uranium are contained in the upper division.

A maximum of 8% of the upper Witwatersrand column comprises conglomerate. Economically exploitable conglomerates make up less than 0.2% of the sequence in most areas, or 1 or 2 conglomerates in a 3,000 m thick section. Even the most proximal facies contain only 2% (by thickness) of payable conglomerate. In areal terms, payable conglomerate sheets have areas of less than 400 km^2 in a basin that was at least 100,000 km^2 in area. The vast bulk of the upper Witwatersrand is thus a monotonous pile of coarse- and very coarse-grained quartz wacke.

Upper Witwatersrand deposition was terminated by several pulses of Ventersdorp volcanism. Episodes of block faulting separated some of the units. Graben valleys were filled by thousands of meters of volcaniclastic conglomerate and some playa lake stromatolitic limestone. Starting some 2,300 million years ago, large parts of the Kaapvaal Craton were covered by up to 10 km of Transvaal Supergroup sediments. Exploration for new Witwatersrand goldfields is proceeding through thick cover of Ventersdorp lava and Transvaal sediments, as well as Paleozoic Karroo strata.

UPPER WITWATERSRAND DEPOSITIONAL ENVIRONMENTS

Thousands of man years of geological observation have been devoted to a study of the thin, economically-important Witwatersrand conglomerate sheets. Very little work has been done to define the depositional paleoenvironment of the vast bulk of the upper Witwatersrand sediments.

To give an idea of background conditions of upper Witwatersrand sedimentation, an example from the East Rand area is given. The Main Reef Leader is the major conglomerate sheet in this area, being developed over an area of about 400 km^2 at the unconformable contact between the upper and lower divisions of the supergroup. The plan view of mined-out areas shows a radial pattern suggestive of a braided fluvial fan. The conglomerate is overlain by 65 meters of impure, quartz wacke arenite, followed by 20 meters of a clean quartzite, followed in turn by a few hundred meters of quartz wacke. The depositional paleoenvironments of these two facies, quartz wacke and quartzite, are analyzed by studying the vertical sequence of sedimentary structures and the orientation of paleocurrent vectors.

The quartz wacke facies is coarse- or very coarse-grained, poorly sorted and comprises angular to subangular sand-sized clasts in a sericite groundmass. The succession is characterized by a series of meter-scale, erosive-based sedimentary cycles. A heavy-mineral-rich small pebble or granule lag is developed at the base of many of the cycles. The lag surfaces are usually overlain by tabular-planar cross-bed sets up to a meter thick. The upper parts of the cycles comprise either:

(i) stacked, tabular-planar cross-bed sets;

(ii) stacked trough cross-bed sets; or,

(iii) crude plane-bedding.

Silty arenite drapes up to a few centimeters thick are preserved at the tops of some cycles.

The cycles have been interpreted as representing the product of Platte type braided stream sedimentation. In such stream systems the dominant bedforms are transverse or linguoid bars, bounded by avalanche faces. Since the avalanche faces are sand traps, the floors of the pools ahead of the bars are usually swept clean of sand, and are floored by a heavy-mineral-rich lag. Advancing avalanche faces cover the lag with tabular-planar cross-bedded arenite. Shallow water areas of bar tops were probably structured by long-wavelength, low-amplitude "ripples", which migrated down current to produce crude plane-bedding. The bar tops were apparently sometimes winnowed to produce heavy mineral concentrations.

Some bars were composite, being composed of a series of avalanche-face bounded sub-bars, which produced a series of stacked tabular-planar cross-bed sets. Lunate megaripples are developed in fairly swiftly flowing water on the backs of bars and result in trough-like cross-beds.

A rose diagram based on 100 measurements indicates transport to the south. The two modes are thought to represent the opposing avalanche edges of linguoid bars.

The fan-delta of the Witwatersrand conglomerate sheet is thus believed to have been overlain by a braided distributary system dominated by Platte-type streams.

A 20-meter interval of pure quartz arenite is developed within the quartz wacke of the beds above the Main Reef Leader. The quartz arenite is texturally mature. It occurs over an area of a few thousand square kilometers and rests with a sharp, slightly undulating contact on the underlying quartz wackes.

The quartzite often shows what could be waning water-velocity cycles. Large, tabular-planar cross-beds (often convoluted) are truncated by trough cross-bed sets. The upper surfaces of the latter are frequently ripple-marked (both current and symmetrical types), and the rippled surfaces are often overlain by silty drapes. Superjacent cross-bed sets are often arranged with azimuths separated by 90 or by 180 degrees. The latter are sometimes arranged in a classical "herring-bone" pattern.

A rose diagram of sediment transport azimuths shows a trimodal pattern. The main transport direction was toward the west-southwest, and is opposed by some 10% of the vectors, oriented to the east-northeast. The third mode is toward the south-southwest.

The interpretation placed on these observations is that the 20-meter thick clean quartzite was deposited on a marine shelf off an unbarred coastline. Dominant transport on the shelf was parallel to bathymetric contours, and showed periodic tide-induced reversals. The vector to the southeast could indicate drainage of water off the shelf. The waning-velocity cycles are thought to indicate individual storm events on the shelves.

The shelf co-existed with the braided fan-delta. During periods of low sediment supply, the fan-delta was transgressed and reworked by the currents on the shallow marine shelf. This pattern is not restricted to the example just given. Several conglomerates (Vaal Reef, Kimberley Reef) are directly overlain by supermature quartz arenite sheets with paleocurrent patterns indicative of shelfal or tidal reworking.

THE CONGLOMERATES

Mineralized Witwatersrand conglomerates are usually thin (few meters to few centimeters) and are composed mainly of clasts of quartz and chert. They are rich in undoubted detrital grains such as zircon and chromite and also carry much leucoxene (altered ilmenite). Economically important metals are gold, silver, uranium and, in some conglomerates, platinoids. These metals, along with some nickel, copper, cobalt, lead and zinc occur with pyrite in the matrices of the conglomerates. Silicate constituents of matrices include quartz, sericite, pyrophyllite and chlorite.

Without exception, major conglomerate sheets rest on gentle angular unconformities or disconformities. The reasons for this control will be explored in the concluding statement.

Gold and uranium-bearing gravels were deposited in a number of different types of deposit:

(i) meandering channels, relatively rare, only when cut on a shale footwall;

(ii) entrenched linear channels, which contain variegated sediments including banded pyritic quartzites;

(iii) braided networks on erosion surfaces;

(iv) carbon seams ranging from thin seams to thicker seams;

(v) gravel sheets on braided, erosion-etched surfaces, with maximum conglomerate development along channel axes. These sheet gravels are by far the most important. An excellent example is the Basal/Steyn reef of the Welkom goldfield, which is preserved over an area in excess of 200 km^2. Pebble sizes decrease from over 40 mm to less than 20 mm down the paleoslope. Pyrite mean grain-size decreases from over 1.6 to less than 3ϕ units, in sympathy with pebble size. High gold (15-60 ppm) occurs in the proximal facies of the sheet. High uranium (150-500 ppm) occurs in the distal parts of the sheet, as does the high U/Au ratio (less than 5 in proximal parts to over 40 in distal parts).

These relations are quantified in graphs showing grain sizes of pyrite, zircon and the U/Au ratio. While the former decrease down the paleoslope, the latter increases. Regional payability patterns are thus primarily controlled by distances down the paleoslope. On a more local scale, the distribution of gold and uranium is controlled by flow patterns. Areas of convergent flow, for example, are likely to be ones flushed of silicate grains where heavy minerals will be concentrated.

Within a Witwatersrand placer, heavy minerals are concentrated in conglomerate rather than arenite. Where carbon seams are present, they often contain the highest concentrations of gold and uranium. In a single placer, those phases containing pyritic foresets are usually enriched in gold and uranium relative to those without pyritic foresets.

Gold occurs as minute grains (0.005 to 0.5 mm) in the matrices of the conglomerates. Some gold is contained within "porous" pyrite. Gold extracted from carbon seams sometimes occurs as micro-nuggets, and in some cases, occurs as a fibrous network. Uraninite occurs as rounded to subangular detrital grains (high in Th) or as clouds of tiny crystals (low Th) in carbon. It is now believed that the uranium in Precambrian conglomerates could have dual source. High-thorium uraninites are relatively stable and can survive detrital transport even under contemporary oxidizing conditions. Low-thorium uraninite was probably precipitated from solution by reduction where it came into contact with carbonaceous matter.

RECONCILIATION OF FACTS

The process-response model (Figure 1) accompanying this summary attempts to show the inter-relationships of the variables controlling deposition of payable conglomerate sheets of the Witwatersrand. An essential step is the development of an erosion surface and an associated water table. Fluvial processes operated on this surface to produce a mechanical concentration of heavy mineral grains. Groundwater played a role in mineralization; there is evidence that some uranium was precipitated by reduction reactions involving Witwatersrand carbon. Reducing paleogroundwater was also probably responsible for the destruction (by iron-leaching) of ilmenomagnetite and ilmenite grains. This process resulted in the formation of very porous titania skeletons with a very high surface area per unit volume. These titania aggregates were very efficient traps for uranium in solution in paleogroundwater. The economically-important brannerite and uraniferous leucoxene of the Witwatersrand (and similar) Basin may well have formed by diagenetic and low-grade metamorphic transformation of the uraniferous titania aggregates.

REFERENCES

1. Button, A., "Houtpoort Quartzite Member: Marine-reworked Facies of a Witwatersrand Fan-delta." Inf. Circ., Economic Geology

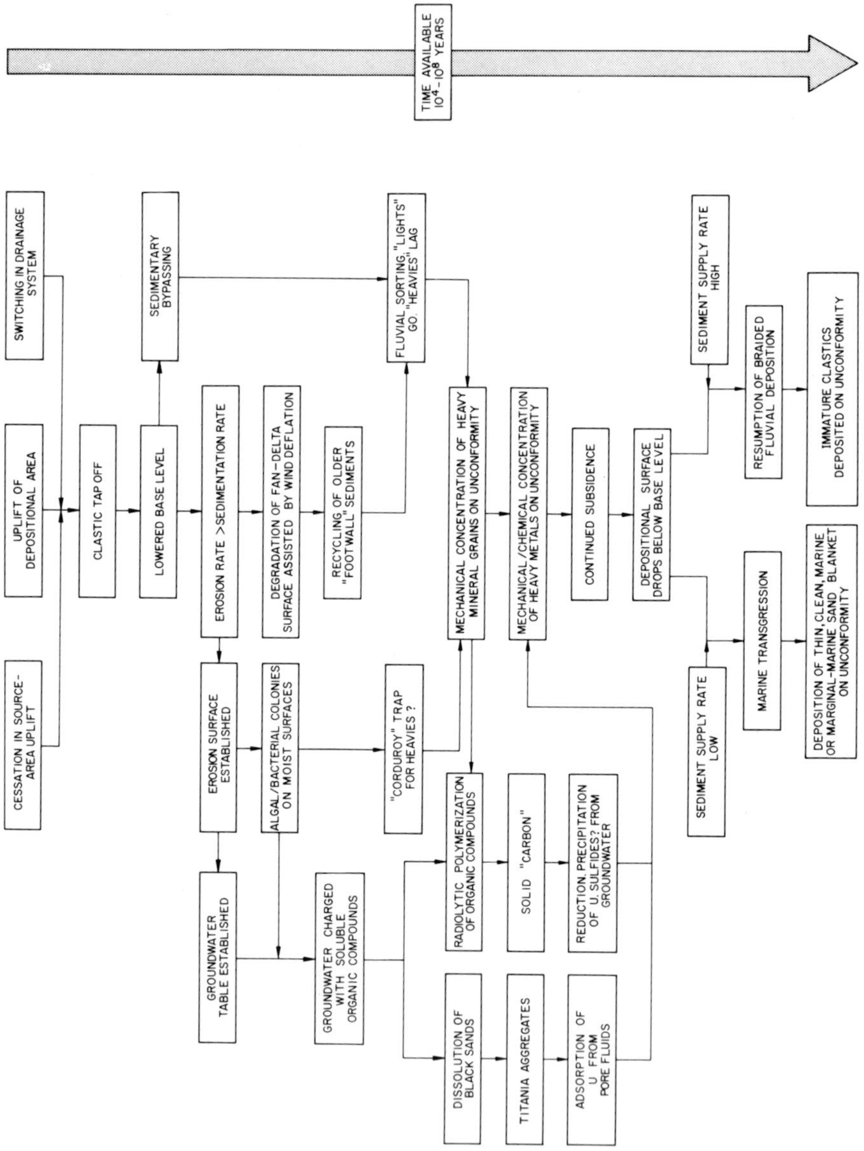

Figure 1. Schematic process-response model.

Research Unit, Johannesburg, In press.

2. Minter, W.E.L., "A Sedimentological Synthesis of Placer Gold, Uranium and Pyrite Concentrations in Proterozoic Witwatersrand Sediments," Canadian Assoc. Petroleum Geol., Mem. 5, 801-829, 1978.

3. Pretorius, D.A., "The Nature of the Witwatersrand Gold-Uranium Deposits." In, Wolf, K. H. (Ed.), Handbook of Stratabound and Stratiform Ore Deposits. Elsevier, Amsterdam, 29-88, 1976.

4. Smith, N.D. and Minter, W.E.L., "Sedimentological Controls of Gold and Uranium in Two Witwatersrand Paleoplacers." Econ. Geol., 75, 1-14, 1980.

Chapter 3

VERTICAL CRATER RETREAT STOPING AS APPLIED AT THE HOMESTAKE MINE

Steven T. Mitchell

Homestake Mining Company
Lead, South Dakota

INTRODUCTION

Mill production from the Homestake Mine in 1979 amounted to 1.29 Mt (1.43 million st) of ore containing 5.9 g/t of gold (0.172 oz per st) compared with 1978 production of 1.44 Mt (1.59 million st) containing 6.2 g/t of gold (0.180 oz per st). The average cost per ounce of gold produced in 1979 rose to approximately $250 contrasted with $176 in 1978.

This increase in unit cost is the result of inflation, a diminishing ore grade, and productivity losses related to a shortage of experienced miners as well as compliance with federal regulatory requirements. Together, these factors prompted the investigation of several large-scale, highly mechanized underground mining methods. This analysis indicated Vertical Crater Retreat mining to be one of the safest underground methods known today, yet competitive with all other forms of stoping in terms of cost and productivity. Therefore, the feasibility of applying the VCR stoping method at the Homestake Mine was explored.

Vertical Crater Retreat stoping is a recent innovation in large-scale underground mining based on C. W. Livingston's crater blasting theories. Figure 1 illustrates the method which consists of drilling 165.1-mm-diameter (6.5-in.) holes from a topsill down-dip to an undercut on the level below. The bottom of each hole is then loaded from above with a predetermined charge weight and horizontal slices up to 4.27 m (14 ft) in thickness are blasted into the undercut. Broken ore is extracted from drawpoints located on or below the undercut level.

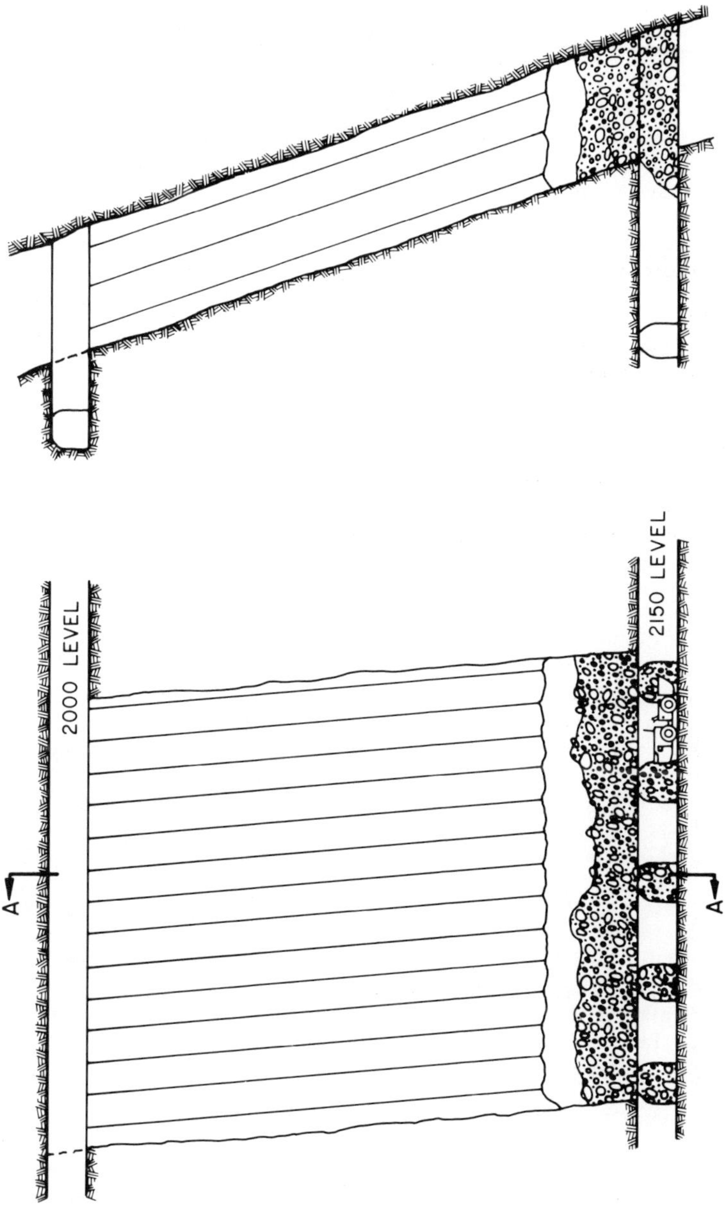

Figure 1. Idealized vertical crater retreat stope.

MINE GEOLOGY

The Homestake Mine is situated in the northern portion of the Black Hills of South Dakota in a series of steeply dipping and highly folded Precambrian rocks. Three well-defined Precambrian formations are exposed in the mine. These formations from oldest to youngest are known as the Poorman Formation (pmf), Homestake Formation (hf), and the Ellison Formation (ef). Intrusive rhyolite dikes and sills believed to be Tertiary in age cut across all of the formations. The gold-bearing ore occurs almost exclusively within the Homestake Formation, which has an unconfined compressive strength of about 275.8 MPa (40,000 psi).

The ore areas are localized along plunging anticlinal and synclinal folds locally referred to as "ledges". Individual ore bodies plunge 30 to 75° from horizontal and usually dip at 50 to 80° to the east. Typical ore zones range from 3.0 to 30.5 m (10 to 100 ft) wide and up to 121.9 m (400 ft) in length with a strike approximately 35° west of north.

Gold values are erratically distributed in a gangue consisting of cummingtonite schist, sideroplesite schist, chlorite schist, quartz, pyrite, pyrrhotite, arsenopyrite, and ankerite. Ore grades are determined by extensive diamond core drilling, dry drilling and secondary wet-sludge sampling for each stoping block.

OVERVIEW OF LIVINGSTON CRATER THEORY

Several mathematical relationships have been derived by C. W. Livingston (1) describing the behavior of rock during the crater blasting process. These relationships attempt to correlate variables involving the mechanics of rock materials, explosive properties and shot geometry. A basic understanding of the cratering phenomenon as learned from theory and experimentation is a prerequisite to the successful design and evaluation of full-scale VCR production blasting.

STRAIN ENERGY EQUATION

Livingston established empirically a strain energy equation relating the energy of the explosive to the volume of material affected by placement of the explosive charge.
This he expressed as:

$$N = E\ W^{1/3} \qquad (1)$$

where N = critical depth or the charge depth at which the explosive begins to fracture rock at the free face (as measured from the free face to the center of the charge)

E = strain energy factor - a dimensionless factor which varies according to the explosive and rock type.

W = charge weight

Also,

$$D_c = \Delta E W^{1/3} \quad (2)$$

where

D_c = charge depth as measured between the free face and the center of gravity of the charge.

Δ = depth ratio - a dimensionless number which equals D_c/N

The charge depth at which the explosive maximizes the crater volume within acceptable fragmentation limits is called the optimum charge depth. Therefore,

$$\Delta o = Do/N \quad (3)$$

where

Δo = optimum depth ratio

Do = optimum charge depth

N = critical charge depth

MAXIMUM EFFECTIVE CHARGE WEIGHT

Experimentation has shown that as the explosive charge weight is increased in a blasthole of constant length and diameter, a point is reached where no increase in crater volume results for a corresponding increase in explosive charge weight. In recent years Lang et al. (2) have theorized that with a cylindrical explosive column, the gases of detonation and resulting borehole pressures are directed laterally outward from the longitudinal axis of the borehole. The magnitude of the force vectors which are directed parallel to the borehole axis toward the ends of the charge is relatively small.

The force vectors produced from a spherical charge (length-to-diameter ratio = 1), however, are directed radially outward along all theoretical planes passing through the center of the charge. Grant, et al. (3) have found that as long as the length-to-diameter ratio of a cylindrical charge does not exceed 6:1, the results of the crater breakage mechanism will approximate those of a true spherical charge.

MATERIAL BEHAVIOR IN THE BREAKAGE PROCESS

Before full-scale production blasting using the cratering phenomenon can be successfully implemented, a series of small-scale crater shots are performed in order to gather information about the effects of different explosives on the material behavior of a given rock type. Since the strain energy factor (E) is a dimensionless constant dependent on the explosive's properties and rock type, it is beneficial to test all of the explosives which later might be used in full-scale production blasting. This data is then incorporated into Livingston's strain-energy equations to generate valid blasting relationships.

Figure 2 depicts an idealized curve for shock-type rock failure expressed as a function of the charge depth. On this graph, energy levels (V/W) versus depth ratios (D_c/N) are plotted from crater test shots to help identify a range of material behaviors which can occur in the crater breakage process.

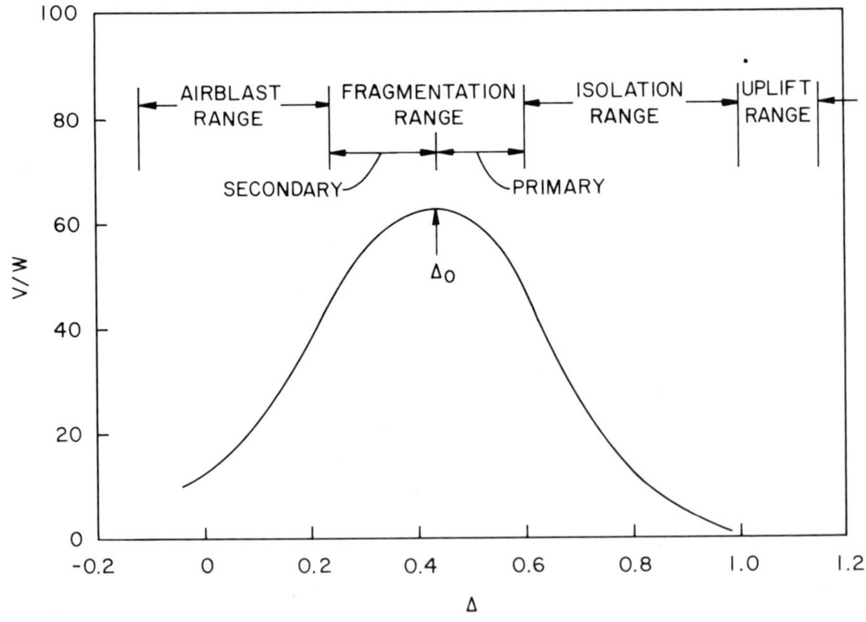

Figure 2. V/W vs Δ curve for shock-type rock failure.

At a Δo value for a given explosive and rock type, flyrock is moderate and crater volume is maximized. As Δ values decrease from Δo, the amount of explosive energy available for the fracturing process decreases and is directed instead toward more flyrock, noise, and airblast pressure. At Δ values greater than about 0.7, the rock material begins to become isolated from its surroundings by spalling only, and true fragmentation does not occur.

BREAKAGE PROCESS EQUATION

From test data Livingston derived an equation empirically to better describe the rock breakage process. This he called the breakage process equation or

$$V/W = E^3 \ A \ B \ C \tag{4}$$

The variables A, B, and C were introduced to describe the utilization of the explosive energy (A), the behavior of the material (B), and the effect of the charge shape (C).

The energy utilization number can be expressed as A = V/Vo. In other words, A is the ratio of the volume of a crater at any charge depth D_c to the crater volume at the optimum charge depth Do. A reaches a maximum value of 1.0 when D_c = Do and V = Vo. The energy utilization number is a measure of that part of the explosive energy devoted to the fracturing process. For most rock materials, graphic illustrations of A values versus Δ values closely resemble the curve shown as Figure 2.

Since $N = EW^{1/3}$, the breakage process equation may be rewritten so that

$$V/E^3W = A \ B \ C \tag{5}$$

and

$$V/N^3 = A \ B \ C \tag{6}$$

Therefore, the materials behavior index (B) can be expressed as

$$B = \frac{V/N^3}{AC} \tag{7}$$

When a spherical charge is placed at Do, V = Vo and A = C = 1.0. Then,

$$B = V/N^3 \tag{8}$$

The materials behavior index, according to Livingston, is a measure of that part of the explosive energy devoted to processes such as crushing, compaction, and plastic deformation, all of which precede the fracturing process.

The value C is a dimensionless ratio of the energy level expressed as a crater volume produced under any conditions to the energy level of a single-shot crater blast under prototype conditions. This number approaches a maximum value of 1.0 when prototype conditions prevail. Obviously, parameters associated with multiple hole blasts, changing geologic conditions, and the manner of detonation of the shot can all create deviations from single shot conditions.

From the mechanics of the breakage process equation, then, it becomes evident why small-scale cratering experiments are a prerequisite to successful full-scale crater stoping. The Livingston approach to crater blasting can be adapted successfully to a given ore body once the empirical data have been evaluated in terms of the explosive energy and the material behavior characteristics.

SMALL-SCALE CRATER TESTS

Test Procedure

In November 1977, small-scale crater testing was initiated in a typical Main Ledge ore body on the 2,073-m (6,800-ft) level. This test consisted of drilling five vertical 101.6-mm-diameter (4.0-in.) holes of varying length into the back of a stope still cut. These holes were drilled with an Ingersoll-Rand JR-300 jackleg drill and were subsequently measured and individually blasted.

The loading procedure for the 101.6-mm-diameter (4.0-in.) holes began by inserting one 76.2 x 406.4 mm (3 x 16 in.) cartridge of Tovex 650 water gel (one-half the charge weight) into the hole. A 0.45-kg (1.0-lb) cast primer with E-Cord Primacord as a detonator was then buried inside the second cartridge and inserted into the hole. Cardboard map tubes filled with sand were utilized as bottom stemming and were inserted into the hole behind the charge. This explosive-stemming sequence was then pushed to the point of the hole and tamped with the aid of 50.8 x 50.8 mm (2 x 2 in.) wooden loading poles. The hole was then blocked with two wooden wedges approximately 0.30 m (12 in.) long. Prior to blasting, a series of concentric circles were painted around each hole so that the radius of the resulting crater could be more easily and accurately determined.

After firing, the crater area was barred down so that the crater depth, radius, and volume could be measured safely. Fragmentation, unusual geologic conditions, and the shape of the crater were also recorded. Figure 3 shows a plot of data obtained from 101.6-mm-diameter (4.0-in.) hole test. Energy levels (V/W) are plotted against depth ratios (D_c/N). The results of this testing indicated that with the explosive-rock combination involved, N=3.35 m (11 ft) and $\Delta o = 0.45$.

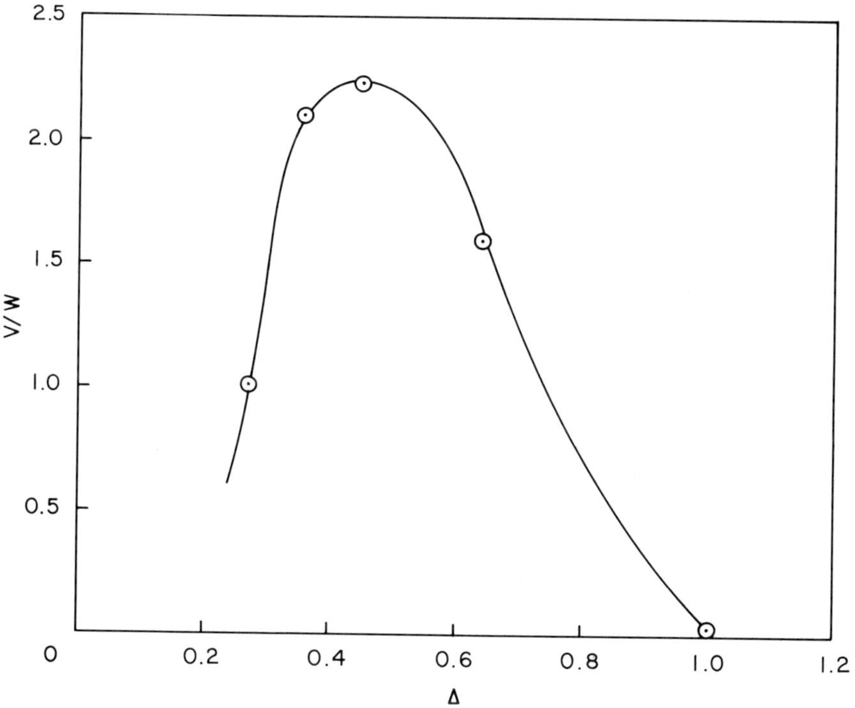

Figure 3. Cratering tests using 10.2-cm-diameter holes.

Once this curve was established, the data was incorporated into Livingston's strain energy equation to determine data useful for scaling purposes. From the 101.6-mm-diameter (4.0-in.) hole tests, then

N = 3.35 m (11 ft)
(shot #4)
W = 6.94 kg (15.3 lb) Tovex 650
(including a 0.45 kg (1.0-lb) cast primer)

Since

$$N = E\, W^{1/3} \tag{1}$$

$$E = 4.43$$

From Figure 3,

$$\Delta o = 0.45.$$

$$Do = \Delta o N \tag{3}$$

$$Do = 0.45 \times 3.35 \text{ m (11 ft)}$$

$$Do = 1.51 \text{ m (4.95 ft)}$$

In March 1978, it was decided that further crater testing should be performed with 152.4-mm-diameter (6.0-in.) holes, since the 101.6-mm-diameter (4.0-in.) holes obviously would not be practical for production blasting. By scaling the 101.6-mm-diameter holes using the strain energy equation, it was calculated that the optimum charge depth (Do) for a 152.4-mm-diameter hole using Tovex 650 should be 2.23 m (7.3 ft).

To corroborate the results of the initial test, then, a second crater blasting test was conducted. This test consisted of drilling five 152.4-mm-diameter holes up-dip (70° from horizontal) into the back of a stope sill cut in a 9 Ledge ore zone on the 975-m (3200-ft) level. A Gardner-Denver PR-123 longhole drill mounted on a bar-and-column arrangement was used to drill the holes. The loading procedure for these holes was essentially the same as before with exception that Tovex Extra was substituted as the explosive and rough-cut wooden 101.6 x 101.6 mm (4 x 4 in.) spacers were used for bottom stemming material.

The results of the 152.4-mm-diameter crater shots are plotted as shown in Figure 4. Data from Figure 4 indicate a Δo value equal to 2.21 m (7.24 ft). It should be noted that although shot #4 produced the maximum volume of rock, the degree of fragmentation was unacceptable. In all instances, visual inspection showed that the single-shot craters tended to be elongated parallel to the strike and bedding planes of the ore zone.

Figure 5 depicts a graph of the energy utilization number "A" versus the depth ratios for data obtained from the 152.4-mm (6.0-in.) crater tests. Livingston defined "A" as a measure of the energy of the explosive devoted to the fracturing process. Generally, an increase in the energy of the explosive and an increase in the scale of the experiment yields a higher Δo value.

CONCLUSIONS OF SMALL-SCALE CRATER TESTS

The main conclusions that are derived from the small-scale crater tests are:

1. Strain energy factors, optimum depth ratios, and crater

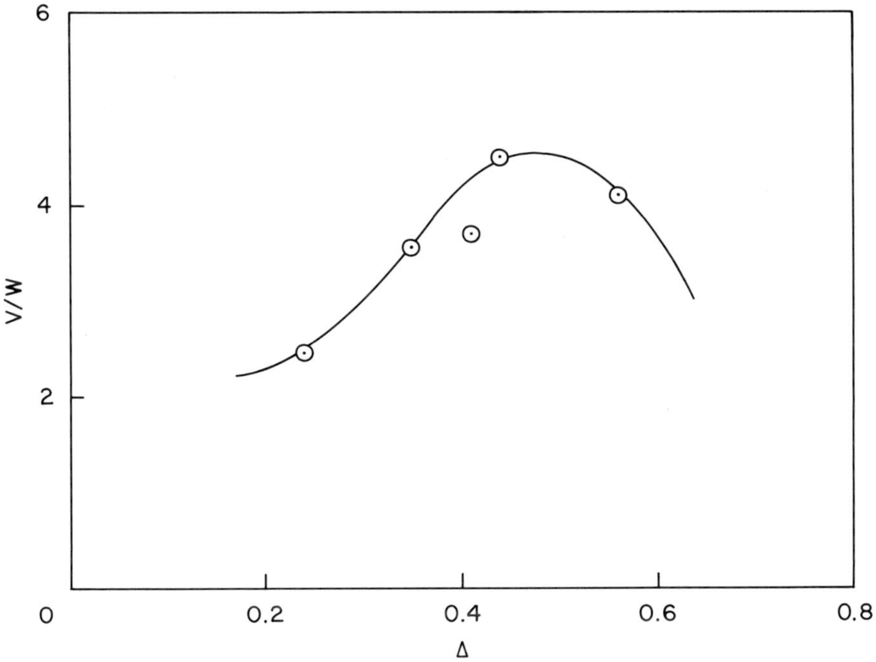

Figure 4. Cratering tests using 15.2-cm-diameter holes.

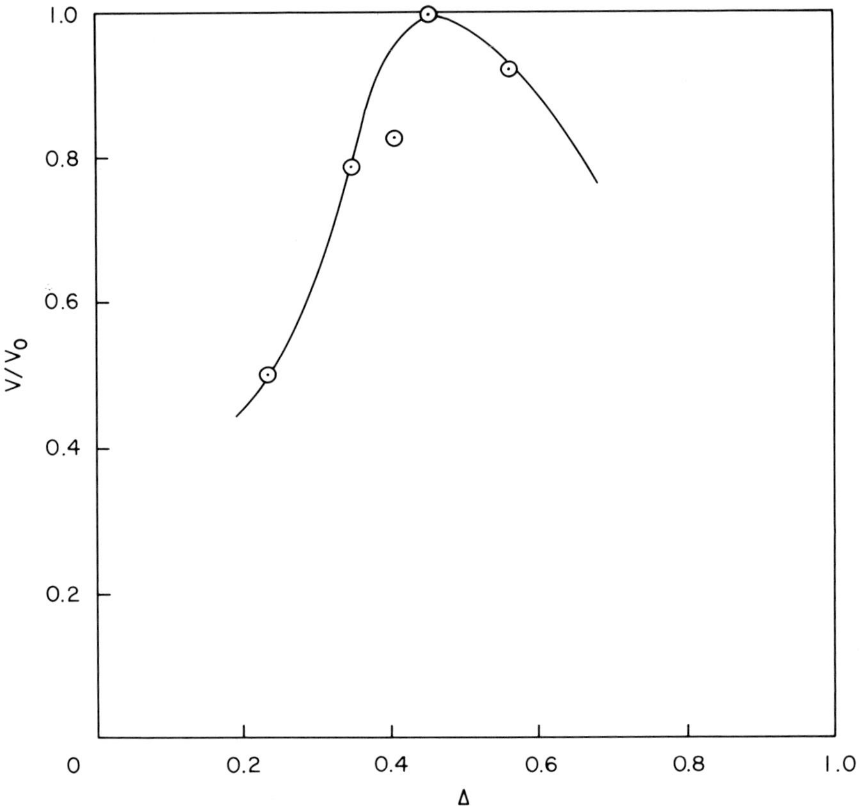

Figure 5. Energy utilization vs. depth ratio for 15.2-cm-diameter crater tests.

volumes are dependent on the explosive-rock combination and must be determined by experimentation.

2. In hard rock formations, there is a need to select explosives which exhibit relatively high densities, high detonation velocities, and high bulk strengths.

3. ANFO blasting agents are not suitable for true cratering applications due to the relatively low density, detonation velocity, and bulk strength values.

4. Based on results of the small-scale crater tests and known powder factors, a 2.74 x 2.13 m (9 x 7 ft) pattern was selected for initial production blasting. The Do value was reduced to 2.06 m (6.75 ft) to improve fragmentation.

IN - THE - HOLE DRILLING

Drill Rigs Used

Four Gardner-Denver ATD 3100 Crawler Carriers each equipped with a TRW Mission Mast and Rotating Head Assembly are currently used at the Homestake Mine. The Gardner-Denver unit utilizes air-powered hydraulics for thrust and rotation. Total hoisting capacity for the mast assembly is 5,670 kg (12,500 lb). Air tramming motors are used for carrier mobility.

The Mission Megadrill B53-15 bottom-hole drill is used in conjunction with 1.83 m (6.0 ft) long x 127-mm-diameter (5.0 in.) drill pipe with 85.7-mm-diameter (3.375 in.) Mission Modified Buttress thread. Operating air pressures at the hammer vary between 1551 and 1724 kPa (225 and 250 psi). A six-month test involving the Ingersoll-Rand DHD 360 hammer indicates a 15% increase in hole penetration rates under similar conditions.

To complement the ITH drill, a skid-mounted booster compressor package was designed by Gardner-Denver for transportation and operation underground. This unit consists of two air-cooled Dearing Model AOL 1000 booster compressors driven by two 460 V, 30-kw (40-hp) electric motors. The 17-m^3/min (600-cfm) compressor unit will boost mine air at 620 to 690 kPa (90 to 100 psi) to pressures as high as 1896 kPa (275 psi). Mine air is furnished by 101.6-mm-diameter (4.0-in.) level lines to a 63.5-mm-diameter (2.5-in.) bullhose at the compressor. Each Gardner-Denver ATD 3100 Carrier coupled with an AOL 1000 booster compressor package constitutes one ITH drilling unit.

Hole Accuracy

Maintaining hole accuracy is important in VCR stoping, since deviations from a desired hole pattern almost always produce substandard blasting results. Improper drill alignment (azimuth and dip angle), the attitude and folding of the ore zone, too much feed pressure, and the weight of the drill string itself are the most probable causes of excessive hold deviation. Hole deviation resulting from the static load of the drill string becomes much more pronounced when drilling at dip angles less than 70 degrees from the horizontal.

Drill operators position the crawler along stringlines that define a drilling grid as marked on the stope walls by mine survey crews. Once the crawler has been situated, the mast is plumbed in the drilling plane and then set on the required dip angle with the aid of a magnetic angle finder (clinometer).

Apparent hole wander with 48.8 m (160 ft) holes averages 3% but has approached 6-8% on occasion necessitating fill-in drilling. Generally, the first 30.5 m (100 ft) of drilling yields little or no deflection, whereas most of the deviation appears to accumulate in the last 15 m (50 ft) of the hole. Once drilling has been completed in a section of the stoping area, mine survey crews establish a transit tie on all holes in the undercut to identify any hole deviation.

Bits

Empirical tests were conducted to determine which type and make of bit would maximize longevity and productivity in ITH drilling. Button bits manufactured by Mission, Dresser, Numa, Epley, Gault, Dalby, and Hughes were tested in the concave, drop center, and flat-face designs. Presently, the Numa bit is preferred in the concave design.

Numa bit life averages 427 m (1,400 ft) but varies between 152 and 975 m (500 and 3,200 ft) depending on drilling conditions. Bit costs alone approximate $1.46/m ($0.46/ft) of 165.1-mm-diameter (6.5-in.) hole. Periodic inspections indicate that most of the button bits should be resharpened after every 30.5 m (100 ft) of drilling. However, simple cost analysis shows it is more economical to sacrifice 15-20% of the possible bit life in order to eliminate up to one hour of raising and lowering rods per shift. Button bits are sharpened underground with an air-powered die grinder using silicon carbide wheels of 60 to 80 grit and J or K in hardness.

Drilling Productivity

Blastholes in all stoping areas are presently drilled on a 2.4 x 2.4 m (8 x 8 ft) pattern which yields about 19.3 t/m (6.5 st/ft) of hole. The majority of all holes are drilled to the undercut sill at dip angles ranging from 45 to 90° from the horizontal. Typical hole lengths vary from 42.7 to 53.3 m (140 to 175 ft) depending on the attitude of the hole.

Since mid 1979, penetration rates have averaged about 7.3 m (24 ft) per drill hour, or 47.5 m (156 ft) per drill shift. Optimum penetration rates occur at 14 to 22 rpm, depending on rock conditions. Total ITH drill footage for 1979 amounted to 25,526 m (83,747 ft).

LOADING AND BLASTING PROCEDURE

Measuring Blastholes

The procedure for loading and blasting begins by measuring the length of each blasthole using either a plastic pipe or wooden loading stick about 0.76 m (30 in.) long fastened to a calibrated rope chain. The rope chain is separated from the end of the pipe by a rope 0.38 m (15 in.) long (exactly one-half the length of the pipe). The "zero" end of the rope chain is then temporarily fastened to the center of gravity of the pipe by a plastic cable-tie wire. When the stope-back measurement has been determined, the rope chain is given a sharp jerk to break the cable-tie wire, thereby allowing the pipe to hang vertically so that it can be pulled back up the hole.

Hole measurements are obtained by lowering the pipe and rope chain down the hole from the top sill until the pipe hits the muck pile. This measurement is recorded on a data sheet for the specified hole. The rope chain is then pulled up until the pipe lodges across the bottom of the hole at the stope back. This measurement is similarly recorded. The amount of relief available for the subsequent blast, then is the difference between these measurements and is also recorded on the data sheet. This procedure is repeated for the remaining holes in the immediate blasting section.

After reviewing the data sheets, a blasting technician prepares a delay pattern according to the stope-back profile. A "plug-set" or blocking height as measured between the stope back and the blocking location based on the angle of the drill hole and the results of small-scale crater tests in similar rock types is then entered onto the data sheet. Although the block height may deviate from normal due to special rock conditions particular to one stoping area, the values shown in the table below are generally valid for most stoping applications at the Homestake Mine:

Hole Angle	Blocking Height
90–80°	1.2 m (4 ft)
79–57°	1.5 m (5 ft)
56–50°	1.8 m (6 ft)
Less than 50°	2.1 m (7 ft)

The required blocking depth is calculated for each hole by subtracting the blocking height from the stope-back measurements. These figures are entered onto the data sheet along with the proper delay pattern. The completed data sheet is then returned to the loading crew so that blocking of the blastholes can begin.

Blocking Blastholes

Initial VCR efforts at Homestake made use of 102 x 102 x 203 mm (4 x 4 x 8 in.) rectangular wooden blocks suspended from the top sill by 6.35-mm-diameter (0.25 in.) polypropylene rope. Improvements in the blocking process have progressed to the point where two wedge-shaped blocks 102 x 102 x 254 mm (4 x 4 x 10 in.) are suspended with 40 grain Strip Mine Special Primacord, which also acts as the initiator for the down-the-hole cap and primer.

By incorporating 40 grain Primacord in lieu of 6.35-mm-diameter (0.25 in.) polypropylene rope, the total loading time is reduced approximately 20% since blocking and priming now become a one step process. Furthermore, elimination of the rope results in a net cost savings of about $0.06/m ($0.02 ft) of blasthole. This savings becomes significant considering the load-blast cycle ordinarily requires a minimum of 381 m (1,250 ft) of Primacord and 381 m (1,250 ft) of rope per hole to completely mine through a 45.7 m (150 ft) high stope block in successive 3-m (10 ft) lifts.

Blocking begins by tying one wedge block onto the 40 grain Primacord and lowering the block down the hole slightly ahead of the rope chain. After the proper depth has been reached, the rope chain is pulled out of the hole and a second wedge block is dropped down the hole alongside the Primacord. After the second wedge block hits the first, the Primacord is tugged to ensure that the blocks wedge together against the wall of the hole.

An alternative method of blocking involves measuring and recording all of the holes in the blasting area. A wooden stick approximately 0.76 m (30 in.) long is tied to one end of the Primacord. Next, a wedge block is tied to the Primacord at a distance equal to the blocking height above the wooden stick. This assembly is then lowered down the hole to the stope back. The Primacord is then pulled upward until the stick catches crosswise at the stope back. This procedure ensures that the wedge block is positioned at the proper elevation above the stope back, even if the hole has been chimneyed out. Finally, a second wedge block is dropped down the hole to lock against the first block.

Next, the Primacord is temporarily tied off to a convenient large rock and a piece of friction tape is wrapped on the Primacord at the collar of the hole to act as a check for block slippage. Approximately one foot of small rocks and cuttings is shoveled into the hole to completely seal and lock the two wedge blocks together. The remaining blastholes are then blocked in the same manner.

Loading Blastholes

The amount of water gel put into each hole depends upon the density of the product. For example, using the 6:1 charge length to hole diameter ratio, approximately 27.2 kg (60 lb) of DuPont's Tovex 650 (1.35 g/cm^3) or Trojan's WS-7F (1.40 g/cm^3) are loaded into the hole. CIL's Aquamex requires approximately 31.3 kg (69 lb) per hole due to the relatively high density of the product (1.54 g/cm^3).

Single Deck Blast:

In loading a single deck, the explosive bag is cut lengthwise and each "sausage" is shucked and dropped into the hole. Experience has shown that this procedure ensures a high degree of loading compaction and coupling within the hole which is a requisite to the cratering process. After half of the charge has been dropped down the hole, the loading data sheet is consulted to determine the proper down-the-hole delay for that hole. Presently, booster primers manufactured by Austin Powder Co. (ADP-1 primer) and Atlas Powder Co. (Deckmaster primer) are used as slider-type primers with nonelectric delay inserts.

Once two adjacent holes have been primed with the slider primers, the Primacord downlines are either tied off separately or permanently tied together with a square knot located midway between the two hole collars. This procedure serves as added insurance in case the wedge blocks fail and the Primacord must support the entire column load. Next, the two holes are loaded with the remaining half of the explosive charge and stemmed with either 2.1 m (7.0 ft) of dry sand or two 1.1 m (3.75 ft) long water bags, followed by filling the remainder of the hole with free-standing water. The process, as shown in Figure 6, is then repeated for the remaining holes.

Double Deck Blast:

The procedure for loading a double deck blast is quite similar to that of a single deck. Generally, in a multiple deck blast, 6.35-mm-diameter (0.25 in.) polypropylene rope is used to hold the wedge blocks, since the static column load could exceed the 136-kg (300 lb) tensile strength of the Primacord downline if the blocks weren't securely wedged within the hole.

After the wedge blocks have been set, bottom stemming is added, followed by the bottom charge and primer. Next, 152 mm (6.0 in.) of cuttings are shoveled into the hole and one wooden spacer 102 x 102 x 762 mm (4 x 4 x 30 in.) is dropped down the hole for decking material as shown in Figure 7. Fine-grained sand or coarse aggregate is normally preferred as decking material, but the bulk handling involved becomes quite time consuming, more expensive, and ultimately requires frequent and careful hole measuring to determine proper decking height.

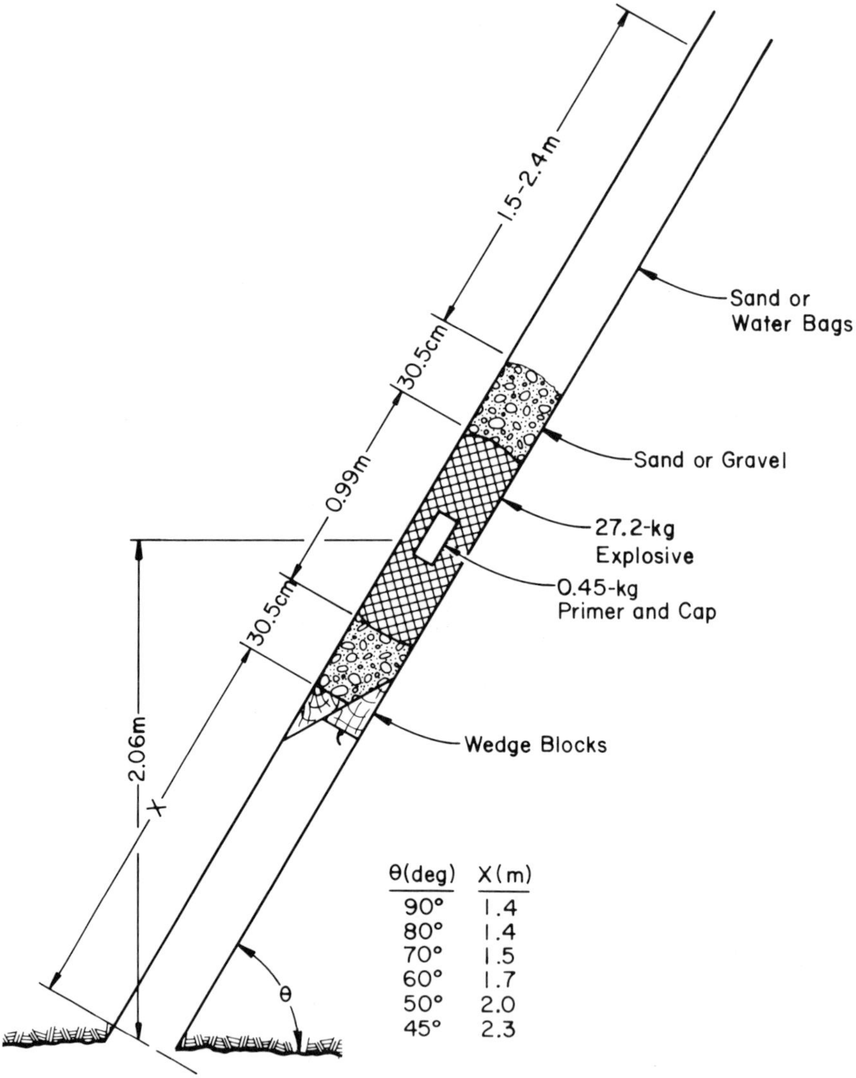

Figure 6. VCR single deck loading procedure.

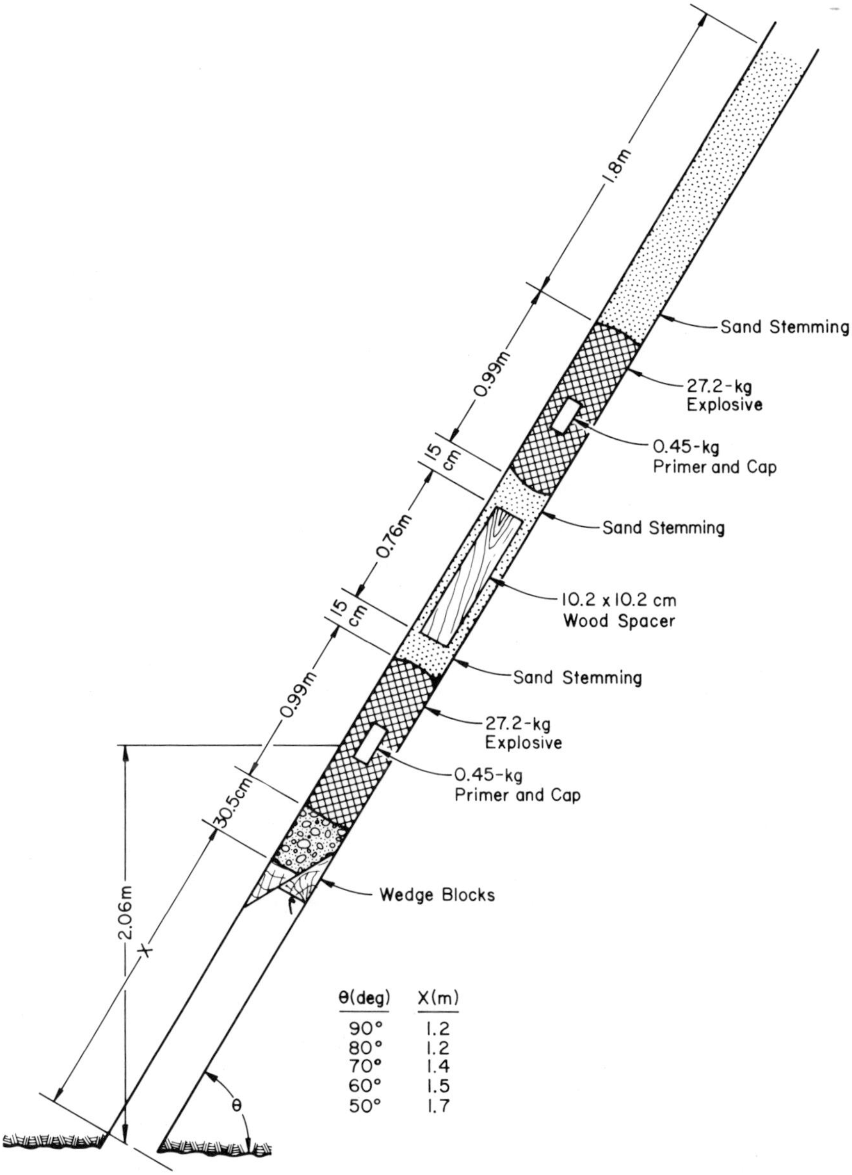

Figure 7. VCR double deck loading arrangement.

After adding the spacer and more sand, the top charge and primer are loaded into the hole. By using the slider-type delay primers, only one downline is required for each hole regardless of the number of decks to be loaded. Approximately 1.8 x 2.1 m (6 to 7 ft) of the hole is then stemmed with sand, and the remainder of the hole is filled with water up to the collar.

Trunkline Tie-In:

Figure 8 illustrates how E-Cord Primacord (25 grain) is tied in to form a trunkline between rows. Plastic cord connectors manufactured by Austin Powder Co. are used to connect the downlines to the trunkline. Experience has shown that use of these connectors speeds up tie-in time and minimizes the possibility of cut-offs due to improper knot-angle connections. Cross ties are utilized every 9.1 to 12.2 m (30 to 40 ft). Once final tie-ins have been made, two DuPont electric "o" Acudet Delay caps are taped onto the end of a Primacord pigtail for trunkline initiation.

Delay Patterns:

Several delay patterns have been tried in true crater blast situations. The preferred method is shown in Figure 9 and consists of shooting a burn out of the middle of the pattern on the first delays. The remaining holes are capped in a concentric or diamond-shaped fashion so that each hole has two free faces to break to, i.e. the horizontal stope back and a resultant vertical interior face. In this circumstance, each hole incorporates a combination slashing action and cratering action. This pattern usually yields the best fragmentation and maximum volume broken.

Generally, a pattern consisting of 30 to 40 holes can be loaded and blasted by an experienced loading crew in one full shift. A typical stoping block 61 m (200 ft) long and 10.7 m (35 ft) wide is divided into three loading sections. Ordinarily, blasting rotates from section to section, thereby maintaining uniformity in stope-back elevations (4). Continued blasting in only one stoping section usually results in excessive sloughing of the brows (5).

Once one end of a stope has been cratered upward one or two lifts, the adjacent section is either slashed into the end section or crater blasted downward into the undercut. A combination slashing and cratering delay pattern is commonly utilized using one or more decks of spherical charges. However, when multiple-deck slashing patterns are used, an excessive amount of rock must be drawn out of the stope so that sufficient relief is available for slashing in the adjacent section. Consequently, dilution in the form of slough rock from the hanging wall almost always occurs. For this reason, multiple-deck slashing is no longer practiced.

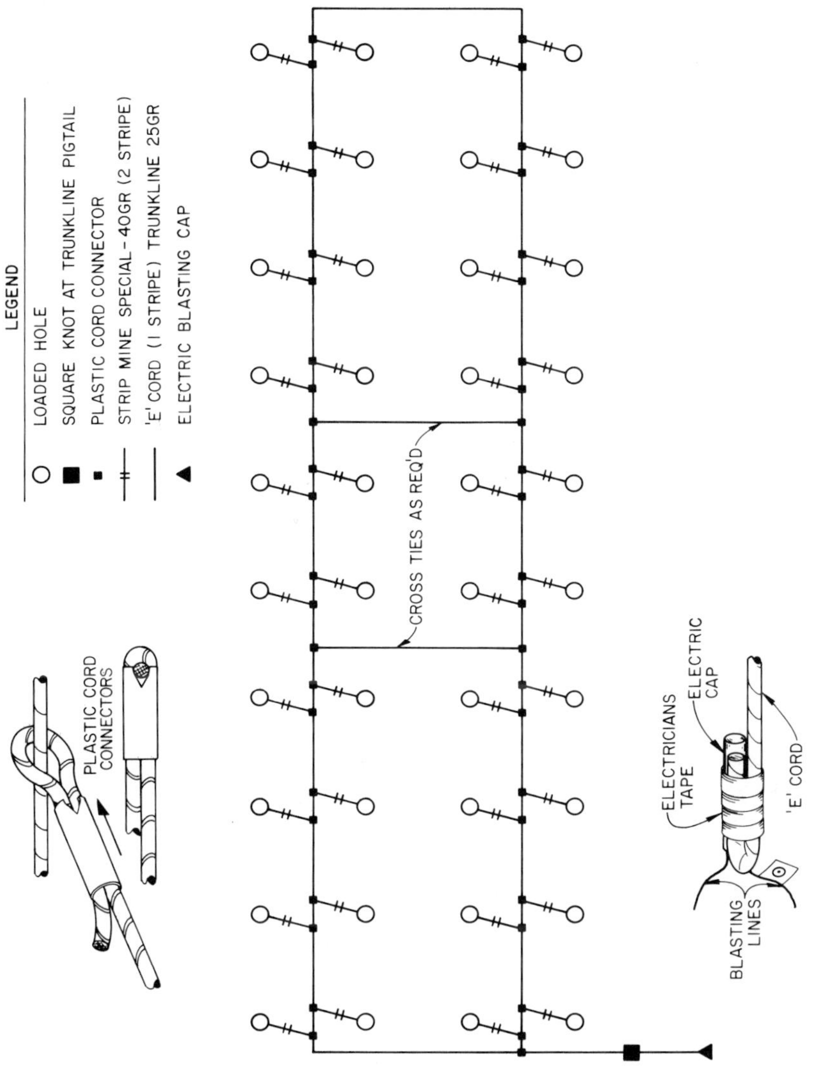

Figure 8. VCR trunkline tie-in procedure.

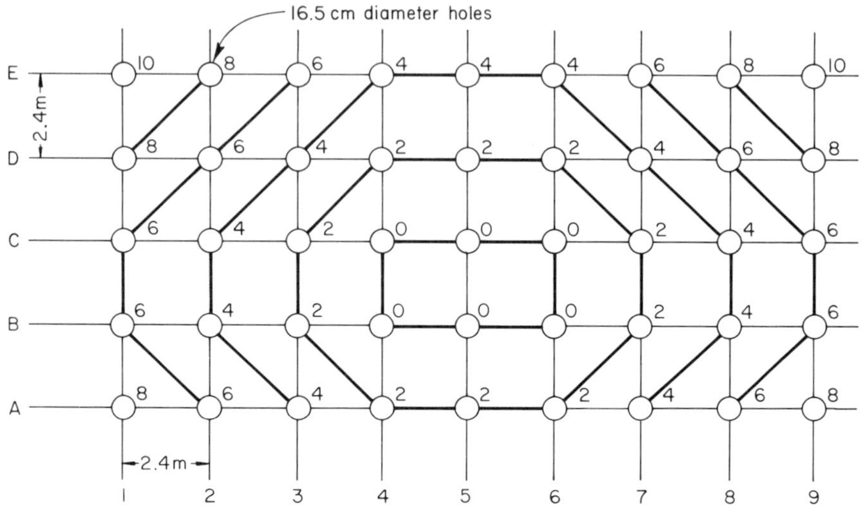

Figure 9. Typical VCR blasting delay pattern.

Explosive Selection:

Successful crater blasting applications are dependent on maximizing the amount of explosive energy per unit length of charge. This requirement is fulfilled by selecting an explosive which exhibits relatively high values of detonation velocity, density, and bulk strength. Also, the explosive should have sufficient consistency so that the cartridge can be shucked to increase coupling ability and fill the entire cross-sectional area of the blasthole.

To date, the most widely-used explosives in VCR applications are TNT-based water gels, emulsions, and nitrocarbonitrate water gels. ANFO blasting agents are unsuited for true crater applications because of the low values mentioned above. Experience at the Homestake Mine indicates that the degree of fragmentation and volume of rock broken is proportional to the relative bulk strength of the explosive.

Observations at other mining operations have shown that blastholes which do not break out into the undercut or later become plugged must be treated with caution, since flashes of post-blast gases may occur. It is theorized these incidents have occurred due to the generation of combustible gases in blastholes which fail to break out upon detonation of the explosive charge. It appears that this phenomenon can occur with any explosive although high-energy aluminized products show a more frequent rate of occurrence.

VCR STOPE DESIGN

LHD Versus Slusher Extraction

VCR extraction systems are variable depending on the size of the ore body, stability of the surrounding country rock, and availability of equipment. Experience at the Homestake Mine has shown that extraction systems utilizing load-haul-dump equipment can achieve two to three times the production rates obtained by slushers. With multiple drawpoints, large rock can be sorted by LHD equipment for blockholing, thereby minimizing the amount of lost mucking time due to oversize rock.

Figure 10 shows an idealized VCR stope using slusher extraction and illustrates how ground control can be a continual problem due to the required close proximity of the slusher drift to the drawpoint hoppers. A LHD haulage drift, on the other hand, can be driven a greater distance from the stope and is less affected by blasting. The equipment needed for LHD extraction requires a relatively high initial capital investment, but the associated high production rates enable the equipment to be cycled among stopes which ordinarily would not start producing until a later date.

Drilling Considerations

The top sill of a VCR stope block is usually a sill cut slashed out to the full width of the ore body. A top sill drift can be used for fan drilling with the ITH drill providing the drift is stripped high and wide enough to facilitate the drill. Typical parallel-hole patterns range from a 2.4 x 2.4 m (8 x 8 ft) pattern to a 3 x 3 m (10 x 10 ft) pattern depending on the rock type and extraction system used.

Figure 11 shows how certain stope blocks can be brought into production sooner by utilizing a combination VCR and blasthole-longhole method. If, for example, manpower is not available or ground conditions will not allow a full undercut for a VCR stope, a ring drilling drift can be driven instead of cutting out the full undercut. The bottom 15.2 m (50 ft) of the ore body is then drilled out using longhole drills and production blasting begins while ITH drilling progresses at the opposite end of the stope block.

Early Crown Pillar Test Stopes

By mid 1978, a decision was made to experiment with the VCR method on a larger scale. The method could easily be adapted to

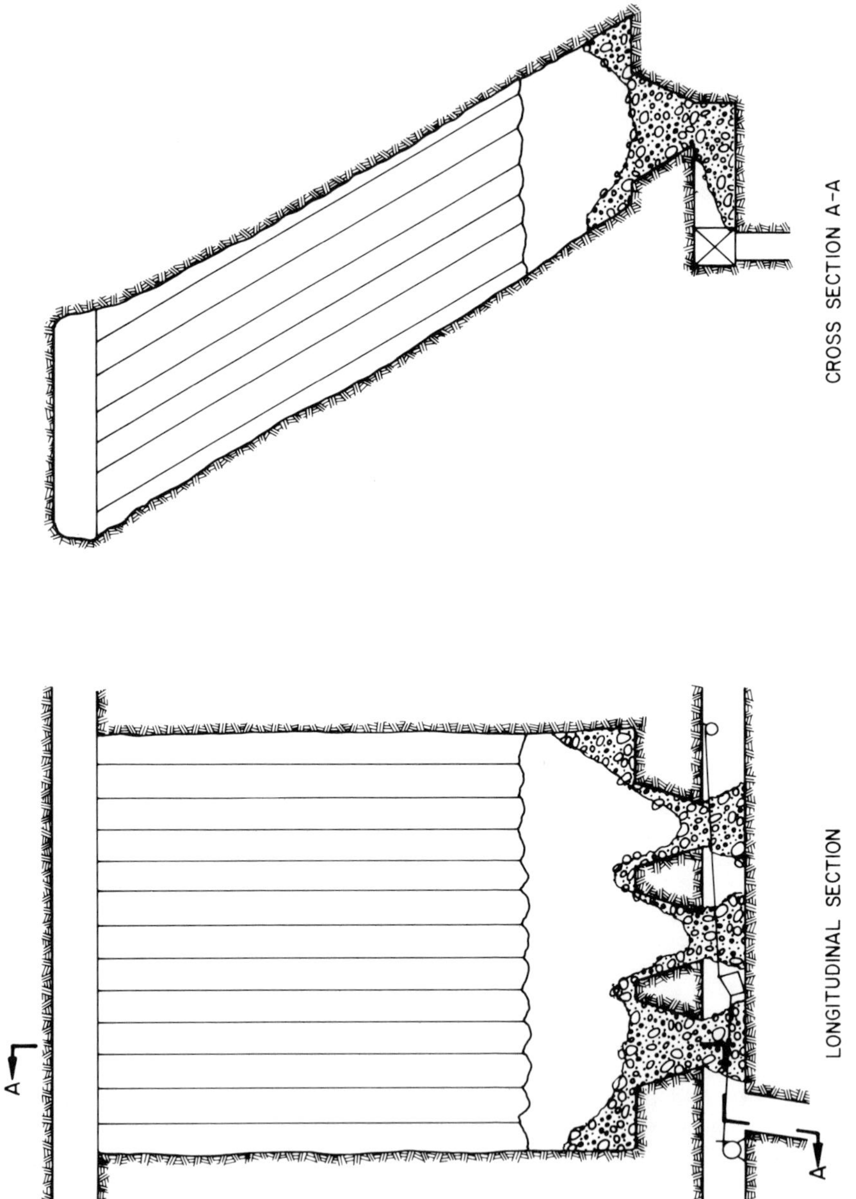

Figure 10. Idealized vertical retreat stope with slusher extraction.

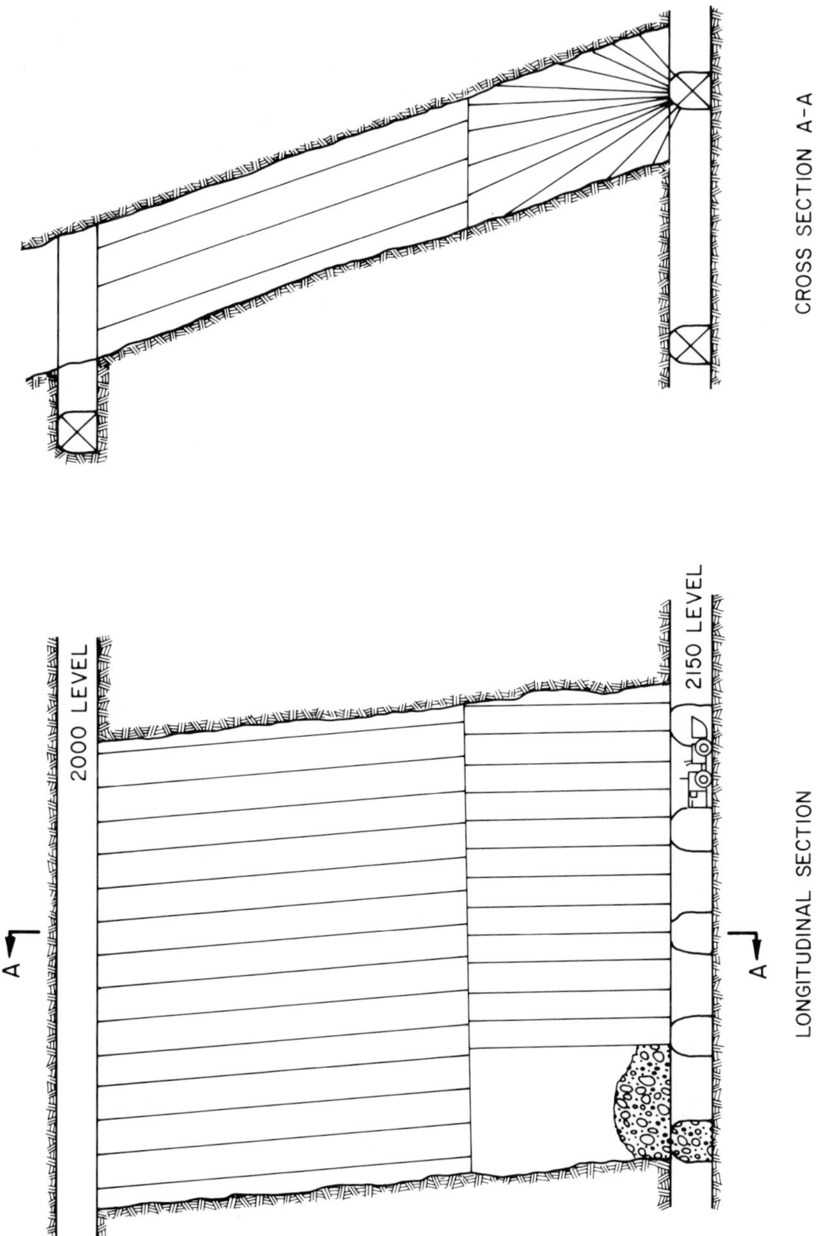

Figure 11. VCR stope with ring-drilled bottom sill.

existing shrinkage stoping areas by cutting the top sill to facilitate an ITH drill. This was tried with some degree of success in 66-67 D Stope 9 Ledge, on the 1,707 m (5,600 ft) level where some forty-three 165.1-mm-diameter (6.5-in.) holes were drilled on a 2.1 x 2.7 m (7 x 9 ft) pattern in a 15.2-m (50 ft) thick ore area over the shrinkage stope. After making two crater blasts, an attempt was made to double-deck the remaining 7.6-m (25 ft) thick crown pillar using a combination of electric and non-electric initiating systems. Unfortunately, the electric circuits were not properly balanced, and several of the electric caps failed to detonate causing the round to freeze up.

In October 1978, a second crown pillar approximately 7.3 m (24 ft) thick was drilled out with 41 holes on a 2.1 x 2.7 m (7 x 9 ft) pattern, as shown in Figure 12. This shot was double-decked with Tovex Extra water gel using 6.1 m (20 ft) long Nonel MS Delay caps as in-the-hole delays in conjunction with HDP-1A cast primers. ANFO was used where necessary to bring the charge height of the top deck up to a 1.2 m (4 ft) stemming distance. Approximately 4,717 t (5,200 st) were broken in the 68-69 D stope with excellent fragmentation (6).

By early 1979, Austin Powder Co. was marketing the ADP-1 booster primer with a 4 grain Primaline Primadet MS Delay blasting cap. This priming system requires only one detonating cord downline, regardless of the number of delays employed in each hole. Austin ADP-1 primers were used in conjunction with Ireco's Iregel 382 Emulsion in a triple-deck shot in a 9.1 m (30 ft) thick crown pillar as shown in Figure 13.

Experience in other crown pillar blasts indicated that a concentric or diamond-shaped delay pattern most often yielded the greatest crater heights and best fragmentation. Because of the limited number of MS delays available, two "breaks" were incorporated in the 42 D crown pillar blast as illustrated in Figure 14. Decking material consisted of a combination of sand and wooden 102 x 102 m (4 x 4-in.) spacers. The estimated powder factor for this 7,620 t (9,400 st) blast was 0.75 kg/t (1.5 lb/st).

Two unsuccessful crown pillar blasts were made in mid-1979 (74-75 B 9 L 6050 and 38-43 F 11 L 5000), both of which left a ledge of solid ground between the cratered areas of the top and bottom charges. The main conclusions reached from these double-deck crown pillar blasts were:

1. All holes in a double-deck cratering pattern should measure less than 7.3 m (24 ft) between the top sill and the stope back.

2. The effect of placing charges at a distance less than the optimum charge depth will reduce the depth and volume of the resulting craters.

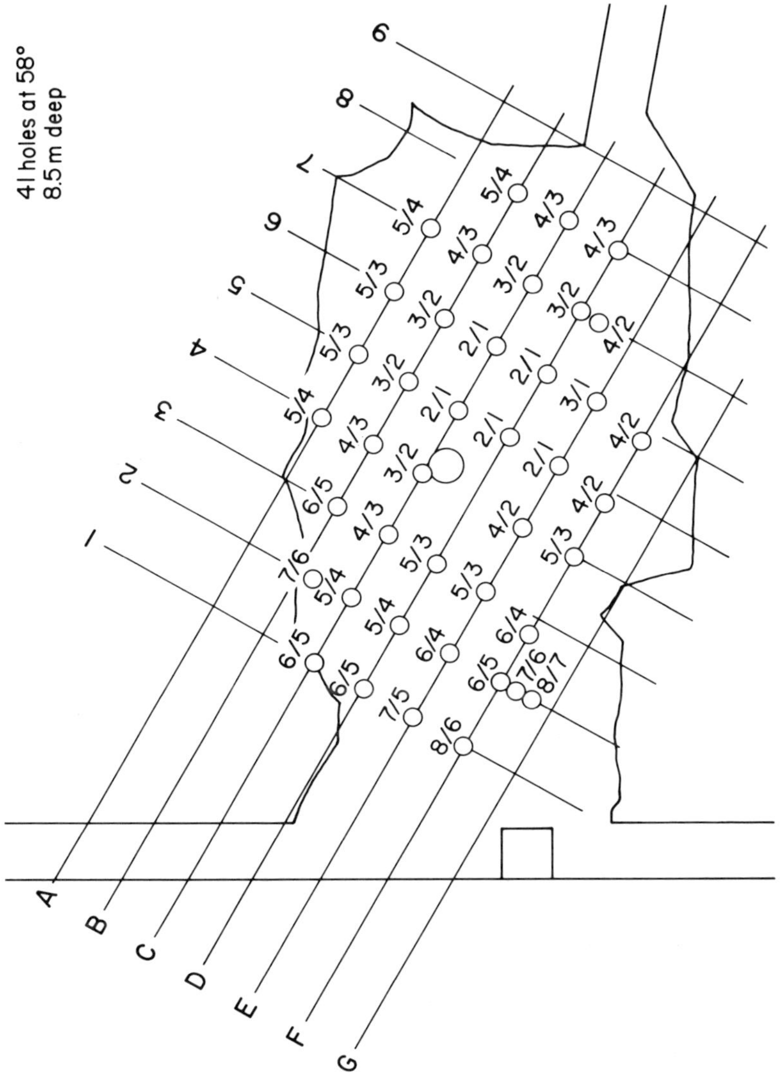

Figure 12. Double-deck crown pillar blast in the 68-69 SD stope, 9 ledge, 5900 level. Values are MS period (upper deck/lower deck).

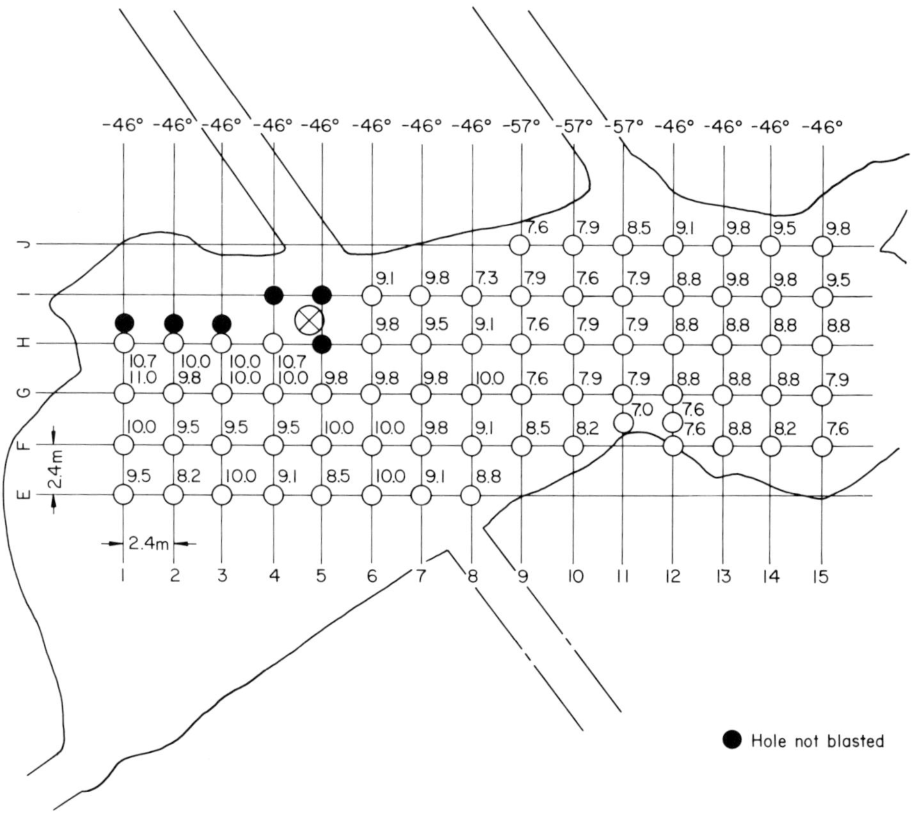

Figure 13. Drill data for 42 D stope, 9 ledge, 4700 level crown pillar blast. Drill hole depths expressed in meters.

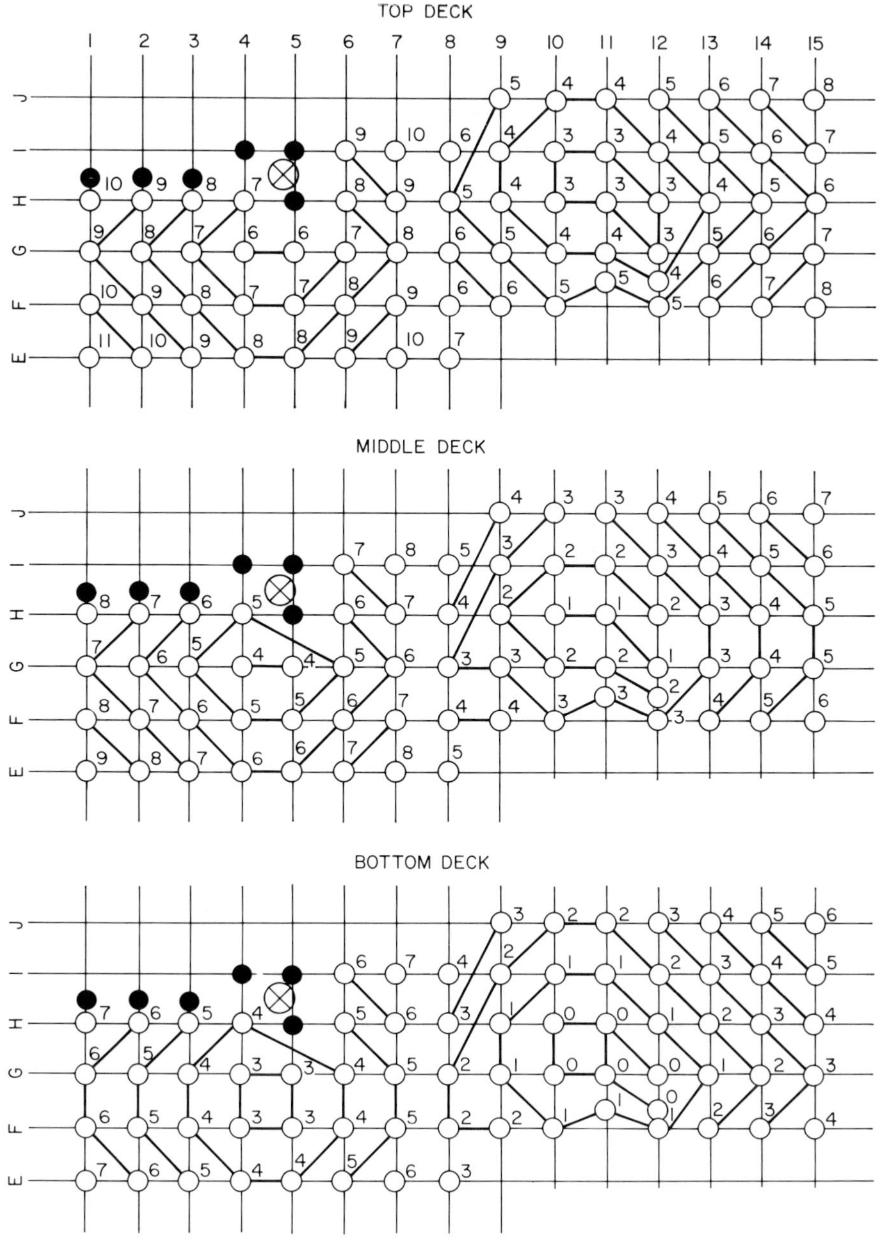

Figure 14. Delay pattern for triple-deck, crown-pillar blast in 42 D stope, 9 ledge, 4700 level (solid holes not blasted).

3. Interdeck stemming in a true crater delay pattern should not exceed 1.2 m (4 ft).

In November 1979, a successful double-deck crown pillar blast was made in 10-15 EA Stope 9 Ledge, 2,750 Level incorporating the above considerations.

Blasting Problems - Solutions

As of July 1980, three VCR stopes had reached full-scale production (38-43 F 11 L 5,000, 53-55 F 9 L 2,150, and 10-15 EA 9 L 2,750). All of these 45.7-m (150 ft) high ore blocks exceed 61 m (200 ft) in length and average 10.7 m (35 ft) in width.

In 38-43 F 11 L 5,000, it became necessary to drill out a 2.1 x 2.7 m (7 x 9 ft) drop raise after a double-deck loaded crown pillar blast failed to break through in an adjacent blasting section. The 19.8 m (65 ft) drop raise was cratered through to sill in about fifteen blasts. Considerable difficulties were encountered with plugged holes.

Two of the most prevalent problems experienced in all VCR stopes have been hole collar damage and plugged holes. Originally, it was thought that setting 152.4-mm-diameter (6.0-in) steel pipe in concrete in the top 1.5 m (5 ft) of the 165.1-mm-diameter (6.5-in.) holes would eliminate most of the collar damage. Results of the initial blast showed, however, that the collar casings apparently created too much restriction, since all of the casings were blown out of the holes. Future tests are scheduled using a 0.9 m (3 ft) length of 203.2-mm-diameter (8-in.) collar casing in conjunction with the 165.1-mm-diameter (6.5-in.) hole.

The occurrence of plugged holes has been indirectly attributed to excessive charge heights in chimneyed holes and excessive broken rock being left on the floor from cutting-out of the top sill. It has been found that lowering wooden blocks 101.6 x 101.6 x 762 mm (4 x 4 x 30 in.) to the chimneyed area is an effective means of blocking chimneyed or bulled holes. The hole is then filled with broken rock and cuttings up to the proper charge depth.

Much of the difficulty experienced with collar damage and plugged holes has been minimized by using sufficient amounts of stemming material. Current practice involves shoveling 0.3 m (1 ft) of drill cuttings directly on the charge column followed by lowering two 1.1 m (3.75 ft) long water stemming bags. Prior to tying in, all of the holes are filled with water from a water hose. All holes measuring less than 18.3 m (60 ft) from the top sill require additional stemming and are stemmed with 2.1 m (7 ft) of fine sand followed by water from a hose.

Figure 15. Gardner-Denver ATD 3100 carrier with a TRW Mission mast and rotating-head assembly drilling a 17 cm-diameter hole.

COMPARATIVE COSTS AND PRODUCTIVITY RATES

A comparison of the year to June 30, 1980 productivity rates at the Homestake Mine shows that on a tons per man shift basis, blasthole sublevel stoping yields 22.8 tons per man shift (25.1 st/ms) while open cut-and-fill stoping averages 15.1 (16.7 st/ms). VCR output has averaged 31.6 tons per man shift (34.8 st/ms) to date in 1980 and is expected to increase as more stopes reach the production phase.

Table 1 summarizes direct stoping costs by mining type for three different methods. Figures shown for the VCR method include development costs associated with mining of the top sill and undercut as well as ITH drilling costs. The average 1980 stoping cost for the VCR method is relatively high, mainly because only four stopes out of a total of seventeen in development had reached the production phase as of June 30, 1980. The VCR stoping costs shown in parentheses in Table 1 were calculated for the four stopes that were producing in the first half of 1980.

Table 1. Comparative Stoping Costs by Mining Type.

	OCF	BH	VCR	
YTD June, 1980 stoping & fringe	$ 9.49	$ 5.31	$ 6.79	($4.91)
Budgeted 1980 stoping & fringe	$12.48	$ 4.97	$ 6.55	
Budgeted 1980 grade (oz/st)	0.221	0.143	0.149	
Budgeted 1980 stoping cost per oz	$56.47	$34.76	$43.96	
YTD June, 1980 stoping cost per oz	$42.94	$37.13	$45.57	($32.95)

As more VCR stopes reach the production stage, the additional tonnage realized is expected to offset development costs and continue to stabilize the overall VCR stoping cost. Although ore grades from producing VCR stopes average only 4.1 g/t (0.149 oz/st), Table 1 shows the VCR stoping method to be more profitable than OCF stopes averaging 6.2 g/t (0.221 oz/st). Ultimately, the greatest percentage of total VCR output will be shifted from remnant low-grade blocks into the higher-grade 19 and 21 Ledge ore areas, thereby making the total stoping cost even more attractive.

CONCLUSION

Applications of the cratering mechanism using Livingston's spherical charge theories became adaptable to underground stoping about six years ago with the introduction of in-the-hole drill equipment capable of drilling large diameter holes at an acceptable cost.

In the late 1970's, substantial increases in the cost of production from traditional stoping methods prompted investigation of the

Vertical Crater Retreat method for possible applications at the Homestake Mine. Experimentation with the VCR method indicated that in many areas of the mine, it would be more economical to mine less selectively using mechanized equipment, with less dependence on the availability of skilled miners.

Most of the VCR production, to date, has been derived from low-grade blocks in the upper levels of the mine. Successful applications in these areas has led mine personnel to plan numerous VCR stopes in the recently developed higher grade areas of the mine. It is estimated that over the next five years, VCR production will account for 35-50% of the total mine production.

REFERENCES

1. Livingston, C. W., U.S. Patent 3,762,771, Oct. 2, 1973.

2. Lang, R. C., Roach, R. J. and Osoko, M. N., "Vertical Crater Retreat an Important New Mining Method," Canadian Mining Journal, pp. 69-76, Sept., 1977.

3. Grant, C. H., "Simplified Explanation of Crater Method," Engineering and Mining Journal, pp. 86-89, Nov., 1964.

4. Crocker, C. S., "Vertical Crater Retreat Mining at the Centennial Mine of HBM&S Co., Ltd.," Flin Flon, Manitoba, 1978.

5. Carter, G. S., "Vertical Crater Retreat Mining at HBM&S Co., Ltd's. Centennial Mine," District 4 C.I.M. Convention, 1977.

6. Ericksen, S. D., "Vertical Crater Retreat Stoping at the Homestake Gold Mine," Homestake Mining Co., Lead, SD, 1978.

7. Monahan, C. J., "The Crater Blasting Method Applied to Pillar Recovery at Falconbridge Nickel Mines Limited," Underground Operators Conference, Feb. 19-21, 1979.

Chapter 4

SILVER MINING IN WESTERN CENTRAL MEXICO

Jack Haptonstall

Homestake Mining Company
Lead, South Dakota

INTRODUCTION

The mines of western central Mexico historically contributed a profusion of gold and silver for the treasury of Spain and later provided sources for financing the Mexican War for Independence and the Mexican Revolution. The western coastal and barranca (cliff) tribes were hunters and extremely hostile, and the Spaniards were not able to penetrate many of the rich districts of their rugged homeland until the mid-1700's. This was more than 200 years after Cortez began the Inquisition by defeating the Aztecs and their allies in a decisive battle in the great city of Tenochtitlan (present day Mexico City). The Spaniards had a voracious appetite for silver and gold, and it was inevitable that they would finally subdue the natives and commence mining activities.

Described in this paper are some of the precious metal mining activities in the southern half of the State of Sinaloa and the western half of the State of Durango (Figure 1), and specifically those mining activities in the area controlled by the company with which the author was associated for 14 years, Minas de San Luis, S. A. de C. V., the parent company of which is now Industrias Luis Min, S. A.

GENERAL DESCRIPTION

Geography, Logistics and Climate

The most spectacular of the physiographic provinces, the western Sierra Madre Occidental is the cliff or "barranca" province, the

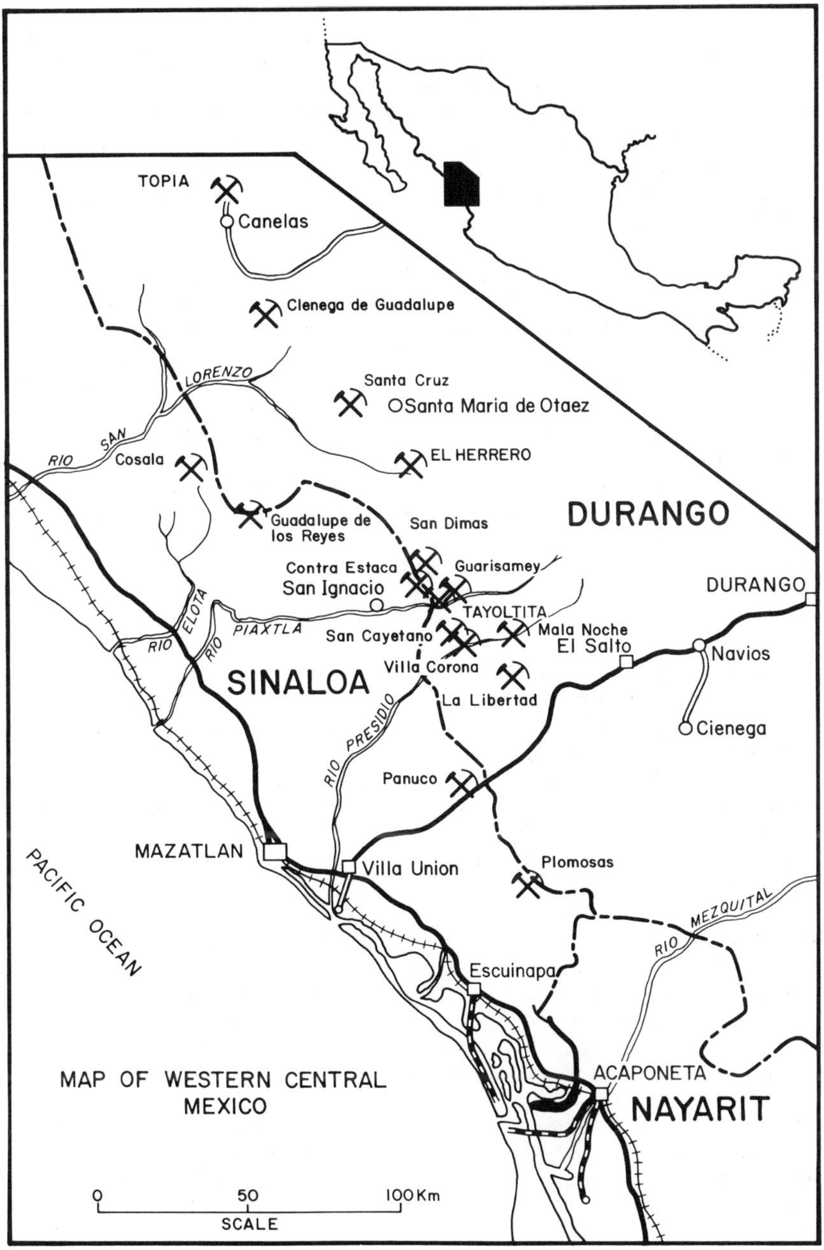
TOPIA
Canelas
Cienega de Guadalupe
LORENZO
RIO SAN
Santa Cruz
Santa Maria de Otaez
Cosala
EL HERRERO
DURANGO
Guadalupe de los Reyes
San Dimas
Contra Estaca
Guarisamey
San Ignacio
DURANGO
TAYOLTITA
RIO ELOTA
RIO PIAXTLA
San Cayetano
Mala Noche
El Salto
Navios
Villa Corona
SINALOA
La Libertad
Cienega
RIO PRESIDIO
Panuco
MAZATLAN
Villa Union
Plomosas
PACIFIC OCEAN
RIO MEZQUITAL
Escuinapa
MAP OF WESTERN CENTRAL MEXICO
ACAPONETA
NAYARIT
0 50 100 Km
SCALE

Figure 1.

provinces having been defined by Henshaw (1). See Figure 2. The barrancas generally parallel the Pacific coast and rise abruptly above the Pacific coastal plain and the valley and ridge provinces. The barrancas grade into the Mexican plateau province. The area embraces the Tropic of Capricorn which passes to the north of the city of Mazatlan. The barranca province is characterized by broad relatively flat-topped mountain ranges, separated by deep gorges cut by the rivers flowing southwest to the Pacific Ocean. Among these rivers are the Rio San Lorenzo, Rio Elota, Rio Piaxtla, Rio Presidio, and Rio Mezquital (Figure 1).

Smith's (2) description of the barrancas is generally applicable to the entire study area. Within the barrancas, steep-sided ridges with gently-sloping crests extend toward the rivers and major arroyos from each side of the canyon rims. These inner-barranca ridges depart the sides of the canyons 300 to 500 m (1,000 to 1,600 ft) below the rims and drop off abruptly to the rivers or arroyos. The inner canyons thus formed are as much as 1,500 m (4,920 ft) deep. Intermittent streams in the arroyos separating the inner-barranca ridges have steep gradients and are typified by spectacular waterfalls. Vegetation of the barrancas is sub-tropical in the river valleys, grading to pines and oaks at higher elevations near the canyon rims. Elevations in the river valleys are typically 400 to 700 m (1,300 to 2,300 ft) above sea level, while the canyon rims rise to 1,500 to 2,400 m (4,920 to 7,870 ft) above sea level.

The Mexican plateau province is a relatively flat-topped mesa with an average elevation of about 2,400 m above sea level. It is typically covered with high altitude vegetation, such as forests of pines and oak. The region is generally hot and humid with rainfall averaging about 600 mm (24 in.) per year. The coldest temperatures recorded in this area are about 10°C (50°F) with the yearly average temperature of about 22°C (72°F); however, in the Mexican plateau province, below zero temperatures are common in winter, and some snowfall is recorded each year.

Presently, one of Mexico's major goals is to expand the road and highway system to open regions of potential economic importance, and several Federal and State highway projects for the barrancas are either in progress or on the drawing boards. Funding is the major problem because of the great amount of expensive rock work necessary in this land of hard rock and bold relief.

Most of the study area is serviced by radio-telephone systems or radio communications. Little effort has been made by the National Telephone Company to install telephones in these remote areas, and it appears that any major program to accomplish this will be far in the future.

The Mexican Federal Power Commission has undertaken an aggressive program to "electrify" the rural areas of the country. Installation of many new lines in the barranca region to stimulate both the agriculture

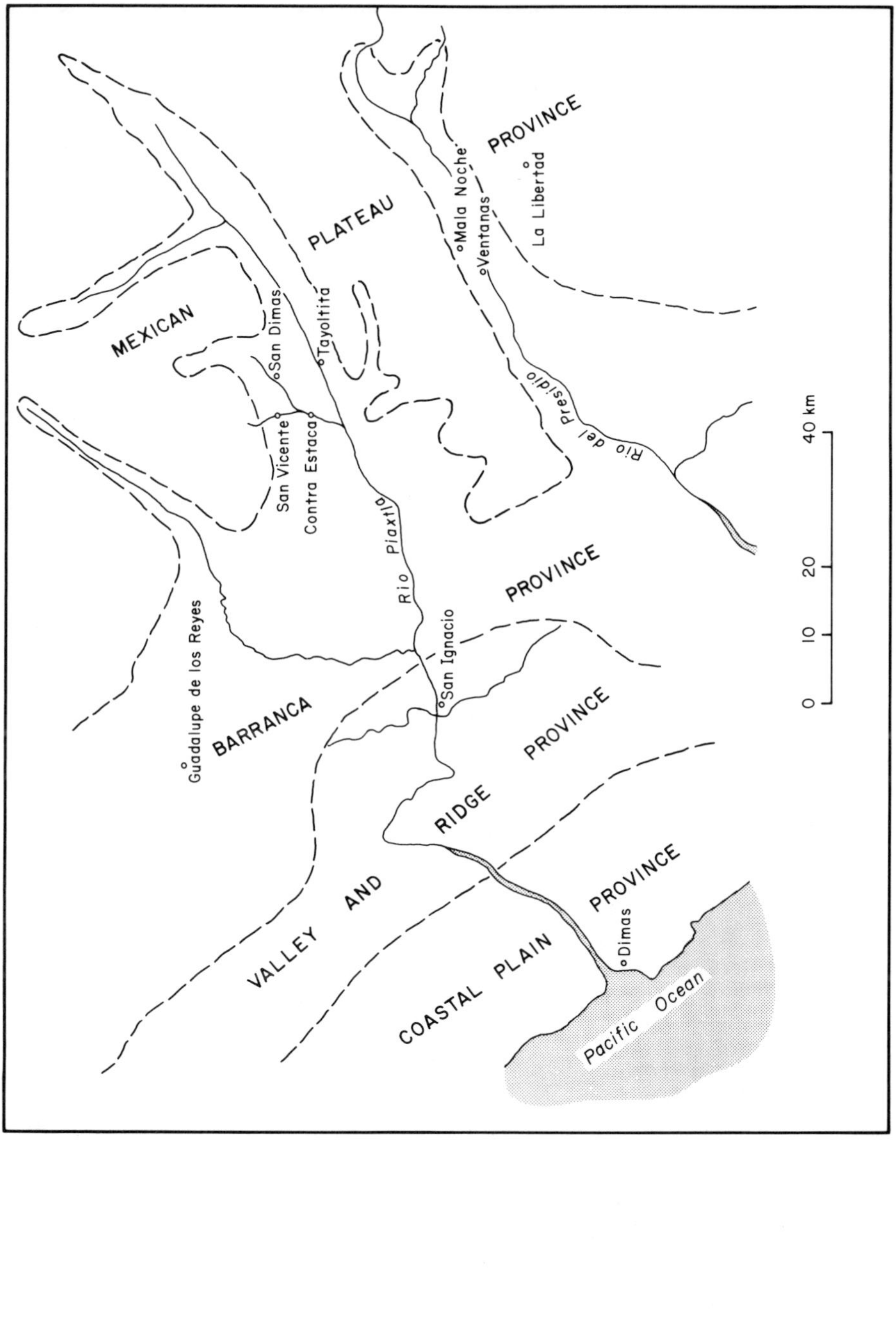

Figure 2. Geomorphic map showing the boundaries of the physiographic provinces of the Piaxtla-Presidio Region, Sinaloa, Durango, Mexico.

and mining industries is the goal. However, this policy is relatively new, and most of the older mining companies were obliged to furnish their own power. Many of the power plants consisted of small hydroelectric units of less than 1,000 kw (1,330 hp) generating capacity; many of these are still in use today, most having been built in the early part of this century.

The inaccessibility of the barrancas prompted the development of small bush-pilot airlines to service them. The inhabitants found that hacking out a short airstrip in a river valley or atop the cliffs was far easier than building roads, and pilots responded by establishing air service throughout the region. Planes are mostly single-engine Cessna's, but both single- and twin-engine STOL aircraft are used.

Some of the barranca airlines were Mexican aviation pioneers. The forerunner of Mexico's national airline, Aero Mexico, was founded around a small company, Lineas Aereos Mineras, S. A., whose sole purpose was flying a route between the sleepy port of Matzatlan, Sinaloa and the mining center of Tayoltita, Durango.

General Geology and Mining

The rock formations of the barranca province generally include a lower volcanic series of upper Cretaceous and Tertiary volcanic flows of rhyolitic, latitic and andesitic composition. The thickness of the lower volcanic series averages about 1,500 m (4,920 ft). The lower volcanics overlie gneises, schists, and sediments of undetermined age, which are interbedded with volcanic formations of Cretaceous (?) age. Much of the area is intruded by members of the Oligocene age Coastal Batholith, which range in composition from diorites to granites. Other Oligocene intrusives consist of rhyolite and andesite dikes, breccia pipes, and stocks.

Topping the lower volcanic series is the tuffaceous Capping Rhyolite of late Oligocene or early Miocene age. This flow rock reaches thicknesses of over 1,500 m and covers most of the higher barranca province as well as the Mexican plateau province.

Granite, limestone, and sedimentary formations are exposed in the coastal plain and valley and ridge provinces. However, most of the area is covered by sediments eroded and transported from the barrancas.

Faulting in the region consists of primary north-south-trending normal faults, and secondary east-west-striking normal faults, all of which tend to divide the area up into great rectangular blocks. Many of these blocks have undergone tilting.

Most of the economic precious metal deposits are found in the lower volcanic series of the barrancas. These generally consist of epithermal open space filling fissure veins, the ore minerals of which are mainly argentite and electrum. There also can be present minor polybasite, minor proustite, native silver, tehrahedrite and base metals in specific locations.

Gangue minerals in the veins are generally quartz, calcite, minor rhodonite, minor flourite, and host rock fragments. In many cases the veins have undergone post mineral flooding by quartz.

In reality, many of the ore deposits found in the barranca province seem to extend under the Mexican plateau, the erosion of the barrancas having provided "windows" for the examination of the lower volcanic series. If the 1,500 m of capping rhyolite could be peeled off the barranca and Mexican plateau provinces, undoubtedly many wonderful ore deposits would be exposed. This is one of the chief problems of the explorationist in Mexico.

Some silver-lead limestone replacement deposits have been mined in the valley and ridge province. The limestones are sometimes highly marbelized, but the formations (of undetermined age) are not extensive, and the deposits tend to be small.

The ore minerals for the limestone replacements are typically galena, sphalerite, minor chalcopyrite and oxides of these. The silver is usually associated with galena.

Examples of the vein deposits are the Tayoltita, La Libertad, and Mala Noche silver and gold mines of Minas de San Luis, S. A. de C. V.; the Santa Cruz silver mine managed by Mr. Phillip Beckley; the El Herrero silver and gold mine of the Gutierrez family; the Topia silver, lead, and zinc mines and the Cienega silver-gold mine of Industrias Penoles; the silver and gold mines of the Panuco Coopertiva, the silver-gold mines of the Guadalupe de Los Reyes Coopertiva; and the La Rastra-Plomosas copper, lead, zinc, silver, gold mines of Industrial Minera Mexico, S. A. All of the above are underground mines, employing either shrinkage or cut-and-fill stoping as the chief methods of extraction.

Base metal-silver limestone replacement deposits are exemplified by La Estrella mine of Sr. Enrique Gaytan and the old Nuestra Senora Mine of the American Smelting and Refining Company. These are located near the town of Cosala, Sinaloa. The principal silver mines in the study area are the Tayoltita, La Libertad, and Mala Noche Mines. The description of these operations is the subject of the remainder of this paper.

TAYOLTITA MINE

Location and Climate

The Tayoltita Mine, located in the San Dimas Mining District (Figure 1), lies about midway between the Pacific coast city of

Mazatlan, Sinaloa and the Mexican plateau city of Durango, Durango; the airline distance from each city is about 125 km (78 miles). The town of Tayoltita is situated on the east bank of the Piaxtla River and is surrounded by the Sierra Madre Occidental barrancas.

The elevation of the town is 485 m (1,500 ft), while that of Bolanos Peak, which rises directly above the town, is 1,905 m (6,250 ft) above sea level. The Mexican plateau commences at about 7 km (4.3 mi.) to the north of Tayoltita.

Until recently, no all weather road had been built into the town; the closest road heads lying in the Sierra at 25 km (15.5 mi.) from the town and at 1,800 m (5,000 ft) above it. A recent combined effort of the Durango State Government and the Mexican Federal Government had resulted in the construction of a road from the plateau to within 12 km (7.5 mi.) of Tayoltita, while Minas de San Luis, S. A. had constructed a second road from the mine to within 10 km (6.2 mi.) of connecting to another Sierra road system.

Since the early 1920's, year round access into the camp has been by light aircraft; this includes transport of personnel, foodstuffs, and most supplies and materials for the mine, mill, and smelter operations as well as the necessities for feeding and clothing over 10,000 inhabitants of the company-sponsored communities. The company supports a three-aircraft airline based at Tayoltita for this purpose.

During the dry season, (usually late in March), an 80 km (50 mi.) road is constructed down the bed of the Piaxtla River to the freight station of San Ignacio, Sinaloa. The road is usually washed out by heavy summer rains by the third week in June, and in the time it is open, all the heavy equipment and many supplies and hand tools for eight months of operation are freighted into the mine. This condition requires the mine to maintain an unusually large warehouse inventory throughout most of the year.

The climate of Tayoltita is sub-tropical with temperatures varying between 10°C (50°F) in winter to 40°C (104°F) in the summer dry season. The average annual rainfall is 700 mm (28 in.). Hurricanes from the Pacific coast will occasionally affect the operation adversely.

Minas de San Luis, S. A. produces all of the power for its mine, mill, and smelter operations, for the three principal towns, as well as for any new prospects or mining operations in the area. The power production amounts to 2,630 kw from five diesel-electric sets and 2,100 kw in from three small hydro-electric plants for a total of 4,720 kw (6,330 hp).

The company maintains about 400 two-, three-, and four-bedroom housing units for both worker and staff employees. These are provided rent-free with full utilities, and water service at no cost to the employee. The company also maintains both a clinic and hospital as well as a dental clinic.

The company operates and maintains the electrical networks as well as the potable water systems in the towns and housing areas. It also provides and supports a limited local telephone service. Minas de San Luis, S. A. also maintains a company store as well as subsidizing three union stores, all of which are operated on a break-even basis. It also maintains and subsidizes three public primary schools and a private primary school, which have a total attendance of over 1,200 children.

Production and Geology

The isolation of Tayoltita as well as the hostility of the natives prevented exploitation by the Spanish until the mid-18th century. Total production from the Tayoltita Mine since Spanish Colonial times is conservatively estimated at 12.5 million tons (13.8 million st), from over 150 veins. Bullion production is calculated at 9,800,000 kg (315,000,000 oz) of silver and 193,500 kg (622,200 oz) of gold. Current yearly milling is about 165,000 t (182,000 st) producing 61,200 kg (1,900,000 oz) and 950 kg (29,500 oz) of silver and gold, respectively. The mine is still Mexico's largest single gold producer after 79 years of uninterrupted operations.

The mine is operated on a 6-day work week, and the mill is on a 7-day basis. Mine production averages about 560 tpd (616 dry stpd).

The Tayoltita ore deposit is classified as an epithermal complex fissure vein system, hosted in bedded Cretaceous and Tertiary volcanics. These rhyolite and andesite flow rocks have been intruded by a premineral batholith complex and andesite stocks, as well as postmineral rhyolite and andesite dikes. The rock units are underlain by the Tertiary Piaxtla granite batholith, which is believed to have been the source of the precious metal mineralization for the veins. The mine's principal ore-bearing structure is the northeast trending Arana normal fault-vein. Most other economic veins intersect the Arana on it's footwall and lie to the west of it. The principal veins in the mine are the San Luis, 219, Arana, San Rafael, Cinco Senores, 904, 921, Escondida, Culebra, Candelaria and Cedral. Dips of the economic veins are rarely flatter than 60°, and widths vary from a fracture to plus 10 m (33 ft) with the minimum minable width at about 1.3 m (4 ft). The ore is localized in shoots which may vary in length from 15 m (50 ft) to plus 300 m (980 ft). The Arana block has been tilted by rotational faulting, thus as a result, the ore shoots rake at about 35° to the northeast.

The ore minerals consist of banded and/or disseminated argentite and tiny plates of electrum. The gangue minerals are mainly quartz and host rock fragments. Manganese is sometimes present as manganese oxides or the silicate mineral, bustamite.

Present mine ore reserves are in excess of 1.40 million t (1.54 million st) grading 450 of silver and 5.9 of gold per t (13.2 oz Ag and 0.172 oz Au per st). This ore is blocked on about 75 different veins and splits of these. Only chip samples from raises and drifts are used for blocking purposes, and no ore is ever blocked on diamond drill holes.

Mining and Milling

The mine, commenced on the vein outcrops near the top of Bolanos Peak, has been developed down-rake for a total inclined distance of 1,815 m (5,950 ft) or 1,214 m (3,985 ft) vertically below the top of the peak (Figure 3). There are more than 100 km (62 mi.) of accessible horizontal and vertical openings. Both the country-rock and veins are very competent, and only occasionally are shotcrete or roof bolt supports needed. The compressive strengths of the veins and host rocks are in the range of 35,000 to 50,000 psi.

Mining is presently done on 20 levels in an area approximately 5 km long (16,400 ft), 500 m wide (1,640 ft), and 1,240 m high (4,100 ft). This area is divided into four different production sections. Level intervals vary from 25 m (82 ft) in the older sections of the mine to 64 m (213 ft) in the newer sections. Access between levels in each section is accomplished through hoist-equipped vertical service winzes.

Most haulage of ore and waste is done with battery or diesel locomotives pulling 1- to 8-t (8.8-st) capacity cars. However, Wagner Scooptrams (both 0.83 and 1.66 m^3 capy) are used very successfully in some areas of the mine to transfer ore from stope draw holes to the ore passes or for rail car loading.

More than 500 m (1,640 ft) of development and exploration are driven in the mine each month. Main level tracked drifts are fully mechanized, employing Eimco 12-B, 21, or 22 mucking machines and locomotive support; however, the diesel Scooptrams are being used for new level development. Raising methods include mechanical boring, Alimak raising, bird-cage raising, but conventional bald headed raising over stulls is still the most common method employed. Drilling in both development and stopes is with Toyo stopers and jack-legs and integral steel. Blasts are initiated with No. 12 caps as hole primers, capped on a safety fuse. The DuPont blasting agent, "Mexamon," constitutes the rest of the hole charge with dynamite occasionally used in wet holes.

Shrinkage and cut-and-fill stoping are almost exclusively practiced at Tayoltita with 70 percent of the mine production obtained from about 20 shrinkage stopes, 13 percent from about 6 cut-and-fill stopes, and the rest from development. A stope is usually opened in the entire ore shoot, and little panel stoping is done. Stopes are flatbacked with both breasting and uppers drilling used (Figures 4 and 5).

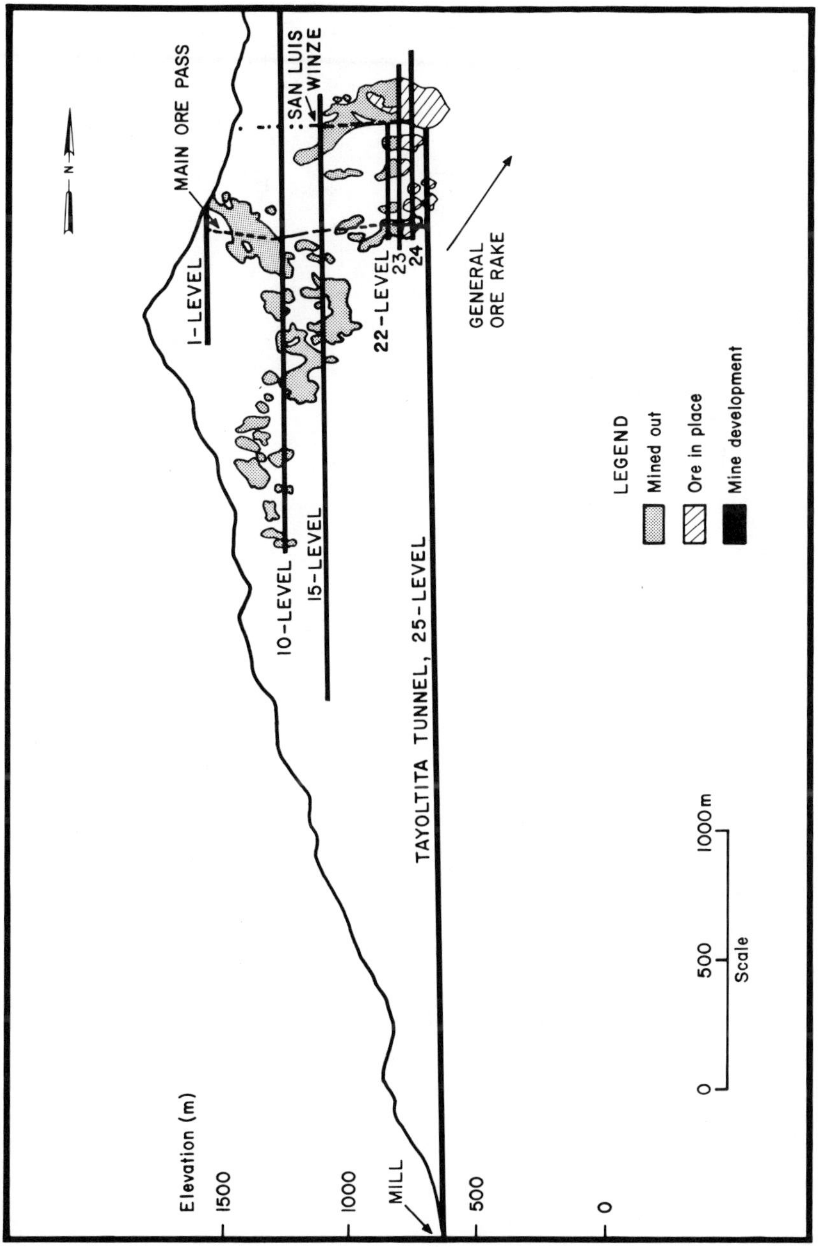

Figure 3. Vertical longitudinal section showing Tayoltita Tunnel related to Arana Ore Bodies.

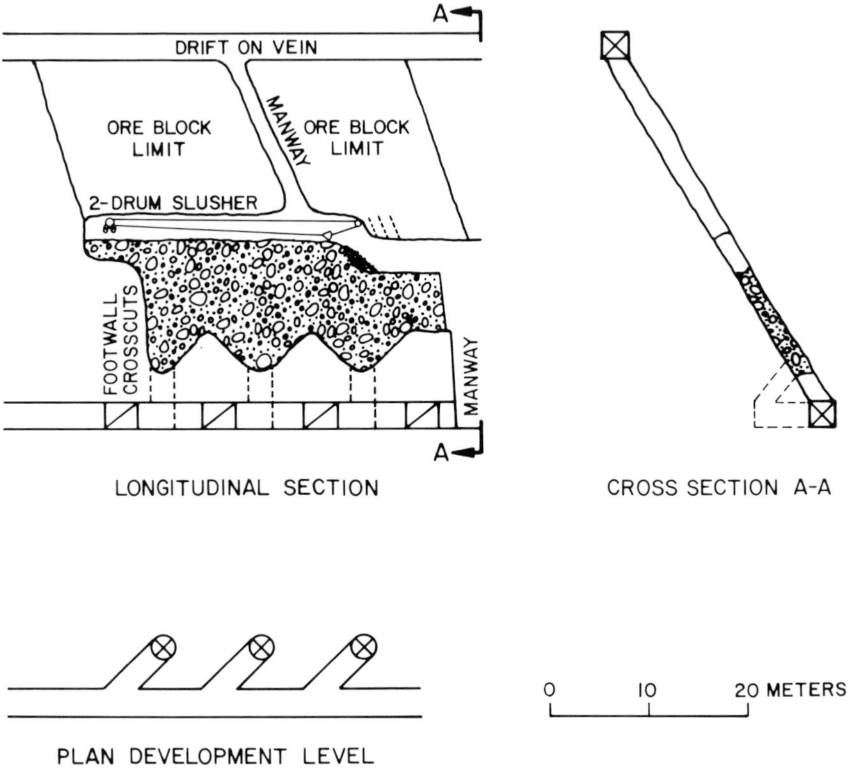

Figure 4. Idealized section showing shrinkage stoping and extraction with LHD's.

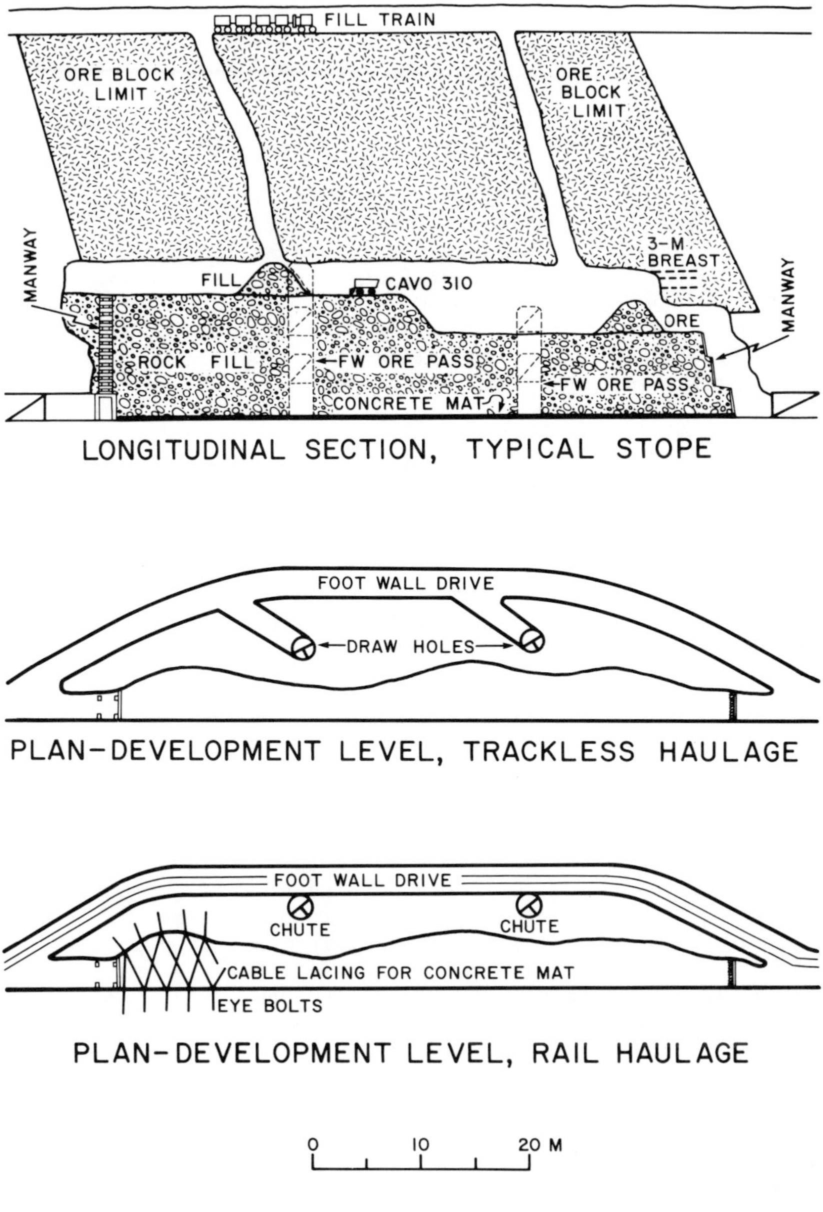

Figure 5. Tayoltita cut and fill stoping.

Shrinkage stopes average approximately 25 m (82 ft) in length and 1.5 m (5 ft) in width. The stopes are prepared over pillars, timber sets, or stulls, with timber chute extractions or are blasted directly into the discovery drift with drawpoint extractions, the latter being the preferred method. Double-drum air slushers are used in the stopes for leveling the pile for drilling. A manway and pipeway are usually carried behind a timber bulkhead on the end of the stope, and a second manway is carried in the development raise from the stope to the next level above.

Cut-and-fill stopes average about 40 m (130 ft) in length and 3.5 m (10 ft) in width. They are prepared over reinforced concrete mats poured in the floor of the original development drifts. A footwall drift is driven parallel to the stope, and extraction raises are driven on 30 m (100 ft) centers from the drift into the stope. These raises are excavated in the footwall as the stope advances upward, leaving alternate pillars and steel gob-fences against the stope fill. Ore is mucked into the footwall ore passes either with two-drum, air or electric slushers, or CAVO 310 air-powered L-H-D units. Fill consists of development waste rock which is also distributed in the stope with slushers or CAVOS. Manways and pipeways in the cut-and-fill stopes are either constructed in the same way as in the shrinkage stopes or are carried through cribbed raises constructed as mining progresses.

The maximum capacity of the mill is 530 tpd (583 dry stpd) on a minus 13 mm (1/2 in.) crushed product. The mill feed, which is 90% quartz, andesite, and rhyolite, is ground in cyanide solution to 70% minus 200 mesh. Lime is used as pH control and is added from a bin directly ahead of the mills.

The process is strictly counter-current cyanidation, employing Aero-brand black cyanide as the leaching reagent. Recoveries are approximately 95% of the silver and 96% of the gold. The total agitation time of the pulp is about 40 hours, and the total retention time in the plant is about 60 hours. Thickening of the pulp and decantation of the rich solutions takes place in three different steps, namely, directly from the ball mills, centrally in the agitation circuit, and on the tail-end of the circuit. The rich solution is filtered in leaf filters, de-aerated and pumped to precipitation. The values are precipitated with zinc dust and filtered in Merrill Crowe presses. Rough 33 kg (1,000 oz) bullion bars are smelted in Tayoltita, and these are flown weekly to the Met-Mex Penoles Smelter in Torreon, Coahuila.

Some of the production and cost figures established in 1980 are presented in Table 1.

Table 1. Data for Tayoltita Mine Operations, January - July 1980 (metric tons and U.S. dollars)

Tons mined per month in stopes	10,530
Tons mined per month in development	3,396
Total tons mined per month	13,925
Tons mined per underground man-shift	2.55
Cost per ton mined	$15.16
Meters development driven per month	561
Cost per meter of development	$130
Total tons milled per month	13,627
Cost per ton milled (including smelting)	$8.04
Tons per man-shift entire operation	1.11
Total cost per ton	$70.50
Bullion bars produced per month	165

Modernization Program Through the 70's

In the early 1970's, the company constructed a 4.4-km (3-mi.) mill-level adit (Tayoltita Tunnel) and primary crushing plant near the adit portal at a cost of $3.4 million. The adit was a single-entry heading and one of the longest ever attempted in Mexico. The 3-by 4-m (10-by 12-ft) adit provides a deep-level rail haulageway between mine and mill and eliminated an aerial tramway and an in-mine hoisting winze bottleneck. Currently 90% of all of the milled tonnage is pulled from three main ore passes which feed through steel chutes into the tunnel level haulage trains. One passway is over 1,000 m (3,300 ft) long.

The tunnel also provided an entry to the main service winze (former San Luis Shaft ore hoisting winze) as well as developing the vein system at the mill level (Figure 3).

The new crushing plant was designed to crush 1,100 tpd (1,210 stpd). This function combined with the elimination of the undersized aerial tramway and the interior winze hoisting bottleneck, as well as the introduction of high-capacity, high-speed rail haulage at the mill level, resulted in an automatic increase in average mill throughput from 400 tpd (440 stpd) to 460 tpd (506 stpd).

Exploration and development at the Tunnel level proved a 170-m (560-ft) long and 6-m (20-ft) wide ore shoot on the main San Luis vein as well as several other shoots on other veins. The ore developed on this level to date has been of a much better quality than average mine grade.

The adit provided a sorely-needed ventilation intake into the mine. About 2,800 m^3/min (100,000 cfm) of additional air enters the mine through tht Tunnel. This has aided in cooling the lower levels where ambient temperatures of 40°C (104°F) combined with 100% humidity are common, and, naturally, the efficiency and productivity of men working in this part of the mine has been improved markedly as a result.

Mine drainage has been channeled into the adit, and the lower levels are actually drying up as the Tunnel-level workings are driven under them. However, water cut on the Tunnel and other low levels of the mine is hot (45°C or 113°F) as is the country-rock, and more heat can be expected at depth. Attempts to deepen the mine below the Tunnel level will be extremely difficult because of the heat and humidity.

Among some of the underground projects undertaken since the Tunnel was placed in service in 1974 are the development of the main San Luis winze to the Tunnel level and conversion of the main service winze for the mine (Figure 6). The winze was extended and stripped (an Alimak raise was used as the pilot hole) between the tunnel and bottom of the winze, a total distance of 112 m (365 ft). Steel sets and an all steel manway were installed in the new section of winze and the timber in the old part of the winze stripped and replaced with steel (306 m or 1,005 ft). Still in progress is the extension of this winze to the surface (437 m or 1,435 ft) and the installation of a rebuilt Vulcan-Denver 1,300 by 2,400 mm (4 by 8 ft), two-drum 300-kw (400-hp) hoist for the shaft. This old two-hoisting compartment winze will then be converted to a single-hoisting compartment shaft upon removal of the center divider to accommodate a large two-deck cage. The inside dimensions of the cage will be 1.8 by 3.0 m (6 by 8 ft) with a 40-man capacity, and this will be balanced with a counterweight running through a 305-mm (12-in.) diameter pipe installed in the manway-pipeway-cableway compartment. The total cost of this project will be plus $2.5 million, and the service winze at 885 m (2,805 ft) in length will be one of the longest in any Mexican mine of comparable size.

An extensive ramp, ventilation borehole, and rock pass system have been constructed in the lower mine between the Tunnel and the 22 level. The ramp gradient averages 15 percent, and its total length is 1,110 m (3,630 ft). Its main purpose is to provide access for rubber-tired L-H-D's, jumbos, and service vehicles between the Tunnel (25) and 22, 23 and 24 levels. The cost of this project was about $200,000 including ancillary ventilation and rock pass raises and boreholes.

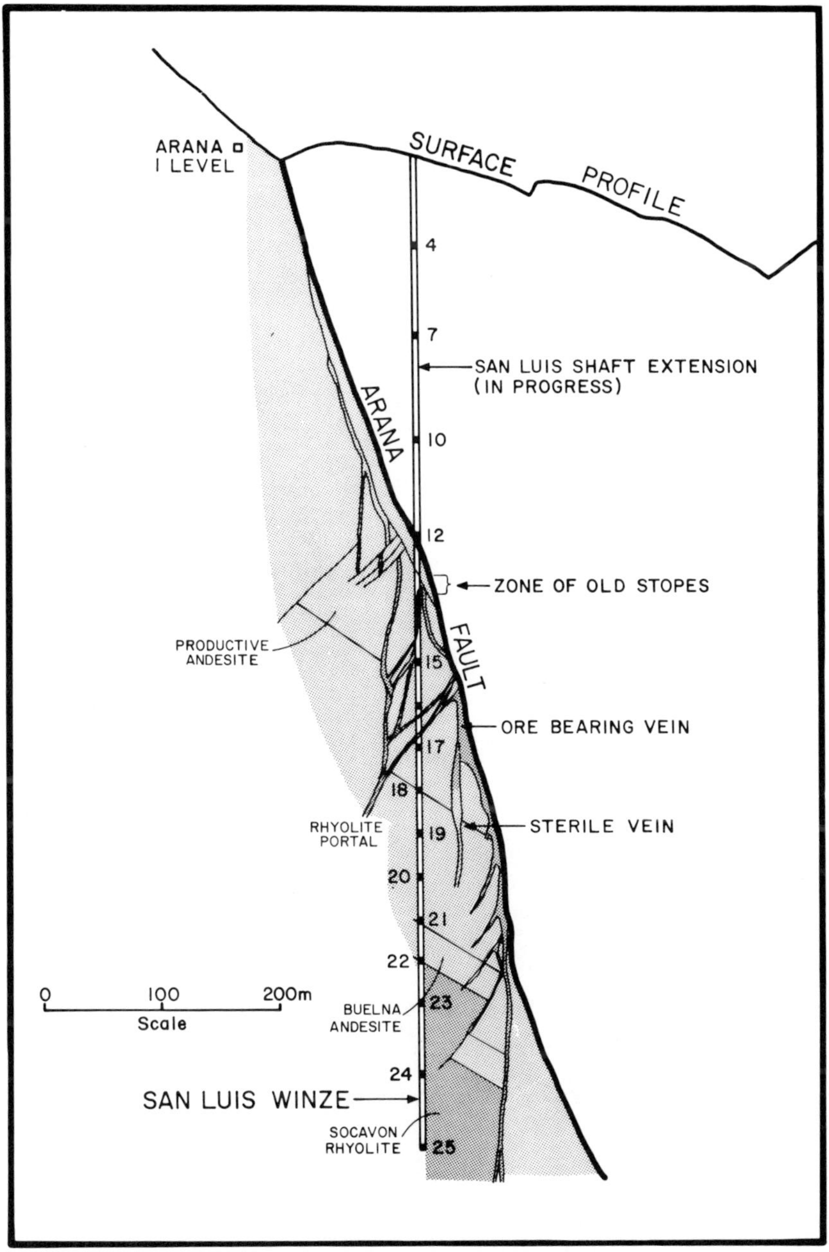

Figure 6. Geological cross section showing San Luis Winze & extension.

At the 10th level of the mine, (about 550 m or 1,760 ft above the Tunnel) a 2,500-m (8,200-ft) crosscut was driven to intersect the most northerly major vein, the Cedral, known in the area accessible from the mine. The crosscut, which cost over $2.5 million, intersected the vein in ore and to date a total of 290,000 tons at an average grade of 350 grams of silver and 2.5 grams of gold per ton have been developed (10.2 oz Ag per st and 0.07 oz Au per st). A second 1,900-m (6,230 ft) crosscut at the 15th level to explore the vein at greater depth is currently underway.

Mine equipment was also upgraded in the 70's. Over $3.0 million has been invested in this program, and some of the major items are shown in the following table:

Table 2. Major Items of Equipment Purchased for the Tayoltita Mine in the Last Decade

1	-	Robbins 61-R, A-C drive raise boring machine
2	-	Alimak STH-5 raise climbers
5	-	Wagner ST-2D, diesel laod-haul-dump units
4	-	Wagner HST - 1A diesel load-haul-dump units
1	-	MM20JB Jarvis Clark, diesel-powered rubber tired pneumatic jumbo
1	-	MM20R Jarvis Clark, rail-mounted pneumatic jumbo
1	-	Ingersoll-Rand 70^3m/min (2,500 cfm) XLE electric compressor
2	-	Atlas Copco Er 42m^3/min (1,500 cfm) electric compressors
4	-	Atlas Copco CAVO 310, air-powered L-H-D units
1	-	15-ton AGV diesel locomotive
1	-	12-ton Plymouth diesel locomotive
3	-	6-ton Hunslet diesel locomotives
1	-	6-ton Plymouth diesel locomotive
300	-	Toyo Ty-24 Jacklegs and stopers
2	-	9-ton AGV diesel locomotives
20	-	10-ton capacity Card Granby cars
1	-	Eimco 950 diesel service truck
1	-	Eimco 965 dump truck
2	-	Diamec 250 diamond drills
		Numerous mucking machines, battery locomotives, etc.

In addition to the projects in the mine, a small satellite operation, El Reliz, lying at 6 km (3.7 mi.) to the west of Tayoltita in the Piaxtla River canyon was brought on stream in 1975. Originally, the reserves of this narrow vein, hosted entirely in the granite basement rocks, amounted to 15,200 t of 313 g of silver and 5.6 g of gold per ton (7 oz Ag per st and 0.16 oz Au per st). However, over 25,000 t (27,500 st) of ore of similar grade have been extracted from the mine and trucked to the Tayoltita mill in the seasons when the short stretch of Piaxtla River road can be opened. The Reliz is still operating, and the current reserve is about 5,000 t (5,500 st). A few other operations of this nature, such as Camichin, are under development and should provide additional mill feed in the near future.

A major project, which can be developed by driving a total of 12 km (7.5 mi.) consisting of an adit and crosscuts into the barrancas to the southwest of the mill and at an elevation about 30 m (100 ft) below the mill level, is planned. This project would develop known ore reserves in the Castellana, Soledad, and Noche Buena Mines, as well as providing an access into the rich San Antonio and Contra-Estaca Mines (Figure 7).

Surface projects have included the construction of more than 66 two- and three-bedroom workers' housing units and 16 three-bedroom staff houses as well as a four-bedroom single-status staff house. A new 18-bed hospital facility was constructed in Tayoltita as was a 10-classroom primary school. A new 37.9 l/sec (600 gpm) potable water system for the towns was constructed and installed. A new 800-seat sports stadium is presently under construction, and a 150-man mine and mine office complex was constructed near the Tayoltita Tunnel portal.

An additional 400-kw (536-hp) hydroelectric generator will soon be installed at the company's Truchas hydroelectric plant. A trunk line from the Federal Power Commission is now under construction and will soon provide 1,000 kw (1,340-hp) to Tayoltita which could be expanded to 5,000 kw (6,700-hp) by changing 26 km (16.2 mi) of undersized power lines. This line passes through the Mala Noche mine area.

The company has invested about $25 million in Tayoltita over the last decade on both operations' improvements as well as projects of public benefit. This does not include investments in normal mine operations nor development and exploration projects in the mine and surrounding area. It appears as though management will be required to invest much more in the coming decade to complete the projects under development and in the planning stage; however, the returns from the effort in this rich old district are considerable.

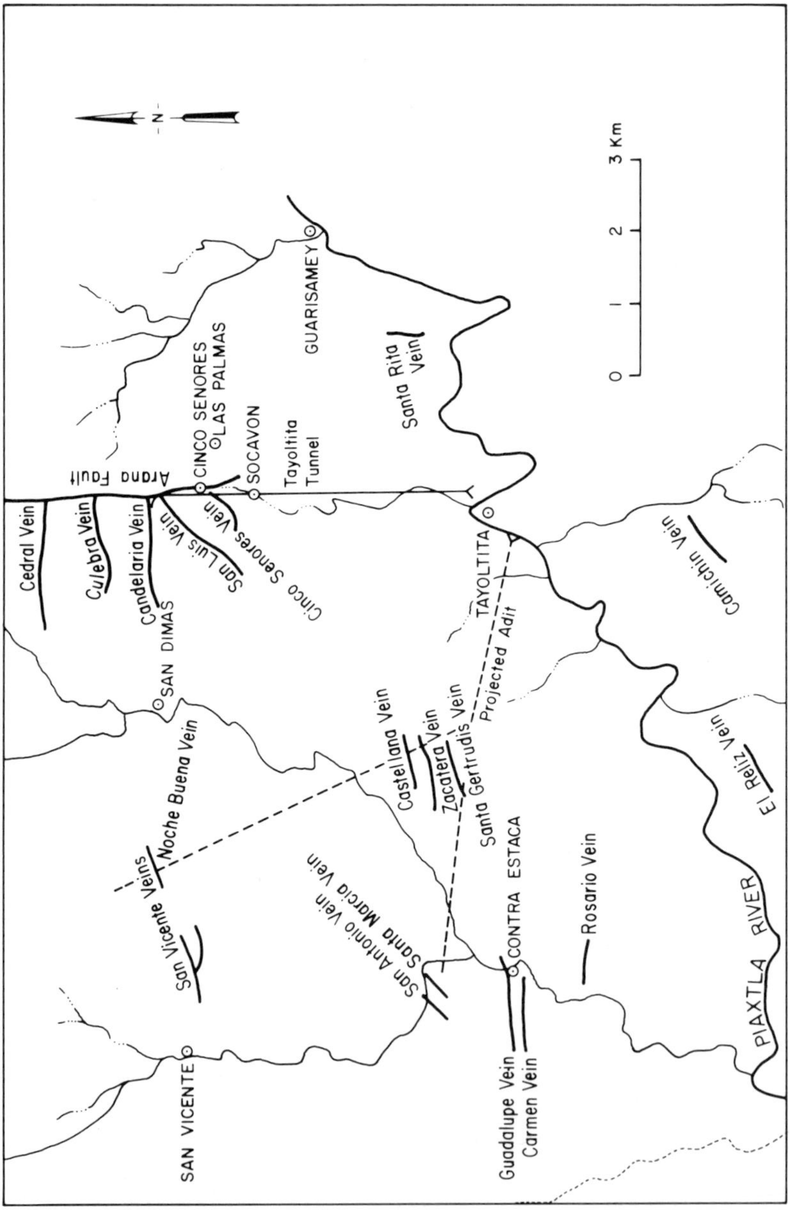

Figure 7. San Dimas Mining District.

LA LIBERTAD MINE

The small La Libertad underground silver mine was brought on stream by Minas de San Luis, S. A., in July 1977. Its planned life is only four years at an annual production of 14,400 kg (447,850 oz) of silver and 30 kg (930 oz) of gold. The mine is located in a deep, vertical-sided gorge at the edge of the barrancas of the Sierra Madre Occidental as they grade into the Mexican plateau of the state of Durango. It lies at 150 km (90 mi.) road distance from the city of Durango (Figure 1).

La Libertad's silver deposits were apparently not known to the Spanish, and the only previously recorded production is from the 1930's. At that time the operators of the Ventanas mining district extracted about 7,000 t (7,700 st) from the mine and shipped hand-cobbed ore via burro-back over 20 km (12 mi.) of steep, mountain trails to the Ventanas cyanide plant for processing.

Minas de San Luis, S. A. explored the mine in the late 60's and early 70's with direct drifting and raising and diamond drilling. A total of 176,200 t (193,820 st) of ore grading 400 g of silver and 0.8 g of gold per ton (11.7 oz Ag per st and 0.023 oz Au per st) were developed (Figure 8). In spite of the small size and low potential for additional ore, management decided to exploit the property, installing a used flotation mill from a defunct fluorite property.

Geology

The silver deposit of La Libertad is essentially one shallow fissure vein, the Santa Rosa, hosted within the lower series of the typical Capping Rhyolite formation covering much of the northern Mexican plateau. The vein splits, forming cymoid loop structures both in the hanging wall and footwall. Two ignimbrige units comprise the wall rock of the Santa Rosa and are known as La Libertad Rhyolite in the hanging wall and the Huizache Dacite in the footwall.

The economic silver mineralization is chiefly argentite, and the small amount of gold present is associated with pyrite. The gangue is chiefly quartz and calcite; however, much clay alteration is found within the vein and on the walls. In the upper 30 to 40 m (100 to 135 ft) of the economic portion of the vein, a mantle of oxidation and enrichment is found, and contaminants range from simple black manganese minerals to limonite, secondary clay, and altered vein material and wall rock. Also found in this zone is a very fine-grained, sooty argentite which is believed to be of supergene origin.

The dip of the Santa Rosa is 70° to the west, and the strike is essentially north-south. The vein tends to be lenticular, both in the horizontal and vertical sense, and width varies from 1.3 to 10 m (4 to 33 ft).

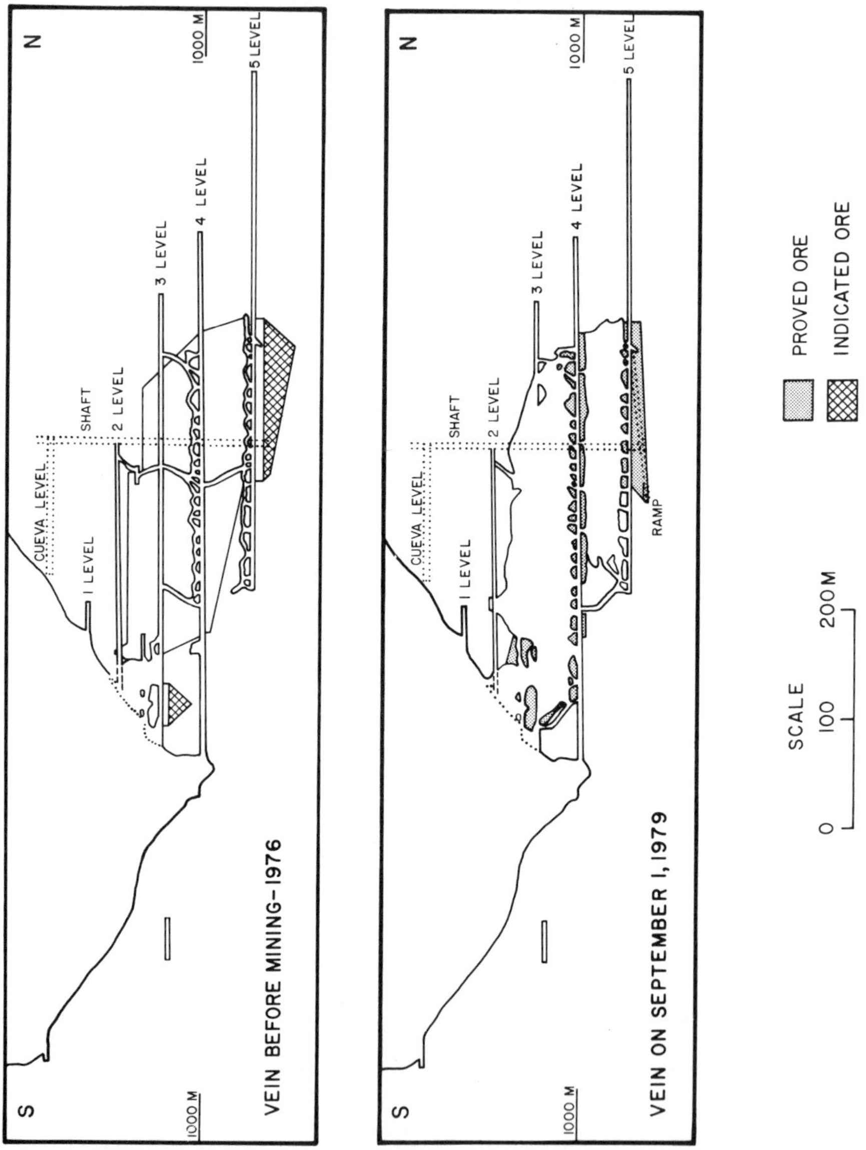

Figure 8. La Libertad, vertical longitudinal section of Santa Rosa Vein.

Mine Operations

The mine was developed originally through five adits from the bottom of La Libertad Arroyo. However, because the nearly vertical cliffs of this gorge made road access to the lowest (level No. 5) adit very difficult, it was decided for production purposes to hoist the ore to a level more easily accessible by road. Therefore, a 2.5-km (1.25 mi.) 10-percent grade access and haul road was constructed from the mill site down the face of La Libertad peak to the site of the new upper adit, baptized "La Cueva."

The excavations for the interior hoist room, rope raise, truck loading and turnaround drifts, and the coarse ore bin are located at the end of the 150-m (500-ft) long, (4.2- by 4.2-m) Cueva level. A vertical, 3-compartment, 280-m (925-ft) deep winze hoisting and service facility is portioned to form two hoisting compartments and a combination manway, pipeway, and power cable compartment. The hoist is a conventional Vulcan Iron Works, 2-drum, 910- by 1,220-mm (3- x 4-ft) unit driven by a 37.3-kw (50-hp) motor. The two skips for rock hoisting are rated at 1.2 t (1.3 st) capacity. One of the skips is removable and is replaced with a cage for raising and lowering men and materials.

The ore is hoisted from a 50-t (55-st) capacity pocket excavated 14 m (50 ft) below the lowest level of the mine. Ore mucked and trammed on level No. 5 is dumped into the pocket, while an ore pass between the 4 and 5 levels also discharges into the pocket via a steel chute. Hoisting is done on both day and swing shift. The ore hoisted to the pocket above the Cueva level is loaded directly into 10-5 (11-st) diesel dump trucks and trucked to the mill. The trucks are fitted with PTX scrubbers to protect the environment of the underground area.

Ventilation of the mine is accomplished with natural airflow. There have been no problems with the system, since all, except the No. 6 level, are adits providing the air intakes, while the shaft serves as the exhaust way. The mine is dry, and the relative humidity is low; thus the working environment is very comfortable.

The level intervals of the mine are 45 m (150 ft) with a total of five complete levels having been developed. Both levels 4 and 5 are tramming levels for the active stopes developed above them. A new 6 level is being developed from a minus 10 percent decline below the 5 level. Its purpose will be to extract the ore block in the floor of 5 level.

The ore body in the principal Santa Rosa vein above the No. 5 level has been essentially extracted in four large overhand shrinkage stopes. Several other smaller stopes have currently been mined on narrow hanging wall structures and remnants and pillars on the south end of the Santa Rosa. The major stopes averaged about 6 m (20 ft) in width. The stope walls, especially the footwall, are blocky, and dilution becomes a problem when the stopes are nearly drawn down.

Drilling is with light Toyo 24-L air leg machines and 22-mm (0.86-in.) hexagonal steel, either of the integral chisel-bit type or with knock-off 4-wing T-C bits, depending on how badly fractured is the face to be drilled. The stopes have been wide enough that breasting is almost exclusively employed, mining them in a rill fashion so that several breasts are available for drilling per shift. Face rounds are usually short because of the generally blocky, ravelly vein material and usually do not exceed 2.4 m (8 ft) in length and about 4.0 m (13 ft) in height. The average hole spacing in the breasts is about 0.8 m (2.5 ft) with the production per round at about 100 t (110 st). Blasting is with the commercial blasting agent, DuPont Super Mexamon D (an ammonium nitrate base blasting agent), primed with a No. 12 cap and standard safety fuse which is ignited by a thermalite ignitor cord. The powder factor is about 0.5 kg per ton (1.2 lb/st).

Slushers are employed in the stopes for leveling the pile for drilling. Manways are carried in the development raise from the upper level or through fenced-off raises on the ends of the stopes and carried up with the advancing stopes.

Muck haulage is achieved with three Eimco 0.8-m^3 (29-cu ft) 911 Diesel L-H-D units on levels No. 4 and No. 5, respectively. These units also do duty in mucking development and exploration headings on the levels and the decline. Development waste is trammed and dumped outside the adit portals on all levels.

Production data for the mine and mill through July 1980 are presented in Table 3.

Table 3. Statistics on Mined and Milled Tonnage at La Libertad in 31.5 Months of Operation (metric tons)

	1977 (5½ Mos)	1978	1979	1980 (7 Mos)	Total or Averages
Tonnage Mines	42,153	80,083	40,633	6,627	169,496
Tons Per Man-Shift UG	9.0	6.4	4.7	1.4	6.4
Tons Milled	15,390	38,882	41,594	22,320	118,186
Kilograms Silver Production	6,339	12,902	15,465	5,296	40,002
Kilograms Gold Production	12.7	27.8	36.0	11.7	88.2

All crushing is done with one 915-mm (3-ft) gyratory crusher which reduces the ore from minus 305 mm (minus 12 in.) to minus 13 mm (minus 1/2 in.). Crushed ore passes into the grinding circuit consisting of two ball mills with a total grinding capability of 224 t (246 st) per day.

The recovery circuit consists of two banks of 2.8-m^3 (10-cu ft) flotation cells. The first bank of 8 cells comprises the rougher cells, while the second bank of 6 cells comprises the cleaner cells. The products of all these cells are pumped to the 9.1- by 3.0-m (30- by 10-ft) concentrate thickener. The thickened concentrate at 18 percent solids is filtered with a leaf filter and subsequently dried for shipment. Concentration ratio varies between 20 and 35 to 1, depending on the contaminants in the ore.

Recoveries are on the order of 80 percent both for gold and silver, depending again on the contaminants in the mill feed as well as the quality of the heads. The La Libertad concentrate retains unusually high moisture because of the clay and insoluble contaminants which report to it. All attempts to reduce the moisture, including trial of different types of flocculants in the thickener and filter, have met with little success. A homemade infrared dryer system was tried, but the unit was dismantled when it was found that it could not handle the wet sticky concentrates. A rotary dryer is being installed at present, and moisture content of the final product is expected to be reduced from the present plus 30 percent to a manageable 10 or 12 percent.

The concentrates, containing about 8.0 kg (280 oz) of silver and 0.1 kg (3.2 oz) of gold per ton, are loaded with a front-end loader on contracted trucks. They are freighted at the rate of 5.5 dry tons (6 st) per trip via 13 km (8 mi.) of steep (plus 10 percent grades) unimproved mountain road over the Central Sierra plateau and over 400 km (248 mi.) of paved highway.

All assays for both mill control and mine sampling are done at the mine. Fire assaying is the only method employed.

The tailings are pumped to a 300,000 t (330,000 st) tailings area to the northeast of the plant. Decant water is returned to the plant for reuse. All other water for domestic, as well as industrial use, is pumped from the bottom of La Libertad arroyo via a 7.6 mm (3 inc.) pipeline to two 200,000 l (52,800 gal) storage tanks above the mill and a 16,000 l (4,230 gal) storage tank above the workers' village. The water is pumped by 150 l per min. (40 gpm) duplex pump against a head of 400 m (1,312 ft) a distance of 900 m (2,950 ft) to the mill tanks.

The tailings contain about 8.0 g (2.5 oz) of silver and 0.6 g (0.02 oz) of gold per ton and recent test work has shown that about 85 percent of both metals can be recovered by cyanidation. The tentative plan for re-treatment of the tailings upon completion of the mine is to install a cyanide circuit in the existing facility. The tailings from cyanidation would be conducted by gravity back into the worked-out mine which would be tightly sealed to avoid the filtration of cyanide solution into the surrounding arroyos.

Infrastructure by Company

The company has constructed 30 frame houses and 25 single rooms for the labor force, as well as 5 frame houses for the married staff and 9 rooms for single-status staff. The company has also constructed and maintained a school for the laborers' children, a meeting hall for the miners' union, a medical clinic, as well as a basketball court, and other facilities for social activities.

The construction of a 13-km (8-mi.) long, 2,500-KVA, 33,000-V power line was contracted with the Federal Power Commission. The line has been instrumental in the electrification of the collective lumber and agriculture communities (ejidos) through which it passes and will be left in place when the mine is finished. All company dwellings are furnished free of charge as is electricity, water, and fuel for cooking and heating.

The company also constructed a 485-m (1,600 ft) long air strip on a mesa overlooking the property. The strip is suitable for light, single-engine aircraft or STOL-type twin-engine planes.

The total number of laborers employed in the operation is 75, while the staff consists of 12 people.

Costs Meet Projections

The total capital investment including property acquisition and exploration costs originally budgeted for La Libertad was $2.8 million, while the actual investment to the date the mill went on stream amounted to $2.9 million, a variation of only 3.6 percent. However, the capital cost per ton of installed mill capacity was $12,900 based on a maximum capacity of 225 tons per day. Needless to say, the profitability of this very timely operation has been far above expectations. At present it appears that over 250,000 t (275,000 st) will ultimately be extracted from La Libertad.

MALA NOCHE MINE

The Mala Noche Mine is an old mine located in the Ventanas - Mala Noche mining district at an airline distance of 150 km (94 mi.) from the city of Durango, Durango. It is accessible by air from the mines of Tayoltita and La Libertad, landing in the Presidio River canyon on the short, 600-m (2,000-ft) long Palmarito air strip and, thence, a mule ride of 1.5 hours to the mine. The mine is also accessible by a 125-km (78-mi.) poor unimproved dirt road from the town of Estacion Coyotes on Mexico's Highway 40 which can take up to 20 hours. The road is narrow and winding with grades up to 20 percent (Figure 1).

The mine is situated at the confluence of the Concepcion Arroyo and one of its smaller tributaries about 4 km (2.5 mi.) to the north of the Presidio River. It is situated in country similar to Tayoltita with the exception that it is somewhat closer to the Mexican plateau, which lies immediately to the east. Again, climate and vegetation are quite similar to Tayoltita, the mine's surface plant lying at an elevation of 800 m (2,500 ft) above sea level in the barranca geomorphic province.

The early history of Mala Noche is somewhat obscure, but undoubtedly the Spaniards did find and exploit it. Early in the 19th century, the mine was worked by an English company but apparently was exhausted by 1906. This company operated a 50-tpd (55-stpd) cyanide plant in the town of Villa Corona bringing the ore from Mala Noche via an aerial tramway.

The mine was then acquired by the Ventanas Mining Company, which optioned the mine to the San Luis Mining Company in the early 1920's. Dr. Hugh McKinstry did a complete study on the mine and surrounding district aided and assessed by Dr. Donald McLaughlin. They made various recommendations for exploration targets in the Mala Noche Mine and for other district mines and prospects. However, economic conditions during this period were not conducive to starting up the mines, and San Luis dropped the project.

The mines were acquired in the 30's by Don Jose Aguilar, who operated Mala Noche until 1968, when a major landslide caused by torrential rains brought by a major cyclone wiped out his main diesel-electric plant and part of the flotation mill he had built at Mala Noche. The Aguilar family worked the mine, upon rebuilding after the 1968 disaster, on a very rudimentary basis. In 1979, Minas de San Luis, S. A. optioned the mine and commenced exploration work.

The historical production from Mala Noche through 1978 has been estimated at 250,000 t (275,000 st) which contained about 112,500 kg of silver and 875 kg of gold (362,000 oz Ag and 28,150 oz Au). Recoveries on this ore would probably reduce this to at best 90 percent of the above.

Geology

The geology of the mine is in a setting very similar to Tayoltita. The host rocks for the veins are extrusive layered rhyolites and adesites and the intrusive Ventanas Porphyry.

The Mala Noche vein is more of a vein system striking N 45° W and dipping 45 to 60° to the northeast. Most of the mine production has been derived from the main hanging wall vein which is in reality an Arana-type fault-vein with a very strong normal fault bordering the hanging wall. Footwall veins and splits have also contributed substantially to the production. The vein has been traced for more than 3 km (1.9 mi.) along its outcrop. The Mala Noche system is also epithermal.

The veins consist mainly of milky white quartz, but at times contain a great deal of breccia derived from the host rocks. Some rhodonite and rhodochrosite have also been identified. Ore mineralization is chiefly fine-grained argentite with minor polybasite and electrum. The silver to gold ratio in the ore is about 120:1.

The ore appears to be localized in shoots, and stopes of up to 220 m (720 ft) in length were mined in the old mine. The width varies between 1 m (3.3 ft) and 10 m (33 ft). Many normal cross faults complicate mining and exploration as well.

Exploration Project

The old mine had been developed down dip to the 8-1/2 level or about 280 m (925 ft) below the outcrop. However, after the 1968 flood, the lower workings below the 7 level became inaccessible. Upon examination of the mine on three different occasions, Minas de San Luis management decided to gamble and drive an 800-m (2,625-ft) crosscut 90 m (300 ft) below the 8-1/2 level and drift the intersected veins for at least 300 m (1,000 ft) both to the north and south, the idea being to develop new ore below the old mine and subsequently establish a mining operation similar to La Libertad (Figure 9).

During 1978 and 1979, the crosscut was driven by company crews and intersected the veins in ore where projected. Drifting on the hanging wall vein to date has developed a large block or ore averaging over 3 m (10 ft) in width beneath the old stopes on the north end of the mine. This ore carries significant values of silver and gold and is still open to the north and downwards. It looks as though San Luis is on its way to making another mine in this isolated rugged region.

ACKNOWLEDGMENTS

Grateful acknowledgment is made to Mr. Robert C. Morel, Director and General Manager of Minas de San Luis, S. A. de C. V., for his permission to publish this paper. Messrs. Carlos Cordova and Mike Clarke of the same company furnished valuable data for the paper.

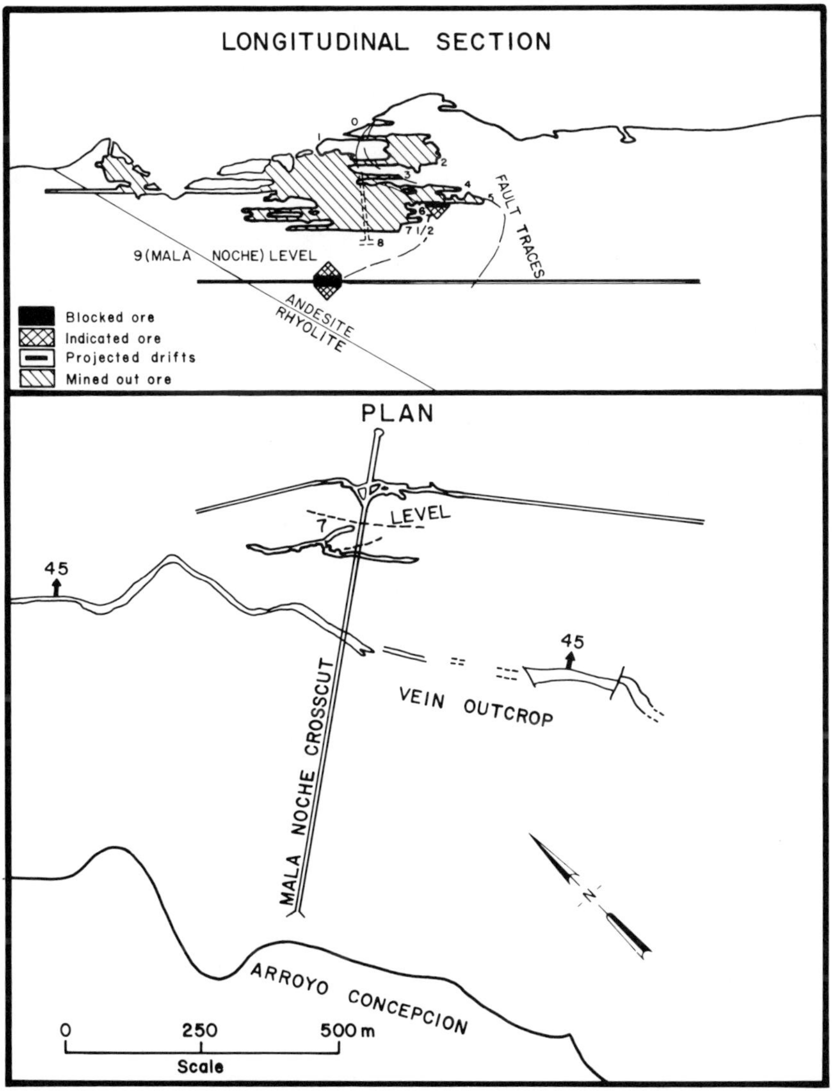

Figure 9. Longitudinal section and plan of the Mala Noche Mine.

BIBLIOGRAPHY

1. Henshaw, Paul C., "Geology and Ore Deposits of the San Dimas District, Sinaloa-Durango, Mexico," Unpublished Private Report, San Luis Mining Company, 1953.

2. Smith, D. M. and Hall, D. K., "Geology on the Tayoltita Mine, Durango, Mexico," Paper presented at SME Fall Meeting, September 22-25, 1974.

3. Bogert, John R., "Tayoltita: Mexico's Most Important Silver-Gold Mining Operation," Mining World, Vol. 25, No. 7, June, 1963, pp. 20-24, and No. 8, July, 1963, pp. 22-25.

4. Haptonstall, Jack C., "Taylotita Mill Level Adit Brightens Outlook for Mexican Silver and Gold Producer," Paper presented at National Western Mining Conference and Exhibition, Denver, Colorado, January, 1976.

5. Haptonstall, Jack C., "Modernization of the Tayoltita Mine, One of Mexico's Major Silver and Gold Operations," Mining Engineering, pp. 171-176, Feb., 1978.

6. Haptonstall, Jack C., "La Libertad: Making A Small Mine Work in Mexico," World Mining, Vol. 33, pp. 42-47, May, 1980.

Chapter 5

WHITEWOOD CREEK PLACER MINING

Fred Mosley and Maurice Hoffman

Strawberry Hill Mining Company
Whitewood, South Dakota

The present operation is located on Whitewood Creek about two miles downstream from Deadwood, South Dakota. Early day placer gold operations took place on this creek from Crook City to an area several miles upstream from Deadwood. Early discoveries were in the Deadwood area.

In this particular zone, much of the early work was accomplished at stream level and at various elevations above. In the original operations, it was not possible to reach bedrock in many places due to the ground water level plus the existence of numerous large boulders in the same area. The bedrock in this area is approximately twenty feet deep, consisting of Cambrian and Ordovician sediments.

Shortly after the early operations began, many small lode mines and several larger mines released mill tailings directly into Whitewood Creek. This practice continued for a full century, thereby limiting the amount of clean water available for sluicing. These tailings in the stream carried large amounts of heavy minerals along with the natural heavier minerals introduced from erosion and were immediately caught in the riffle boxes, resulting in the suspension of many small sluicing operations. No large scale sluicing or mining was ever done at bedrock level in this area, resulting in an almost virgin placer gold deposit. The old tailings cover much of the area, contributing an additional deposit carrying appreciable gold values.

In recent years, a company known locally as New Era Mining Company attempted with some success to exploit the area at bedrock level. The equipment used included a Marian dragline with a three-yard bucket, a washing trommel mounted on a 40 ft by 20 ft floating pontoon barge, together with two banks of Humphrey Spirals and a secondary spiral for collecting concentrate from the two primary banks. The use of the Marian dragline to lift the material from the bottom of the barge pond to the washing section was a dubious practice due to the loss of gold during passage up through ten feet of water.

Four years ago, the present Strawberry Hill Mining Company acquired the property and proceeded to operate the mine in a more efficient manner. The floating barge and the washing plant were removed from the barge pond, and the pond was drained by means of submersible units. The barge pontoon tanks were later used for water storage.

At that time, a D-8 Cat was employed, and an International Harvester front-end loader was put into use to transport the gravel to the washing plant.

The washing and recovery equipment now consists of the following: a Hewlitt Robbins Eliptex vibrating feeder, a conveyor 20 ft long to the washing trommel, a washing trommel and a 100 ft by 2 ft stacker conveyor belt. The washing trommel is constructed of 1/2 in. plate in five sections, each five ft long and a scrubber section five ft in length. The diameter is five ft, and overall length is 30 ft. One-half inch holes have been drilled in the plate to allow the undersize to pass into the sluices. Oversize carries through the trommel and discharges onto the 100 ft stacker conveyor.

The sluices are constructed in a series of shallow steel boxes on each side of the trommel. Ten sluices are on each side. In each sluice is placed a section of Astro-Turf carpeting, covered with expanded metal. These sluices in turn discharge onto a longitudinal riffle on each side of the trommel and collect any gold that by-passes the first riffles.

At regular intervals, depending upon yardage washed, the sluices are taken up and washed clean; expanded metal and Astro-Turf are removed; and concentrate is collected and processed in a Denver mechanical pan. The larger particles of gold are removed, and the finer gold and black concentrate are charged into a small amalgamating barrel. After grinding in mercury for several hours, the amalgam is recovered, retorted and cast into small bars together with the coarser gold.

Tailings are pumped into holding ponds in the vicinity, and the settled clear water is eventually returned to the stream. Large boulders and coarse gravels are conveyed and restored to the area worked previously by means of the D-8 Cat.

Plans are in progress for the construction of a milling facility to process 20-mesh product which can be recovered from the 1/2-in. trommel discharge, carrying from 0.10 oz./T to 0.15 oz./T gold.

Gold and Silver II

Session Chairman
Kenneth B. Hall
Homestake Mining Company
Lead, South Dakota

Chapter 6

METALLURGICAL PRACTICE AT HOMESTAKE, LEAD OPERATION

Harold L. Hinds and Larry L. Trautman

Homestake Mining Company
Lead, South Dakota

The Homestake Ledge or Lead was discovered by Fred and Moses Manuel on April 9, 1876. They recovered a small amount of gold from the quartz outcrop, but, unfortunately, they were prospectors not hardrock miners. Their claim was sold to a group of experienced California mining men in 1877. These mining men incorporated their holdings as the Homestake Mining Company. This was the beginning of one of the greater gold mining enterprises of the Western World. Metallurgical operations progressed over the years from an eighty-stamp grinding circuit using amalgamation for recovery to the more complex circuit of today.

The Homestake Mine, today, is an underground mine with the deepest winze reaching the 8,000-foot level. The ore is hoisted to the surface by two primary shafts, the Ross Shaft and the Yates Shaft. See Figures 1 and 2. It is crushed in three-stage crushing plants located in the headframes. From here, it is trammed to the mill fine-ore bins by electric locomotives.

The Lead grinding plant milling circuit, known as the South Mill, receives ore from the crushing plant at minus 1/2 inch (Figure 3). This material is dumped into overhead fine-ore bins from five-ton rail cars. The fine-ore bins have a live capacity of about 3,700 tons. The South Mill bins can also be fed by a conveyor system which reclaims ore from an 80,000-ton capacity stockpile. This stockpile is built up from peak mining production periods and available surface low grade material. It is then reclaimed during low mine production periods.

The grinding circuit at the South Mill consists of four rod mills in open circuit followed by four ball mills in closed circuit. Three of the rod mills are 6-foot by 12-foot Marcy's. The other is

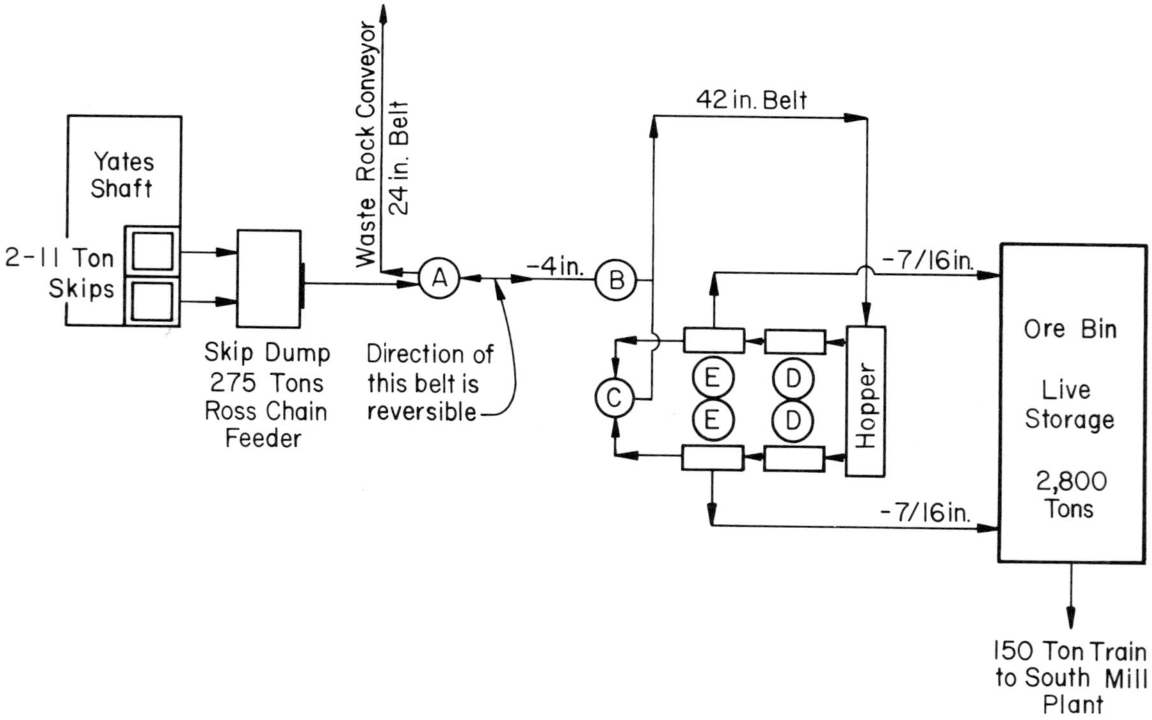

Figure 1. Flowsheet for the Yates Shaft Plant.

A. 30 in. Allis-Chalmers gyratory crusher.
B. 7 ft. standard Symonds cone crusher (2).
C. 7 ft. short-head Symonds cone crusher (2).
D. Jeffrey-Taylor 42 x 120 in. vibrating feeders (2).
E. Tyler 5 x 12 ft., F-600, 2 deck screen (4).

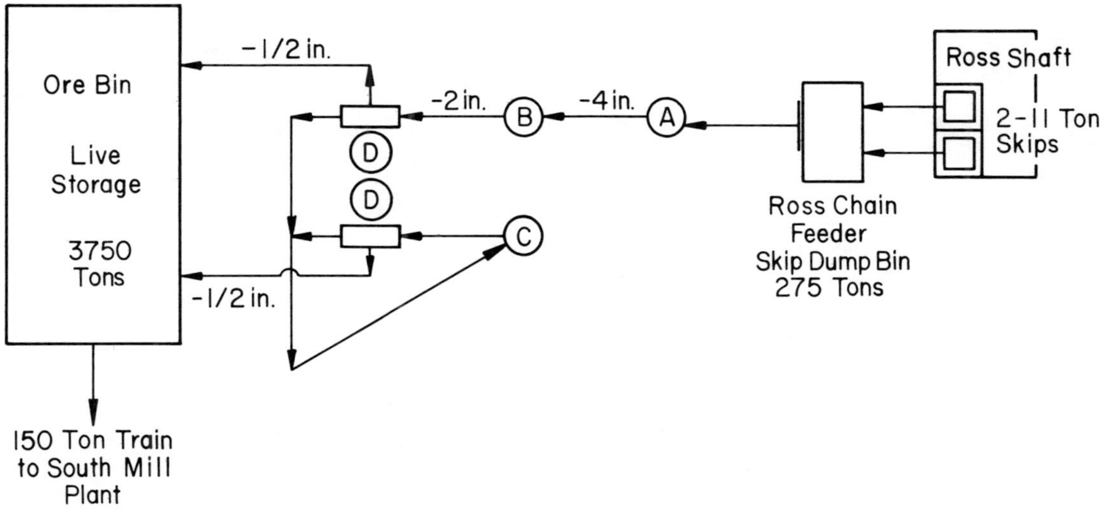

Figure 2. Flowsheet for the Ross Shaft Plant.

A. 30 in. Allis-Chalmers gyratory crusher.
B. 7 ft. standard Symonds cone crusher (2).
C. 7 ft. short-head Symonds cone crusher (2).
D. Jeffrey-Taylor 42 x 120 in. vibrating feeders (2).

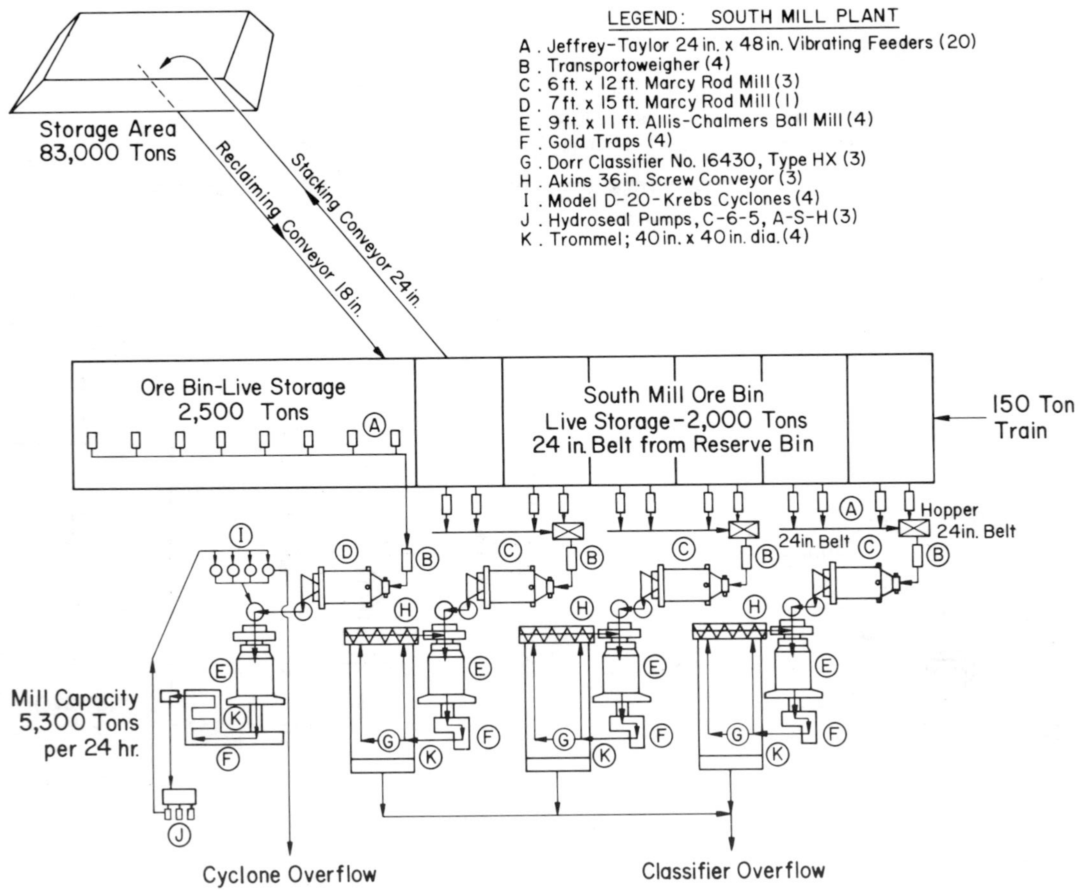

Figure 3. Flowsheet for the South Mill Plant.

a 7-1/2-foot by 15-foot Marcy. All of the ball mills are 9 feet by 11 feet Allis Chalmers.

Ore is fed to the rod mills by means of automatic Jeffrey-Taylor vibrating feeders at the bottom of the ore bins. The feeders are in turn controlled by Transportoweighers. The ore is mixed with water at the rod mill feed spouts. Rod milling is done at 77 to 79 percent solids. No chemicals are added in the milling circuit. The rod mills are in open circuit, and the ore passes through and goes directly to the ball mill scoop. Rod and ball mills are set up as units. Number one rod mill feeds number one ball mill; number two rod mill feeds number two ball mill and so on down the line. Size reduction by the rod mill is given in Table 1. Rod usage in the rod mills is about 0.5 pounds per ton of ore. New rod size is 2-15/16 inch diameter.

Ball milling takes place in closed circuit. Primary classification on the first three units is accomplished with Dorr rake classifiers. The fourth unit primary classification uses cyclones. Circulating loads average 500 percent. Balls used are 35 percent 1-1/2-inch and 65 percent 2-inch diameter. Ball mills are run at 80 percent of critical speed. Ball usage is about one pound per ton of ore ground. Primary classifier overflow from the ball mills is 65 percent minus 200 mesh. This material then goes to secondary classification where it is separated into sand and slime fractions. Secondary classification is done by nine D-20B Krebs cyclones. Sizings of these fractions may be noted in Table 1. The cyclone underflow, or sand fraction, goes directly by gravity to the sand leach plants, and the cyclone overflow or slime fraction goes to thickening and then to the carbon-in-pulp plant. Lead ore contains a certain amount of gold which is liberated by grinding. Until 1970, about 65 percent of the total gold was recovered by amalgamation. When it became necessary to discontinue amalgamation, the total recovery declined. Larger particles of gold were reaching the cyanide circuit and did not have enough contact time to dissolve totally. It was evident that some type of gravity separation was necessary for recovery of these larger particles.

The gravity recovery system in use today consists of a series of sumps and riffles after the ball mill discharge. Riffles are constructed of expanded metal at different heights from the bottom of the launders. Once a week the mill is shut down, and a cleanup is made. This high-grade sand is then run over a half size gravity table whose concentrate will run about ten ounces per pound dore'. Gravity separation now accounts for 28 to 30 percent of total recovery. Testing is continuing for a means of improving gravity recovery.

Table 1. SIZING OF HOMESTAKE MILLING PRODUCTS

SIZE	FEED TO OPEN CIRCUIT ROD MILLS	ROD MILL DISCHARGE UNITS 1 - 2 - 3	ROD MILL DISCHARGE UNIT 4	BALL MILL DISCHARGE UNITS 1 - 2 - 3	BALL MILL DISCHARGE UNIT 4	BALL MILL CLASS.&PRI. CYCLONE O'FLO FEED TO SEC. CLASS No. 1 - 2 - 3	BALL MILL CLASS.&PRI. CYCLONE O'FLO FEED TO SEC. CLASS No. 4	SECONDARY CLASS. PRODUCTS SAND E. Plant	SECONDARY CLASS. PRODUCTS SAND W. Plant	SECONDARY CLASS. PRODUCTS SLIME
	% Wt.	% Wt.	% Wt.	% Wt.	% Wt.	% Wt.	% Wt.	% Wt.	% Wt.	% Wt.
+ 3/4"	2.0									
+ 1/2"	16.0									
+ 1/4"	28.5									
+ 8 M	19.3	9.3	0.6							
+ 10 M	4.2	7.7	1.4	2.0	0.6					
+ 28 M	8.8	27.6	20.9	12.2	7.5					
+ 48 M	3.2	11.2	15.8	16.0	18.5	0.0	0.0	0.0	0.1	
+ 80 M	3.4	6.4	9.6	22.0	25.9	3.1	4.0	1.5	4.9	
+ 100 M	1.1	2.3	3.5	9.8	9.6	5.0	5.5	3.7	5.2	
+ 150 M	2.0	4.3	5.3	10.8	11.8	11.6	11.7	14.6	15.0	
+ 200 M	2.6	5.9	8.4	7.8	10.0	16.0	15.3	25.9	24.7	0.3
+ 200 M	8.9	25.3	34.5	19.4	16.1	64.3	63.5	63.5	50.1	99.7
TOTAL	100.0	100.0	100.0	100.0	100.0	100.0	100.0	100.0	100.0	100.0

Milling rate = 5,350 tons/day (yearly average)
Cyanidation of sand = 59.3% of tonnage - 3,173 tons/day
Cyanidation of slime = 40.7% of tonnage - 2,177 tons/day
Rod mill solids = 74–76%
Ball mill solids = 67–70%
B/M classifier overflow solids = 17–18%
B/M circulating load = 500%

CYANIDATION

Processes used to remove gold and silver from the precious metal-bearing ore and from precious metal-bearing solutions are referred to as cyanidation processes. Sand vat leaching, carbon-in-pulp and carbon cones constitute the cyanidation processes at Homestake.

The sand vat leaching process has been used at Homestake since 1901. Approximately 55% of the gold is recovered with this process. The process involves treatment of the coarse fractions, called sands, from secondary classification with cyanide-bearing solution in a leaching vat. There are thirty-five leaching vats in two plants (Figure 4). Twelve leaching vats are located in the West Sand Plant that treat 800 tons of sand per vat. The East Sand Plant contains twenty-three leaching vats with approximately 775 tons of sand per vat.

Each sand vat has a cocoa matting and canvas bottom stretched over two-inch by four-inch notched support lumber to make a false bottom. The false bottom is needed to allow the percolation of solution through the sand bed and through the cocoa matting and canvas bottom to collection sumps.

The sand plant's leaching cycle consists of filling, draining, aeration, strong solution leaching, washing and sluicing of each leach vat. The total treatment cycle takes approximately eight days.

After a vat is filled with sand, most of the water included in the filling is drained through the bottom. Following draining, compressed air is forced upward through the sand to saturate the vat with oxygen needed in the gold dissolution reaction. Cyanide solution at approximately 0.05 wt.% strength is then added to the top of the sand. The treatment cycle consists of four drain, air and cyanide solution periods. After the cyanidation period, the sand vat is sampled and then sluiced to the sand dam where the sand can be put back underground for backfilling stopes or sent to the Grizzly Gulch tailings disposal dam. A sand plant treatment cycle is explained in detail in Appendix A.

The solutions are collected during the cyanidation periods of the leach cycle in 250-ton pregnant solution storage tanks. The solution is then passed through a Merrill-Crowe zinc precipitation process. The zinc precipitation process consists of adding fine particles of zinc dust to a de-aerated gold-bearing solution to precipitate the gold. The gold-bearing precipitate is filtered from the solution in plate and frame filters. The solution is then returned to storage tanks where the cyanide solution strength is raised back to the 0.05 wt.% starting strength and reused in the leaching cycles. The precipitate is then removed twice a week from the presses and treated in the refinery to remove the gold.

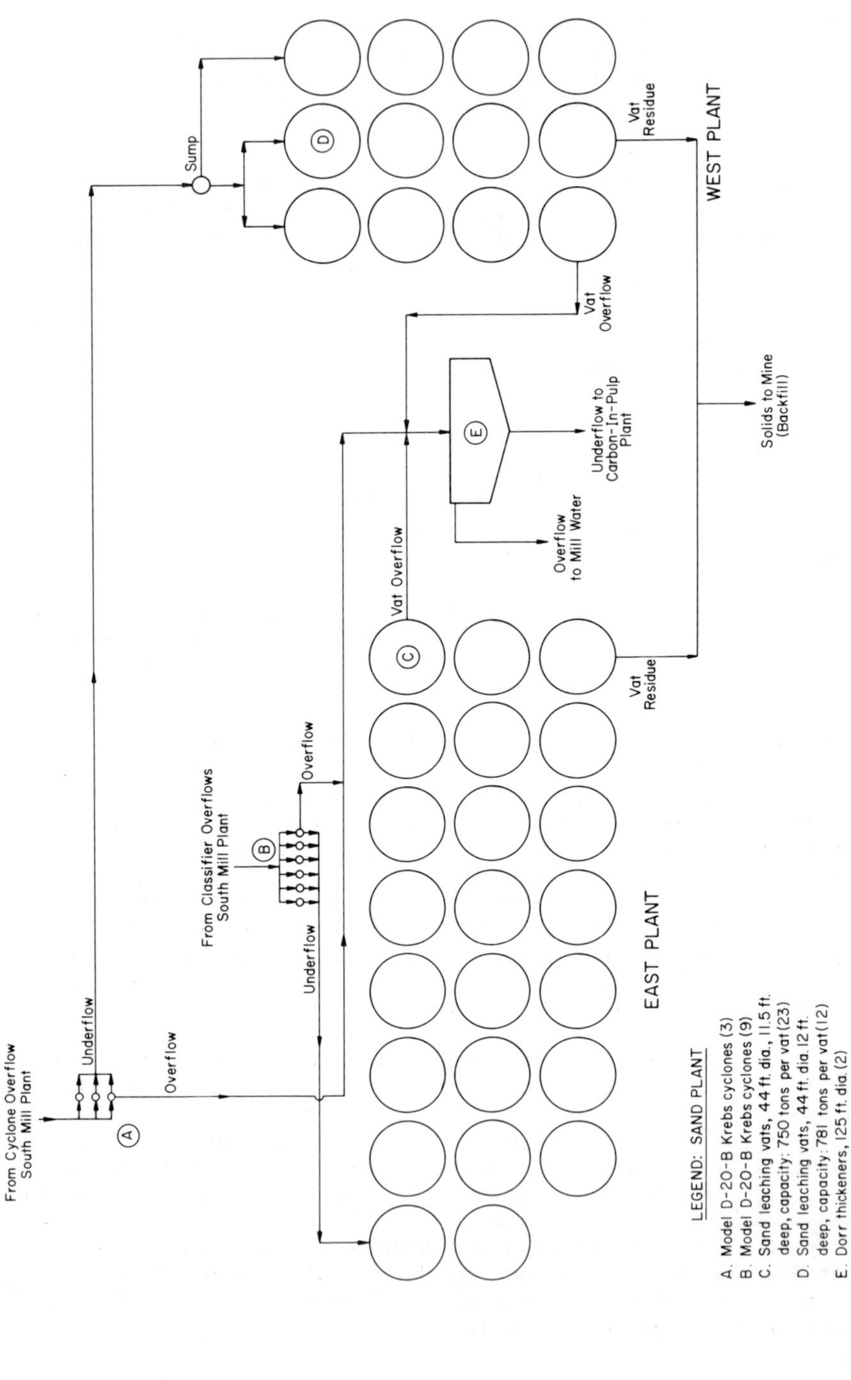

Figure 4. Flowsheet for the East and West Sand Plants.

The second phase used in the cyanidation process is the carbon-in-pulp process which treats the fine fraction, or slime, from the secondary classification.

For many years the slime fraction was treated in the plant in Deadwood, South Dakota. The slimes were pressure leached with cyanide solutions in thirty Merrill filter presses. Gold contained in the pregnant solutions was precipitated with zinc dust. The precipitate was then sent to the refinery.

In August of 1973, the carbon-in-pulp process replaced the seventy year old slime plant. Operation of the carbon-in-pulp plant is on a continuous 24-hour seven day per week basis. Slimes from the grinding circuits which average 0.06 ounces of gold per ton of ore are thickened in two 125-foot Dorr thickeners to approximately 45% solids and then pumped to the carbon-in-pulp plant (Figure 5).

The carbon-in-pulp plant consists of conditioning, dissolution, adsorption, desorption and reactivation circuits. In the conditioning circuit of the carbon-in-pulp plant, slimes are passed over two-feet by four-feet Denver-Dillon vibrating screens with 28 mesh stainless steel cloth to remove wood chips from the slimes prior to treatment. The screen undersize flows to one 18-foot diameter by 20-foot high tank with propeller agitation to condition the slime with air and lime. Lime is added at a rate of about two pounds per ton of ore treated.

Following conditioning, the slime pulp flows to a dissolution feed sump where sodium cyanide is added to a cyanide strength of 0.04% to 0.07%. From the sump the slime pulp is pumped to one of seven 30-foot diameter by 22-foot deep Denver rake type agitators. The slime pulp flows by gravity from tank to tank and is agitated and aerated by sparge rings and center air lift columns. Dissolved oxygen is measured and controlled to satisfy the oxygen demand required in the dissolution of gold. Recovery is attained in approximately 20-hours with closely controlled cyanide strength and pH adjustments. Undissolved gold in the pulp from the circuit averages 0.005 ounces per ton. At this point the gold in solution will average 0.04 ounce per ton of solution.

From the last tank in the dissolution circuit, the pulp is pumped to the first of five adsorption tanks where it comes in contact with activated coconut carbon.

The adsorption circuit at Homestake was modified in recent months from the original design because of adsorption problems and higher tonnages. That is, one of the conditioning tanks was converted to an adsorption tank.

The adsorption problems have been associated with the recycle water system. Homestake experimentation has determined that calcium carbonate loading on the carbon caused gold adsorption problems.

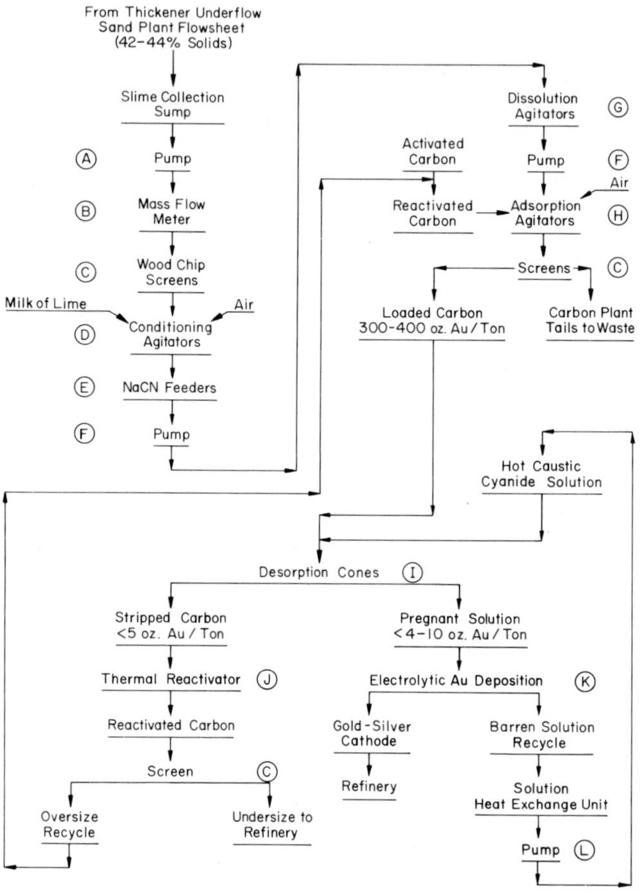

A. Ash 6 in. centrifugal pump with automatic controls
B. Fischer-Porter flowmeter with Chicago-Nuclear density gauge
C. Denver-Dillon 2 ft. x 4 ft. vibrating screens, 24 mesh
D. 18 ft. dia., 20 ft. deep conditioner, propellor type agitator
E. "Vibra-Screw" variable speed belt feeder
F. Ash 6 in. centrifugal pump
G. 1 Tank 35 ft. dia., effective height 16.5 ft. 6 Tanks 30 ft. dia., effective height 20 ft. (with center air lift and nozzle type air sparges located near bottom of the tank)
H. 5 Tanks in series, 18 ft. dia., effective height 13.7 ft.; air lift to screens
I. 3 Zadra cones; 2,000 lb. carbon/charge
J. Gas fired rotary kiln; 12.6 ft. long, 20 ft. dia.
K. 3 electrolytic gold desposition cells in series
L. 1 in. Worthington centrifugal pump

Figure 5. Flowsheet for the carbon-in-pulp plant.

Either periodic acid treatment of the carbon or phosphate additions to the slime pulp has minimized the calcium carbonate problem. Preliminary metallurgical testing has also indicated recycle water temperature could affect carbon loading.

Also, with higher tonnage of lower grade ore being treated, the use of the additional adsorption tank makes it possible to do preventative maintenance on the circuit without gold losses and tonnage losses.

The adsorption circuit consists of one adsorption tank containing propeller and air sparging agitation. The remaining four tanks contain center column airlift and air sparge agitation. Slime pulp is moved through the tanks from No. 1 to No. 5, and the activated carbon is moved counter current to the slime pulp flow. The slime pulp and activated carbon are elevated by outside airlifts from each tank to Derrick double deck vibrating screens which have 22 mesh stainless steel cloth. The slime pulp is separated from the carbon when it flows through the screens, and the carbon is retained on the screens as oversize and circulated back to the adsorption tank. Carbon is advanced on a batch basis from tank to tank when carbon is pulled from the No. 1 tank to the desorption section. Reactivated carbon is returned to the circuit in tank No. 5.

A typical profile from the adsorption circuit is as follows:

ADSORPTION PROFILE

	Loaded Carbon (oz gold/ton carbon)	Solution Values Tank Outflows (MO Au/ton solution)
Adsorption Plant Feed		30.0
No. 1 Adsorption Tank	150	12.0
No. 2 Adsorption Tank	90	6.5
No. 3 Adsorption Tank	55	4.5
No. 4 Adsorption Tank	20	1.5
No. 5 Adsorption Tank	10	0.5

Total carbon losses from the plant average 0.04 pound of carbon per ton of slime treated. The carbon losses are composed of carbon that passed the 22 mesh screens in the adsorption section and carbon removed at the reactivated sizing screen. The fine carbon accumulated over several months is sent to a smelter for recovery of gold that is on the carbon.

After the carbon has been loaded to 150 ounces per ton, it is educted to a 1-1/2-foot by 3-foot vibrating wash screen to remove the slimes that coat the surface of the carbon and prevent the desorption of the gold.

Desorption of the carbon consists of passing a solution containing 1.0 wt.% sodium hydroxide and 0.2 wt.% sodium cyanide at 190°F at a flow rate of 10-12 gallons per minute through three cylindrical stainless steel desorption vessels. The vessels are cone shaped on the bottom and have a capacity of about one ton of carbon. Two cone vessels are used during stripping, while the third vessel is being filled with carbon for later stripping.

Once the solution has passed through the carbon beds the solution will then pass through three stages of Zadra electrolytic cells. The gold is deposited on cathodes containing 8 to 10 pounds of medium grade steel wool per cathode.

After the solution passes through the electrolytic cells, it is returned to the boiler and reheated for reuse in the stripping step. About 50 hours of time are required to strip the carbon from 150 ounces per ton to less than five ounces per ton.

About 600 to 700 ounces of gold are removed from the No. 1 Zadra cell each week. The steel wool in the No. 2 and No. 3 Zadra cells are moved up to the No. 1 and No. 2 Zadra cells. A new cathode of steel wool is then installed in the No. 3 Zadra cell.

During the time the activated carbon is in the process, the carbon adsorbs materials which are not stripped off during the desorption steps. Therefore, a loss of gold adsorption activity occurs if the carbon is sent back to the adsorption step without going to reactivation.

Reactivation is accomplished by heating the carbon in the absence of air to a temperature of 1100°F. Each batch of carbon is reactivated in a gas-fired rotary kiln. The kiln consists of a rotary horizontal stainless steel tube 20-inches in diameter and 12 feet 7-inches in length, which is heated externally by a gas-fired rectangular-shaped firebox. About seventeen hours are required to reactivate one ton of carbon.

Once the carbon has been air cooled, it is passed over a sizing screen of 16 mesh screen cloth. The undersize is collected and sent to a smelter. The oversize is educted back to the No. 5 adsorption tank for reuse.

Carbon cone tanks are being used as the third form of gold recovery system. Homestake is recovering gold from decant water from our Grizzly Gulch impoundment. We have additional gold being dissolved in small amounts during the pumping of carbon-in-pulp tailings and sand plant tailings to Grizzly Gulch. Homestake presently has 8 four-ton carbon cones in use, and four additional carbon cones are in design stage. We are loading the carbon to about 40 ounces per ton before removing it from the circuit. The forty ounces per ton on the carbon is about maximum loading because iron and copper cyanide complexes also occupy carbon adsorption sites. The carbon is presently being shipped to a smelter. Homestake is in the process of designing a carbon strip system to handle the water treatment carbon.

The carbon cone tanks are 8 feet in diameter and 17 feet in height. Each tank has a screen box to dewater the carbon. The flow of water through the carbon is maintained at a rate to very slightly fluidize the carbon bed.

SMELTING AND REFINING

Feed to the Lead Refinery consists of gold-zinc precipitate from the sand leach plants, gold plated steel wool from the carbon-in-pulp plant and high grade concentrate from the South Mill gravity circuit. These three sources are treated as separate entities through the refining operation. See Figures 6, 7 and 8. This is done for calculations of efficiencies and recoveries of each plant.

Gold-zinc precipitate comprises about 55 percent of the refinery feed. It is cleaned from filter presses and transported to the refinery in locked containers. This precipitate contains about four ounces of gold per pound. After being sampled, the precipitate is dried to around ten percent moisture and mixed with flux. Homestake does not acid treat or roast the precipitate but elects to flux off impurities with a strongly oxidizing acid flux. The flux used consists of soda ash, silica, sodium borate and sodium nitrate.

The fluxed precipitate is compressed into ten pound briquets and charged into a natural gas-fired 26-inch Monarch-Rockwell single-chambered furnace. Total capacity of the furnace is 500 pounds. Melting temperatures are 2,000 to 2,200°F. Time for each melt is approximately two hours. During this time the base metals have been oxidized and taken up in the slag. When the melt is complete, the slag is poured off and the dore' bullion collected. This bullion, assaying 800 fine gold and 170 fine silver, is weighed, sampled, and stored for further refining.

Steel wool cathodes from the carbon-in-pulp plant are treated in much the same manner as the gold-zinc precipitate. The cathode consists of eight to ten pounds of medium grade steel wool which is loaded with 600 to 700 ounces of dore' bullion. The plated steel wool is fluxed with sodium nitrate, silica and sodium borate and melted in a Monarch tilting furnace. Melting time is one hour at 2,000 to 2,200°F. The melted charge is poured into a cone mold, allowed to cool, and the bullion recovered. The resulting bullion

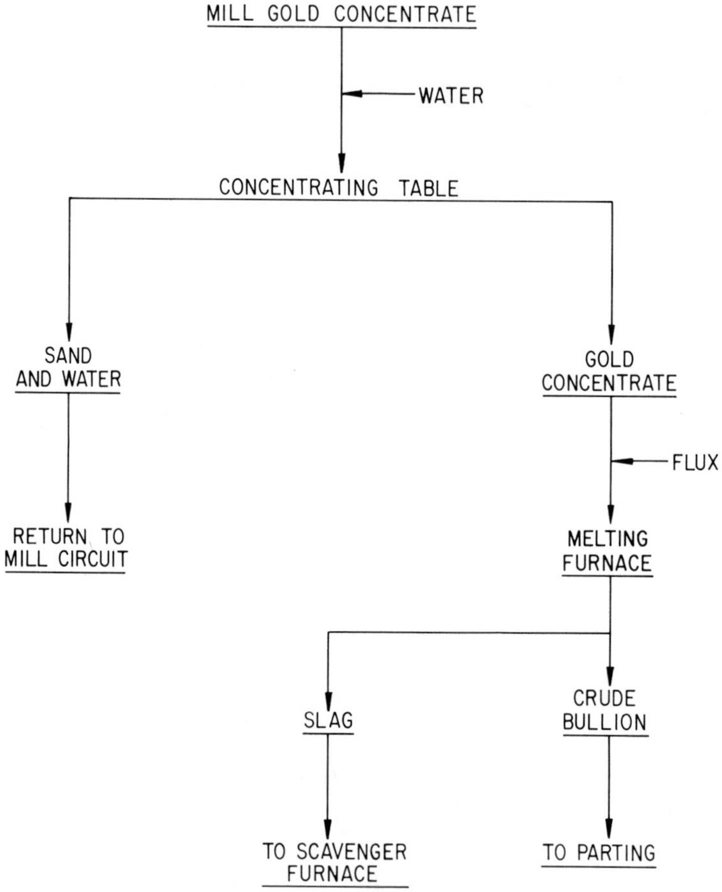

Figure 6. Flowsheet for smelting the mill gold concentrate.

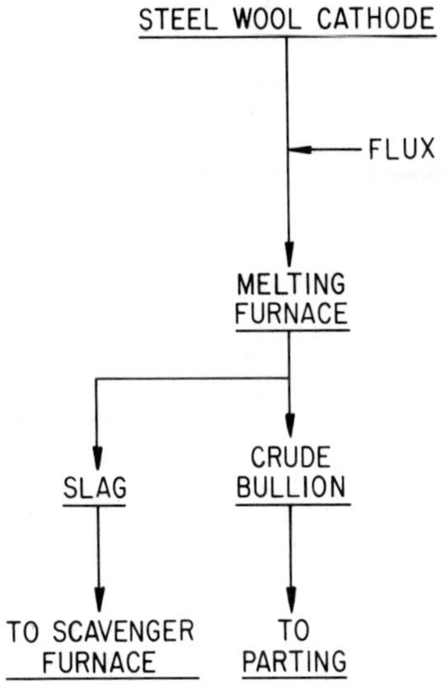

Figure 7. Flowsheet for smelting the carbon-in-pulp steel wool cathodes.

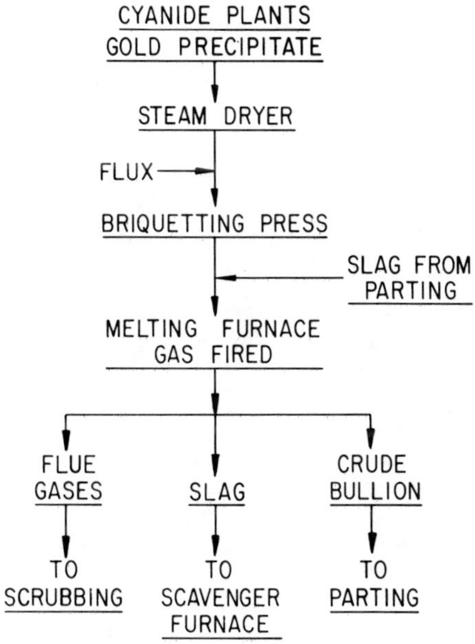

Figure 8. Flowsheet for smelting gold precipitate.

will assay 815 fine gold and 170 fine silver. This bullion is weighed, sampled and stored for further refining.

Concentrate from the South Mill gravity circuit averages ten ounces dore' per pound. Primary impurities are metal sulfides present in the ore. These metal sulfides are fluxed off in the same manner as the steel wool. The concentrate is also melted in a Monarch gas-fired tilting furnace. The resulting bullion assays 780 fine gold and 160 fine silver. This is again stored for further refining. Slags from each operation contain about 20 ounces gold per ton and are eventually run through a scavanger furnace.

All crude dore' bullion, from the three prementioned plants, is upgraded by the Miller process or chlorine parting. The Miller process consists of forcing chlorine through the molten bullion and changing all but the gold to chlorides. See Figure 9.

The Miller process is done in 500-ounce lots in small gas-fired furnaces. The bullion is contained in a clay crucible protected by a larger graphite crucible. A clay tube, with chlorine passing through it, is inserted into the molten bullion. Silver and base metals quickly react with the chlorine and form chlorides which float to the top. The chlorides are skimmed off until no more are formed. At this time sodium chloride and borax are placed on the surface of the molten bullion, the crucible covered to prevent volatilization, and the chlorine flow is continued at a reduced rate for one hour. At the end of an hour, the borax and salt are skimmed off with any other chlorides which may have formed. The refined gold, assaying 997 to 999 fine, is then poured into bars. These bars are cooled, cleaned and charged into a 4,500 ounce capacity induction furnace. Here, they are poured into either anodes for electrolytic refining or good delivery bars of desired weights. About 40 percent of bars sold come directly from the Miller process. Presently, about 60 percent of the gold from the Miller process is cast into 200 ounce anodes for further refining by the Wohlwill process or electrolytic refining.

The electrolytic gold refining section has eight cells in series, each with a capacity of 800 ounces (Figure 10). Anodes from the Miller process are placed in polypropylene tanks containing 90 to 100 grams per liter of gold chloride. The solution also contains 15 percent excess HCl for proper dissolution and deposition. The HCl reacts with the silver, which is the main impurity in the anode, and forms silver chloride, which being insoluble, falls to the bottom of the cell. Cathode starter strips are rolled from previously-produced fine gold. Direct current is supplied by a rectifier at 0.5 volts and 100 amps. A small amount of air is bubbled through the cells to provide a mixing action. Gold deposited on the cathode is washed, dried, and melted into good delivery bars. Bar fineness is 999.9 plus.

A lead blast furnace is used to recover values remaining in

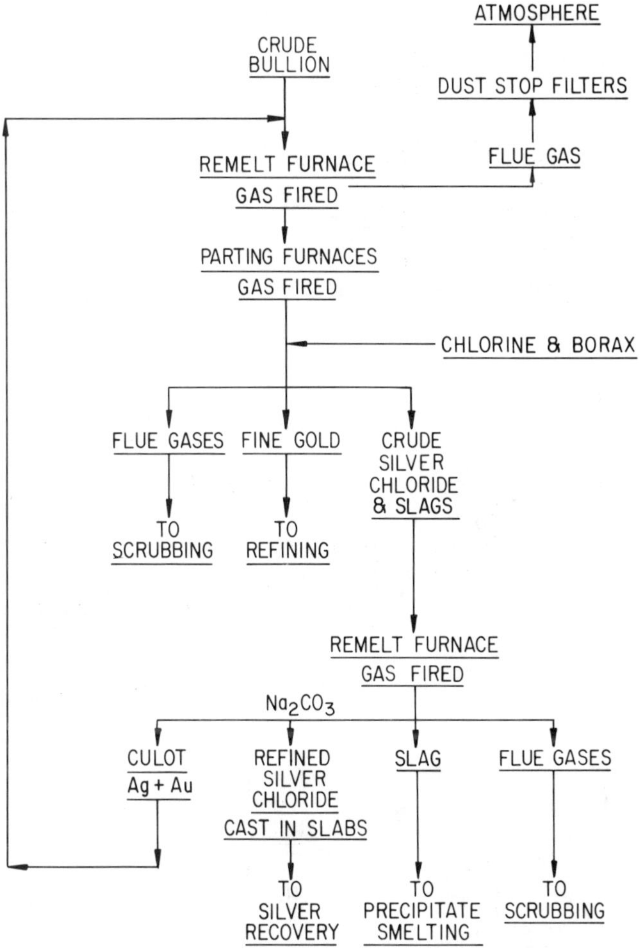

Figure 9. Flowsheet for gold and silver parting.

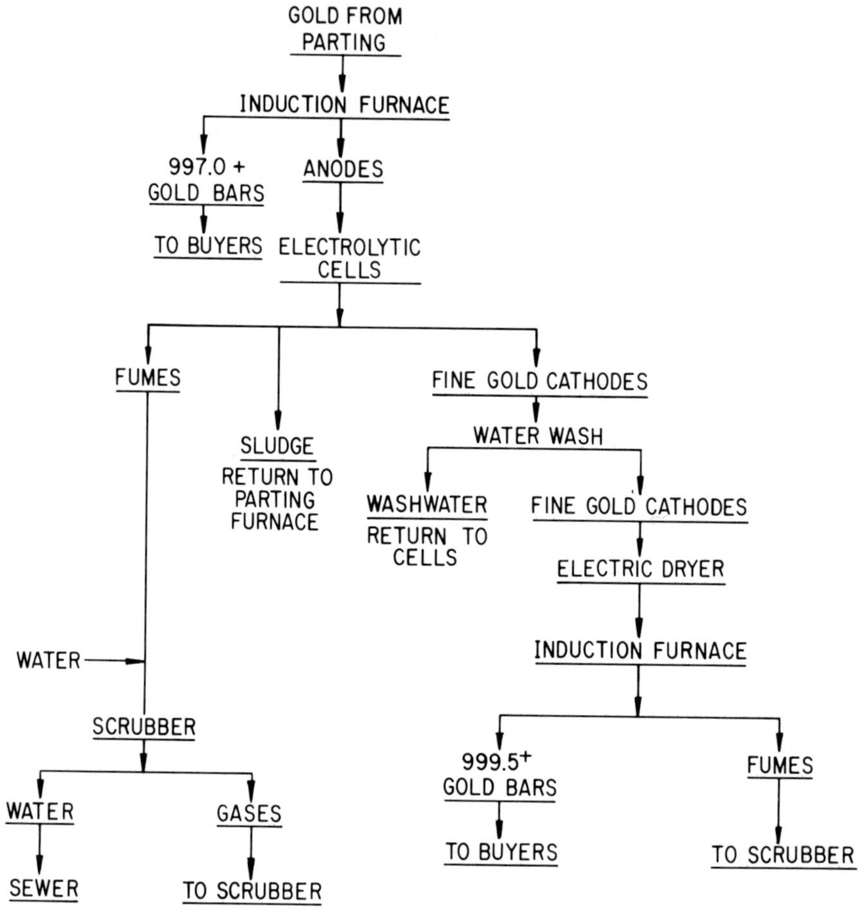

Figure 10. Flowsheet for gold refining.

slags, broken crucibles, used cupels, floor sweepings, and any other source which contains a small amount of gold (Figure 11). The blast furnace is 24-inches in diameter and is operated every two weeks for 12 to 14 hours. Coke is used as a heat source and also acts as a reducing agent for the litharge.

Products recovered from the blast furnace are a high-grade matte, lead bullion and the final slag. The matte is shipped to a smelter once a year; the lead bullion is cupelled to recover precious metals, and the slag is returned to the milling circuit.

Lead bullion from the blast furnace and sludge collected from flue gases are charged into a 50-inch by 40-inch by 9-inch elliptical cement cupel. The cupel is preheated, before charging, in a natural gas-fired Homestake designed furnace. Melting of the sludge takes place in a reducing atmosphere. When this charge is completely fused, an oxidizing atmosphere is used. A stream of air is blown on to the surface of the lead bullion, oxidizing the lead to litharge. The litharge produced is skimmed continuously. Eventually, after eight to ten hours, the lead has been oxidized, and the dore' bullion remains. This bullion is recovered and goes to the Miller process.

Silver chloride from the parting operation is remelted, and a small amount of sodium carbonate is added to the molten chloride. This reduces any gold chloride which may have been formed during parting and also a small amount of the silver chloride. The charge is allowed to cool to a dull red heat and the chloride poured into 1/2-inch slabs six inches square. The reduced gold and silver are solidified and are held for the next parting run.

The silver chloride slabs are reduced to metallic silver in acid-proof stoneware trays (Figure 12). These trays hold 75 pounds of chloride cakes alternated with a layer of iron slugs made from scrap steel. The chloride-iron layers are covered with dilute muriatic acid. The silver is reduced to silver metal, and the iron is oxidized to iron chloride. This reaction is complete in four to five days.

When the silver chloride has been reduced to metallic silver it is melted together with a small amount of borax and poured into 400 to 500-ounce bars. These bars assay 920 to 960 fine with copper the primary impurity. The crude silver is then refined electrolytically using Thumb cells. Final silver purity from electrolytic refining is 999.0 plus. This is poured into forms to suit customer needs.

MINE BACKFILL AND TAILINGS DISPOSAL

Sand residue from the sand leach plants is sluiced from the vats and runs by gravity to two holding areas called sand dams.

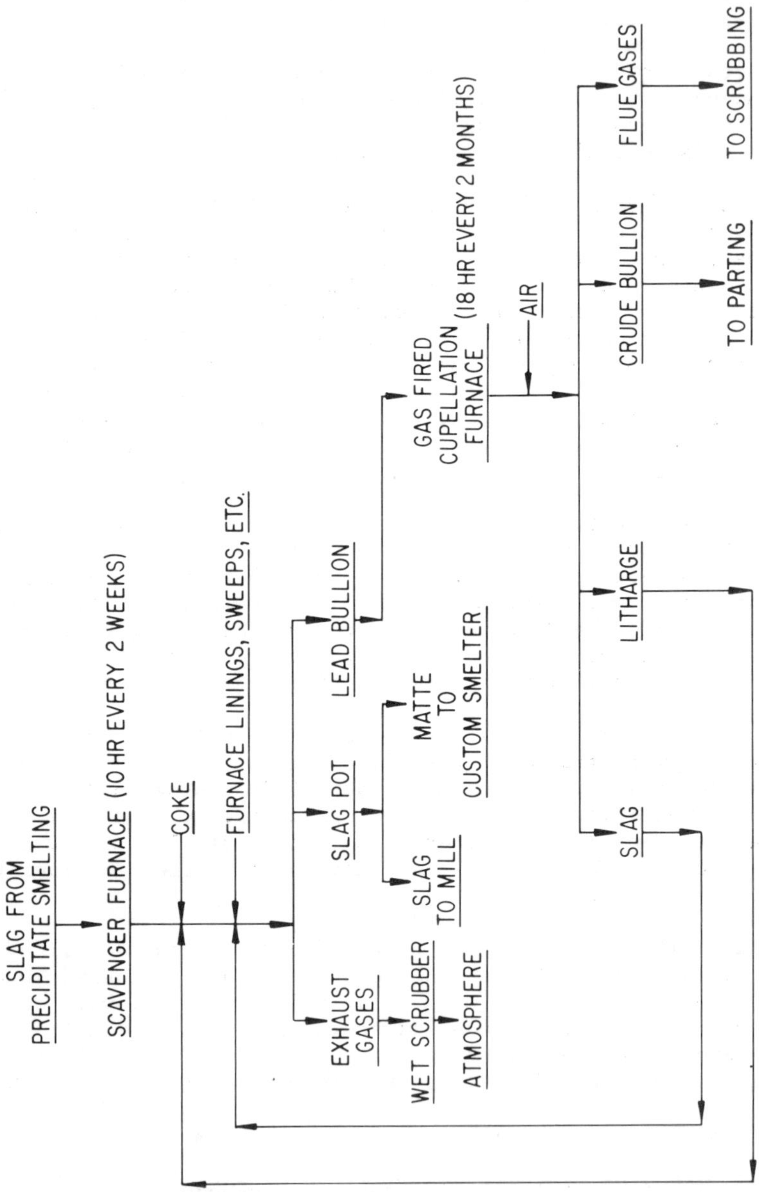

Figure 11. Flowsheet for slag processing.

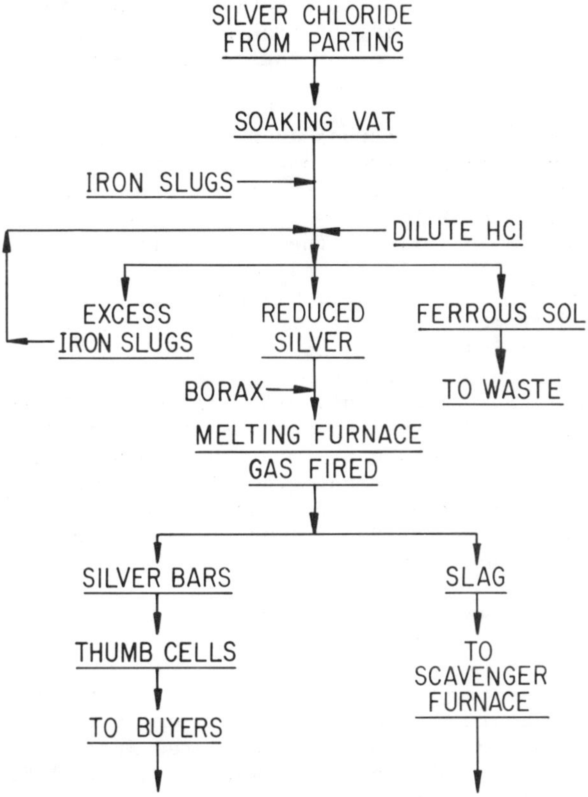

Figure 12. Flowsheet for silver chloride reduction.

Dam capacities are 6,000 tons. Directly below the sand dams is a raise which bottoms at the 1,100 foot level of the mine. The sand in these holding areas is moved hydraulically into a pipeline running down the raise. Normal density of sand slurry is 55 to 60 percent solids. Pipe connections on the 1,100 foot level direct the sand to the level and stope to be backfilled. During the final six to eight inches of stope filling, dry cement at a one to ten ratio is added to the sand slurry. This forms a cement cap on the floor of the stope on which slushing of broken rock can be done. At the present time, about 50 percent of the residue sand is used for mine backfill.

All tailings from the carbon-in-pulp plant and sand tailings not used for mine backfill are pumped to a tailings pond located in Grizzly Gulch which is located three miles from Lead. These tailings are pumped at 35 to 40 percent solids by a bank of eight BC-6-6C ASH rubber lined pumped in series. The system has two banks of pumps and two eight-inch rubber lined tailings lines. Vertical lift of the tailings system from the tailings pump station to the impoundment area is 487 feet. Tailings line length is 15,651 feet. The line is unusual in that it goes up a steep hillside, down a deep valley and back up another steep 15 percent grade to the tailings dam area. The lines are equipped with vent valves to relieve entrapped air and with dump valves for draining. The dump valves, located in Kirk canyon, drain the pipelines to a holding area on the 1,100 foot level of the mine. The lines must be drained when not in use in the winter and also during emergency shutdowns. Pump discharge pressure from the final pump stage ranges from 400 to 500 pounds per square inch. Normal tonnage pumped is 4,000 tons per day.

WASTE WATER TREATMENT

Waste water is comprised of fresh water taken into the water system from mine cooling, gland water for pumps, rain water and springs. The fresh water cannot be separated from internal water, and this creates an imbalance in the water system, which requires a discharge to keep the water system in balance.

Techniques are presently being researched to meet cyanide and heavy metal standards. A Baker pressure sand filter has been installed for suspended solids removal, and an 80 x 150 foot cooling reservoir has been constructed for cooling mine water.

Flowsheets for waste water treatment in the refinery and a flowsheet showing the water circuit for the mine and mill are given in Figures 13 and 14.

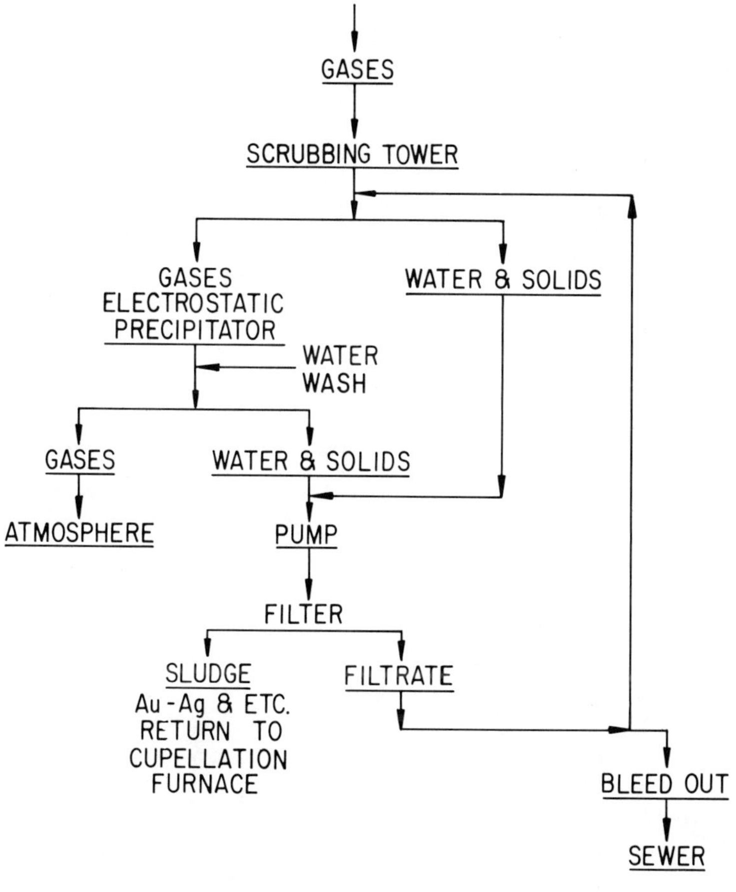

Figure 13. Flowsheet for wastewater treatment in the refinery.

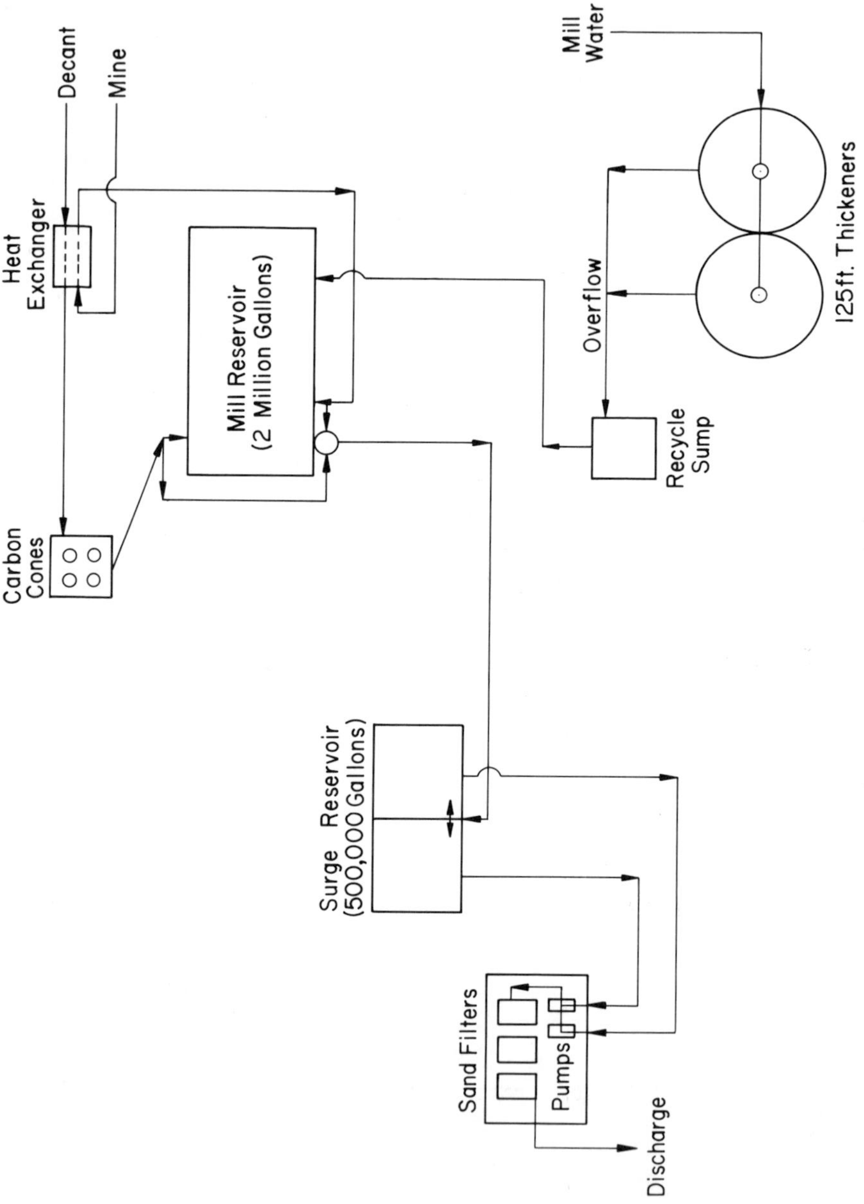

Figure 14. Water circuit for the mine and mill.

APPENDIX A

SAND PLANT TREATMENT CYCLE

Operation	Duration (hr)	Remarks
(1) Filling	7-8	Filling time of East Plant, 23 vats, 7-8 hours; filling time of West Plant, 12 vats, 17-18 hours. Pulp at 25% solids, minus 16 mesh; lime slurry (2.0 lbs/ton of sand) charged to water filled vat through Butters-Mein distributor. Vat overflows pulp deslimed and water reused. Charge, 0.160 oz. gold/ton. Extraction 92.0 to 94.0%.
(2) First drain	19	Effluent to waste.
(3) First aeration	18-32	Gage pressure. 9 lbs. Effluent valve closed. Aeration period varies according to alkalinity determination of sample taken 8 hours after vat filling.
(3a) Alkaline wash	2-4	Clarified filling of water pH 11.5 to 12. Effluent to waste.
(4) First strong solution leach	20	Leaching rate, 2-3/8" per hour. Ongoing solution strength 0.0475 to 0.06% NaCN. Effluent to waste.
(5) Second drain	14	Effluent to waste until trace of cyanide, then to solution collector tanks followed by conventional zinc dust precipitation. Barren solution used as 3rd period leach solution.
(6) Second aeration	6	Gage pressure. 9 lbs. Effluent valve closed.

SAND PLANT TREATMENT CYCLE (Continued)

Operation	Duration (hr)	Remarks
(7) Second strong solution leach	16	Leaching rate 2-5/8" per hour. Ongoing solution strength 0.0475 to 0.06% NaCN. Effluent solution to collector tanks and precipitation. Barren solution reused.
(8) Third drain	12	Effluent solution to collector tanks and precipitation. Barren solution reused.
(9) Third aeration	4	Gage pressure. 8 lbs. Effluent valve closed.
(10) Third strong solution leach	16	Leaching rate 2-3/8" per hour. Ongoing solution strength 0.0475 to 0.06% NaCN. Effluent solution to collector tanks and precipitation. Barren solution reused.
(11) Fourth drain	12	Effluent solution to collector tanks and precipitation. Barren solution reused.
(12) Fourth aeration	4	Gage pressure. 8 lbs. Effluent valve closed.
(13) Weak (barren)	20-W-16E	Solution strength 0.02 to 0.03% NaCN. Effluent solution to collector tanks standardized to 0.0475 to 0.06% NaCN and recirculated for ongoing strong leach solution. Period duration varies according to amount of wash water used.
(14) Washing	19-22	Leaching rate 2-1/2" per hour. Effluent solution to collector tanks, standardized to 0.0475 to 0.06% NaCN and recirculated for ongoing strong leach solution. Normal requirements are 48" to 55" of wash water to maintain solution balance.

SAND PLANT TREATMENT CYCLE (Continued)

Operation	Duration (hr)	Remarks
(15) Sluicing	7	With clarified water through Butters-Mein Distributor. Residue sands returned underground for stope fill.
(16) Refilling of vat with water	2	Clarified water (limed when available).

TYPICAL SAND PLANT CHARGE

Dry Weight - 775-800 tons per leach vat

Assay - 0.150 oz. gold/ton

Extraction - 92.0 to 94.0%

SCREEN ANALYSIS

Mesh	+48	+80	+100	+150	+200	-200
Wt. Per Cent	0.0	6.0	8.3	17.4	22.4	45.9

Total treatment cycle - 196 hours

Chemicals Used:

Sodium cyanide	- 0.30 lbs. per ton of sand treated
Lime	- 1.35 lbs. per ton of sand treated
Zinc dust	- 0.035 to 0.040 lbs. per ton of sand treated

Chapter 7

THE ADSORPTION OF GOLD AND SILVER CYANIDE FROM SOLUTION BY ACTIVATED CHARCOAL

E. H. Cho and C. H. Pitt

Department of Metallurgy
University of Utah
Salt Lake City, Utah

INTRODUCTION

Activated charcoal has been used for several centuries in many capacities including the recovery, isolation and purification of various materials (1). The first use of charcoal in a gold metallurgical system was in the extraction of gold after chlorination (2, 3,4). The gold was dissolved in a chlorine system and subsequently filtered through charcoal.

The application of charcoal with cyanide extraction was patented by Johnston (5) in 1894 and adopted commercially in Australia in 1897. The process consisted of dissolving the gold and silver in a potassium cyanide solution and passing it through a series of charcoal filters. The charcoal was subsequently smelted to recover the gold and silver.

The interest in the use of charcoal diminished in the first part of the twentieth century as the practice of the use of zinc as a precipitating agent increased. However, in 1916, the Yuanmi Mine in Australia replaced their zinc precipitation plant with a charcoal adsorption plant.

In 1926, Scott and Gross (7) of the U. S. Bureau of Mines conducted extensive research on the adsorption of gold and silver from cyanide leach solutions. Among their findings was the result that the most important factor in improving the adsorptive power of charcoal was heat treatment with oxidation by air or steam. The work of Zadra (8) in 1952 stimulated renewed interest in the use of activated charcoal in gold and silver recovery with the development of a method of recovery of the gold and silver from the charcoal other than its destruction by burning. The process involved simultaneous desorption with hot caustic and cyanide and electrowinning of gold and silver. Stripping and regeneration of the charcoal, however, still remains a problem to this day, and researchers continue to attempt to develop better methods for stripping the gold and silver values from the charcoal. Recent work by Potter and Salisbury (9) has utilized high pressure leaching with hot caustic cyanide solution.

Activated charcoal consists of a family of substances. The basis for differentiating between the types of charcoal is usually by their adsorptive and catalytic properties and by their method of manufacture. The basic activation process consists of carbonization and oxidation. The carbonization process is usually conducted in the absence of air at temperature in the range of 600-900°C, while the oxidation process is performed in the presence of oxidizing gases at about 900°C. Different permutations of the basic process produce different types of activated charcoal.

The structure of activated charcoal is that of flat plates in which the carbon atoms are arranged in a hexagonal lattice. A crystallite is formed by two or more of these plates stacked one above the other. The oxidation process erodes the surface so as to increase the surface area, develop greater porosity and also form oxygen containing functional groups on the surface which appear to play an important role in the adsorption processes.

One of the unique characteristics of activated charcoal is its porosity or pore structure. There appears to be a random distribution of amorphous particles which produce a complex network of irregularly shaped and partly interconnected passages between the particles. Dubinin (10) classified the pore size distribution of activated carbon into three size groupings: macropores, transitional pores and micropores. Macropores have an effective radius of 1,000-2,000 Å and contribute very little to the total surface area. Transitional pores have an effective radius of 20-1,000 Å and contribute about 5% of the total surface area. Micropores range in size from 10-20 Å and contribute about 95% of the total surface area. The surface area of a typical steam activated charcoal can be as great as 1,100-1,200 square meters per gram.

Adsorption Isotherms

When a solid and a liquid phase are in contact with each other, the composition of the liquid at the surface is different from that of the bulk composition of the liquid phase. The increase in the concentration of a species at the surface as compared with the bulk solution composition is known as adsorption. Desorption is the release of a species from the surface into the bulk solution. When a solid and a liquid phase come into contact, rates of species' adsorption may be measured which generally vary with the amount of species' already adsorbed. By measuring these adsorption rates, the kinetics of adsorption are determined. When the rate of adsorption is equal to the rate of desorption, a steady state equilibrium is achieved, and the concentration of adsorbed species remains constant with time. Determination of the equilibrium adsorption concentration as a function of the solution species concentration at constant temperature determines an adsorption isotherm. Heat energy may be evolved or consumed as adsorption takes place and generally also varies with surface species concentration. The heat change per mole at a given surface concentration is known as the isosteric heat of adsorption.

Adsorption is generally classified into one of two types. If the adsorbate is bound to the surface by relatively weak forces, it is termed physical adsorption. Here, the binding forces are mainly van der Waals and/or weak electrostatic forces. The second form of adsorption is chemical adsorption where electrons are exchanged or shared with surface atoms. Physical adsorption usually requires energies of less than 10 Kcal/mole, while chemisorption is generally much higher.

EXPERIMENTAL MATERIALS AND PROCEDURES

The charcoal used in the experiment was granular activated carbon obtained from Calgon Corporation. The charcoal was made from coconut shell by high temperature steam activation and was 10 x 30 mesh in size with a surface area of 1,150 to 1,250 m^2/gram. About 80 percent of the surface area was contributed by micropores of 18-21 Å diameter.

Silver and gold cyanide solutions were prepared by dissolving the high purity metal in cyanide solutions and subsequently diluting to the proper concentration (11,12).

The charcoal particles, usually of the order of 1-2 grams, were reacted with the precious metal solutions in a one liter flask containing about 500 ml of solution. The flask temperatures were controlled in constant temperature baths. Solution samples were taken from the flasks at regular intervals to determine the change in solution concentration as adsorption proceeded.

Analysis of the gold and silver values was made using a Perkin Elmer Model 305 atomic absorption apparatus. Cyanide analysis was made by titrating with silver nitrate using an indicator of a rhodaine group chemical.

Zeta potential measurements were made using a Simpson Model 3326 Zeta Meter. Adjustments in pH were made using HNO_3 or NaOH solutions.

Heats of adsorption were measured experimentally using a Trovac Model 850 calorimeter by titrating solutions of metal cyanide into charcoal contained in distilled water.

EXPERIMENTAL RESULTS

Typical adsorption isotherms at various temperatures for silver cyanide on the activated charcoal are shown in Figure 1. Mathematically, the adsorption isotherms can be fit by the Freundlich equation,

$$Q = bC^m$$

which may be derived theoretically if one assumes an exponential variation in the heat of adsorption with surface coverage. Figure 2 shows the adsorption isotherms for silver cyanide on charcoal plotted according to the Freundlich isotherm. By obtaining values for the amount of silver cyanide in solution at constant surface loading of the charcoal at different temperatures, one can determine the isosteric heat of adsorption. Figure 3 shows the isosteric heats of adsorption as a function of surface coverage for silver cyanide on charcoal.

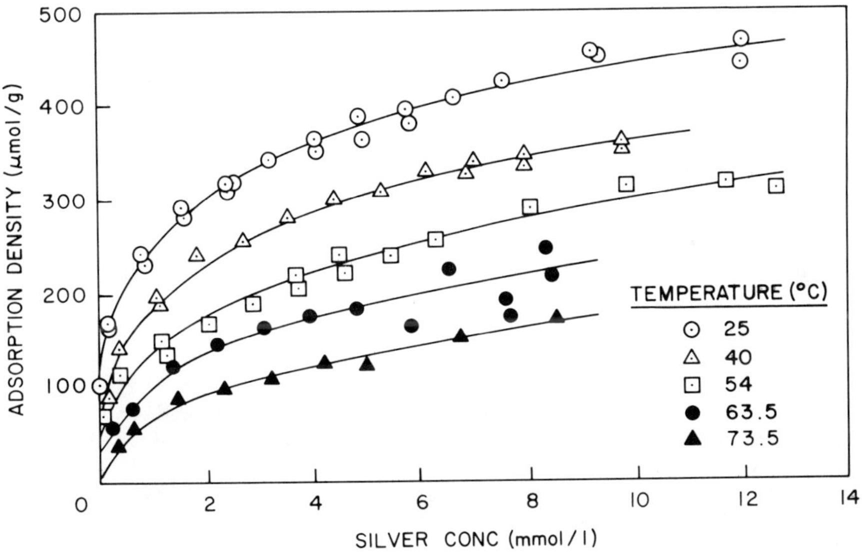

Figure 1. Adsorption isotherms for silver cyanide on charcoal.

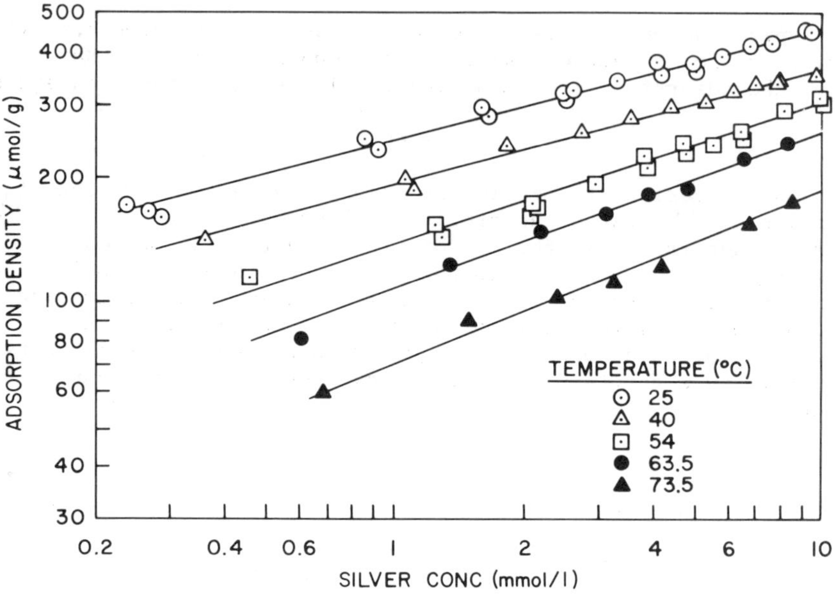

Figure 2. Adsorption isotherms for silver cyanide on charcoal plotted according to the Freundlich isotherm.

Effect of Counter Ions

The effects of sodium ions, free cyanide ions and calcium ions were also determined. Figure 4 shows the effect of sodium ions and free cyanide on silver cyanide adsorption. When sodium cyanide solution alone was placed in contact with the charcoal, it was found that cyanide ion but not sodium ion was adsorbed. It was also found that as the initial sodium cyanide concentration was increased the loading of silver cyanide went through a maximum (13) and that sodium ions also adsorbed in the presence of silver cyanide. This indicates that the sodium ions likely adsorb on the silver cyanide already adsorbed on the charcoal. As calcium ion is added to the adsorption system, results are obtained similar to the case of sodium. See Figure 5. When calcium cyanide only is added, calcium ion is found to adsorb below a pH of about 8. Again, it may be concluded that calcium adsorbs on silver cyanide ions already absorbed on the charcoal.

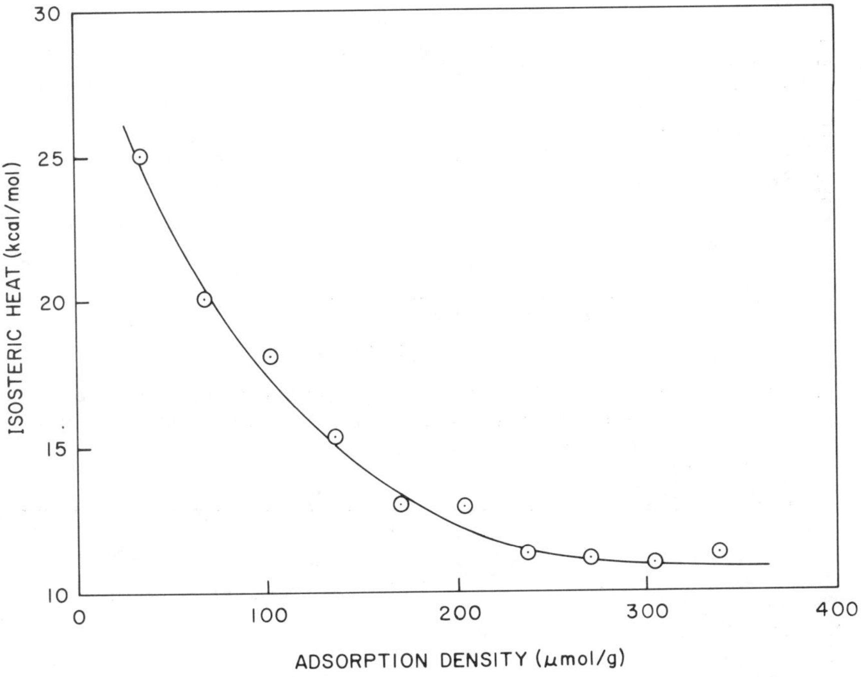

Figure 3. Isoteric heat of adsorption of silver cyanide on charcoal as a function of adsorption density.

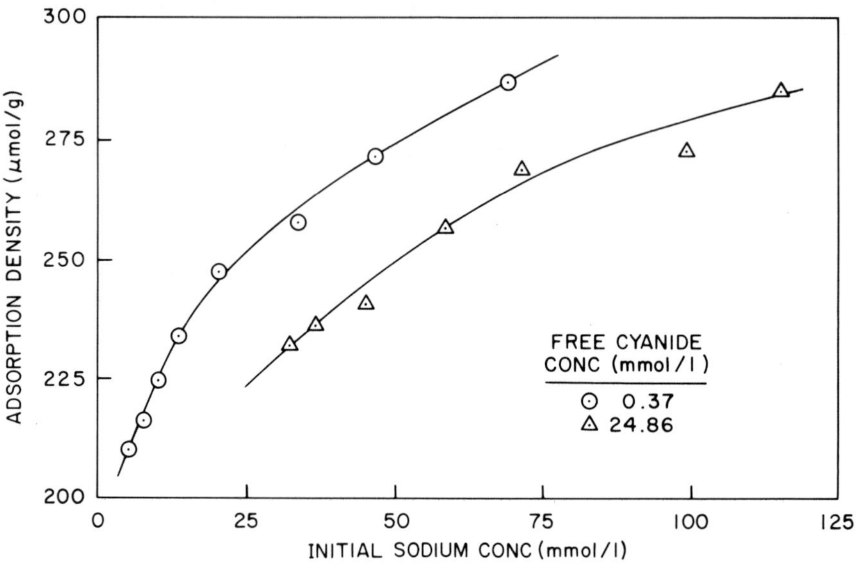

Figure 4. Effect of sodium ion and cyanide ion on the adsorption of silver cyanide on charcoal for an initial silver concentration of 200 ppm.

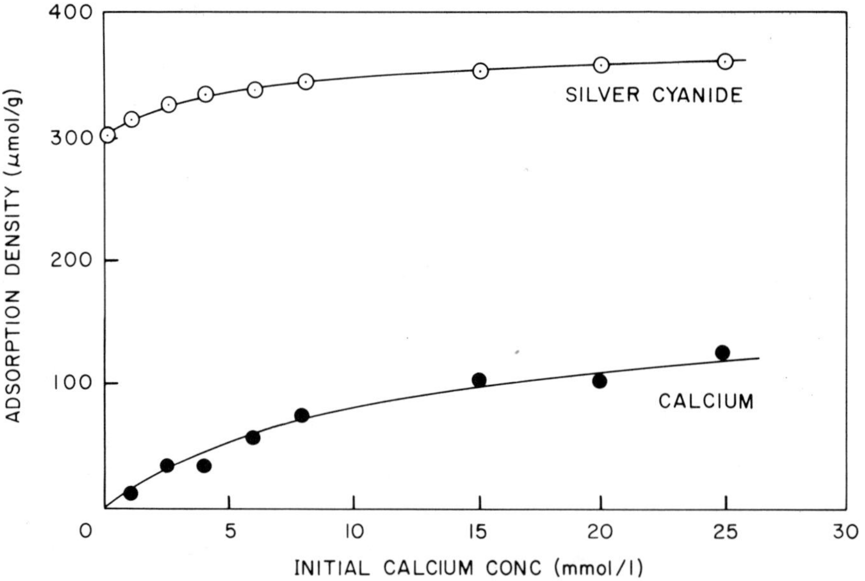

Figure 5. Effect of calcium ion on the adsorption of silver for a sodium cyanide concentration of 0.14%. The initial pH was 7.4 ± 0.2, and the initial silver concentration was 200 ppm.

Zeta Potential

The variation of zeta potential with pH is shown in Figure 6. The pH values were measured after 30 minutes conditioning of the mixture.

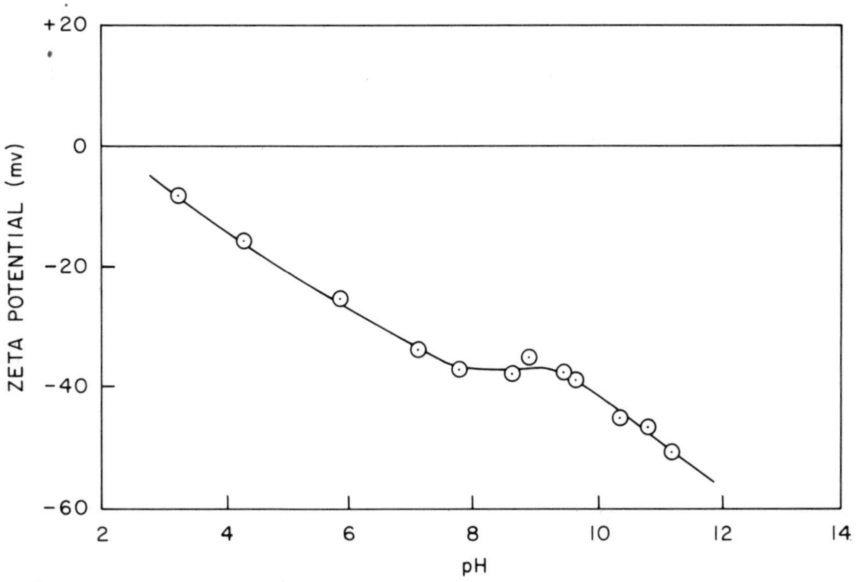

Figure 6. Variation of the zeta potential of charcoal with pH.

Effect of pH

The effect of pH on the adsorption of silver cyanide by charcoal is shown in Figure 7. Decreasing pH causes an increase in the adsorption of silver cyanide with the exception of the pH range of about pH 5 to 7 where no change is noted. The pH curve has a similar shape to that of the zeta potential curve. Since the acidity of the solution is important in the adsorption reaction, the pH change caused by the adsorption of hydrogen ion by the charcoal was determined. The changes are shown in Table 1 for two different strengths of cyanide solution. These data indicate that greater adsorption of hydrogen ion may increase the adsorption of silver cyanide in certain pH ranges, but where the adsorption of silver cyanide remains constant with pH, hydrogen ions still adsorb. It seems likely from the evidence presented that the adsorption of silver cyanide on charcoal is influenced more by the zeta potential of the charcoal particles than by adsorption of hydrogen ions.

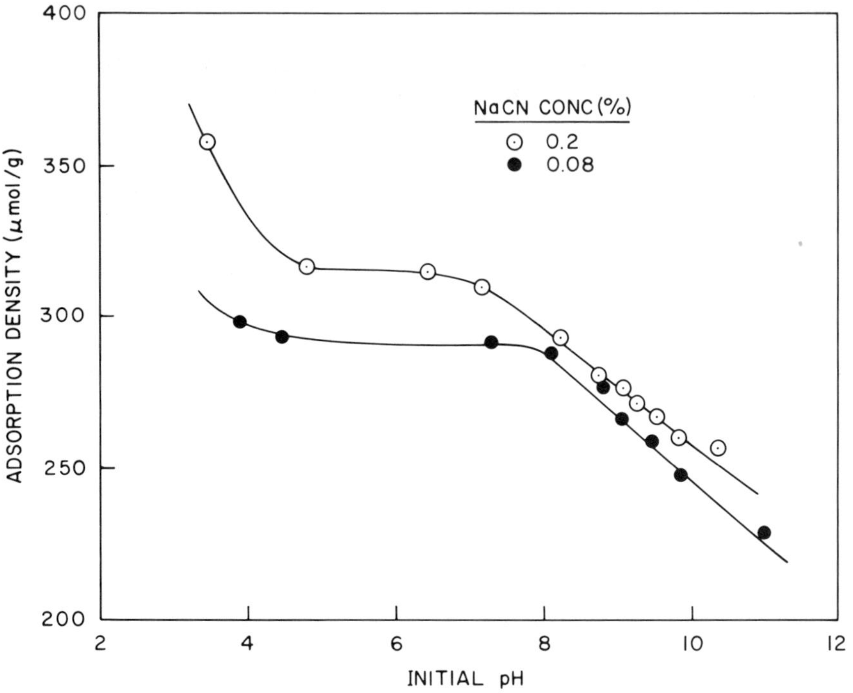

Figure 7. Effect of pH on the adsorption of silver cyanide on charcoal at 25°C for an initial silver concentration of 200 ppm.

Adsorption Isotherms for Gold Cyanide

Adsorption isotherms for gold cyanide at 25°, 40°, and 55°C are shown in Figures 8, 9, and 10. As the temperature was increased, there was more scatter in the data which indicates that the lower energy adsorption sites may be highly affected by slight variations in operation conditions. Isotherms at temperatures greater than 55°C were not attempted because of the large amount of scatter in the data. These isotherms can also be fit to the Freundlich equation.

Heat of Adsorption

As mentioned previously, the heats of adsorption for gold were determined by using a microcalorimeter. Isosteric heats from the adsorption isotherms could not be determined due to the paucity of isotherm data. The results are shown in Figure 11. At low free cyanide concentration, the initial heat of adsorption was about 60

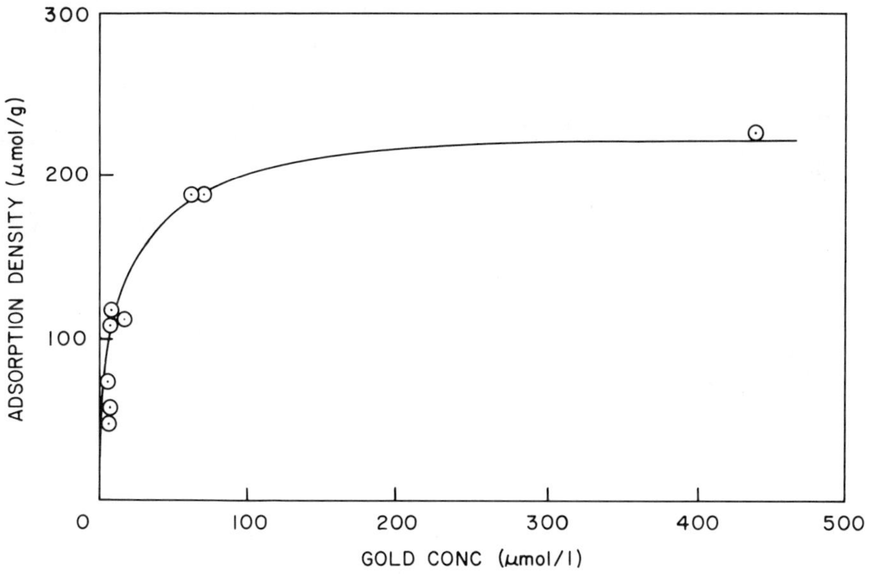

Figure 8. Adsorption isotherm for gold cyanide on charcoal at 25°C.

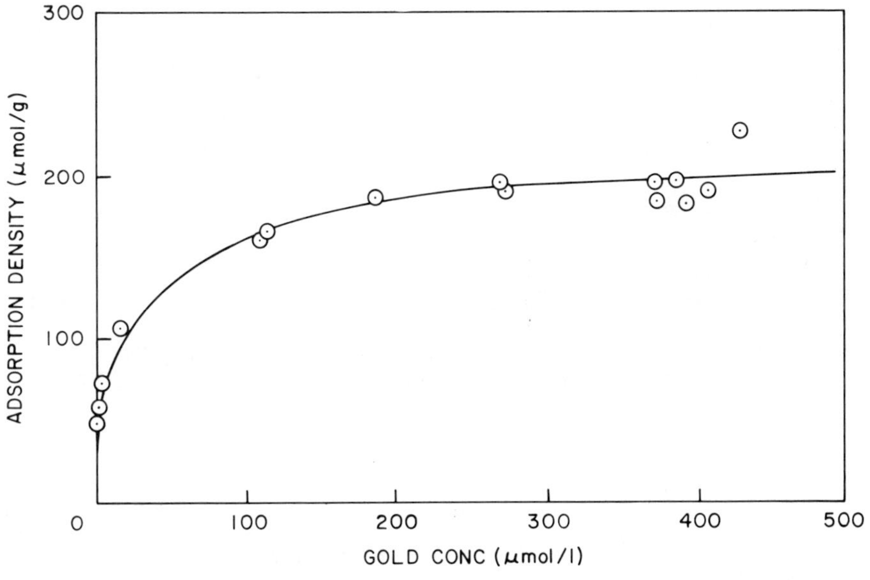

Figure 9. Adsorption isotherm for gold cyanide on charcoal at 40°C.

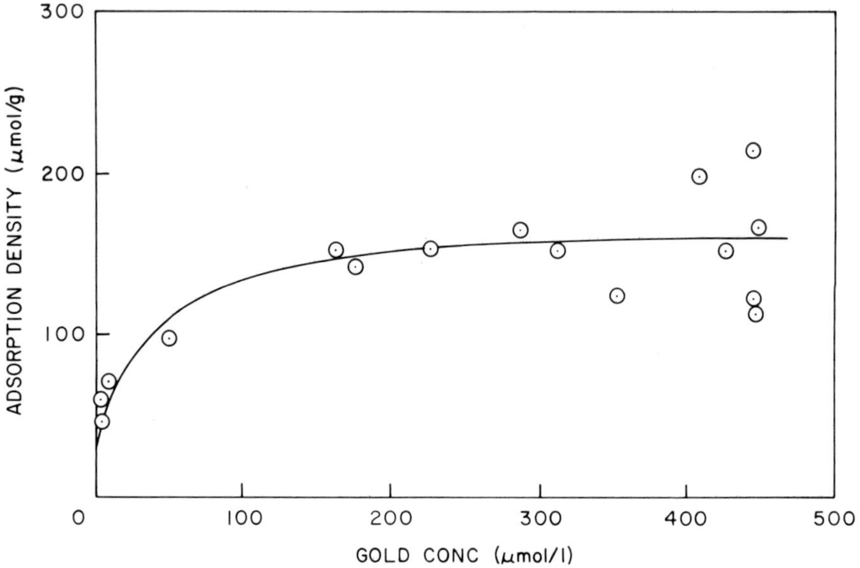

Figure 10. Adsorption isotherm for gold cyanide on charcoal at 55°C.

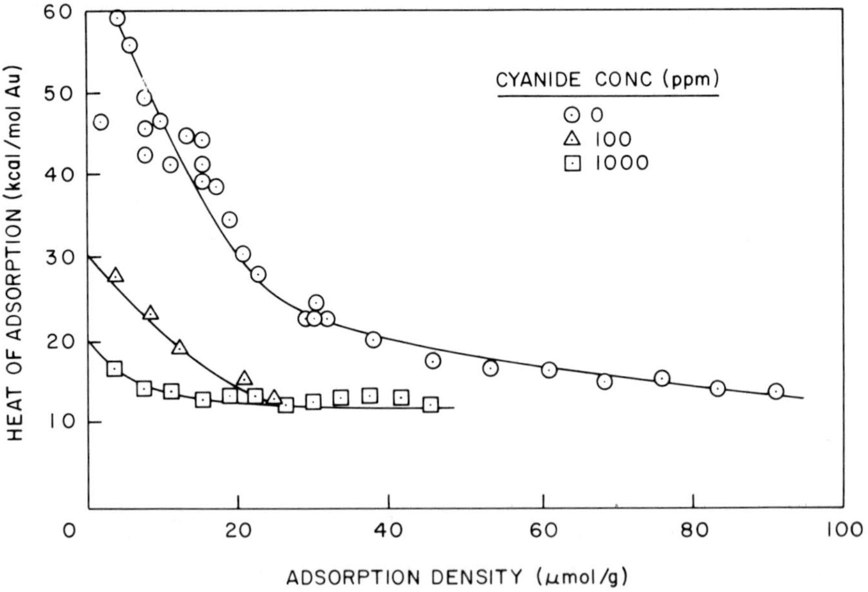

Figure 11. Heat of adsorption of gold cyanide on charcoal as a function of the adsorption density of gold. The data were obtained by microcalorimetry.

Kcal/mole of gold adsorbed. The heat of adsorption decreases exponentially with the adsorption density of gold. The presence of 100 ppm free cyanide reduces the initial heat of adsorption to 30 Kcal/mole of gold, while the presence of 1000 ppm free cyanide reduces the initial heat of adsorption to 20 Kcal/mole of gold.

The heat of adsorption of free cyanide or charcoal after the surface coverage was 40 micromoles of gold per gram of charcoal is presented in Figure 12. At low surface coverage the heat of adsorption was 4 Kcal/mole which decreased to about 1 Kcal/mole above 50 micromoles of CN^- per gm of charcoal. These results indicate that gold cyanide is chemisorbed, while cyanide ion is physically adsorbed.

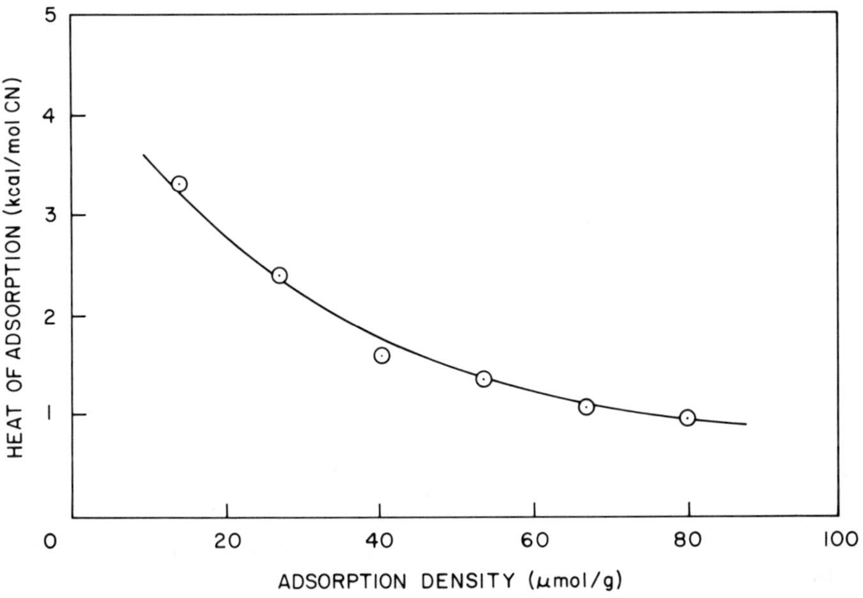

Figure 12. Heat of adsorption of free cyanide on charcoal as a function of the adsorption density of cyanide.

Effect of pH on Gold Cyanide Adsorption

A plot of gold cyanide adsorption versus pH is shown in Figure 13 in the presence of 25 ppm free cyanide. The curve is somewhat different from that for silver cyanide in that there is no plateau at intermediate pH values. The maximum loading value is 575-625 micromoles of gold per gram of charcoal in the pH range 2-3, and the minimum loading is 250-352 micromoles of gold in the pH range 8-12.

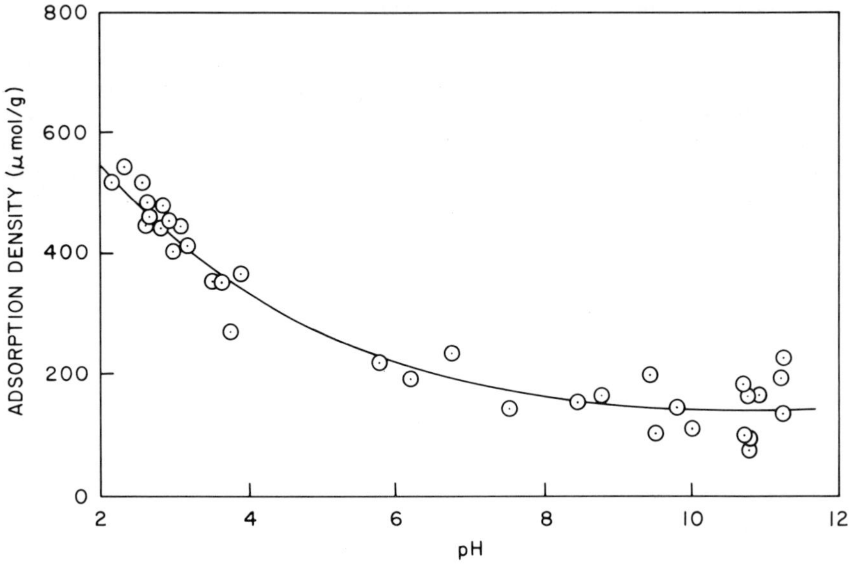

Figure 13. Adsorption of gold cyanide on charcoal as a function of pH.

Selective Adsorption Between Gold and Silver Cyanides

The results of an experiment to determine the selective adsorption between gold and silver cyanides are presented in Figure 14. The initial concentrations of both cyanide complexes were the same in the experimental solutions. The solution pH was 11.2. It may be observed that the amount of cyanide adsorbed is about three times that of silver cyanide.

PROPOSED ADSORPTION MECHANISM

The results of experiments conducted under the natural solution pH showed that the adsorption of the metal cyanide occurs in solutions having pH values of 11 to 11.5. At these pH values, the zeta potential of charcoal particles is negative (50 to 60 mv). Assuming that the zeta potential represents the surface charge, the number of negative sites on the surface is predominantly greater than the number of positive sites at these pH values. Thus, the adsorption of negative ions such as gold and silver cyanide ions cannot be readily explained by electrostatic forces alone. Also, the results shown in Figure 14 show that gold cyanide absorbs more than silver cyanide. Since the predominant ionic species of gold and silver cyanides are $Au(CN)_2^-$ and $Ag(CN)_2^-$ and the ionic radii of gold and silver are 1.37 and 1.26Å, respectively, the size of the gold cyanide ion will be larger than the silver cyanide ion.

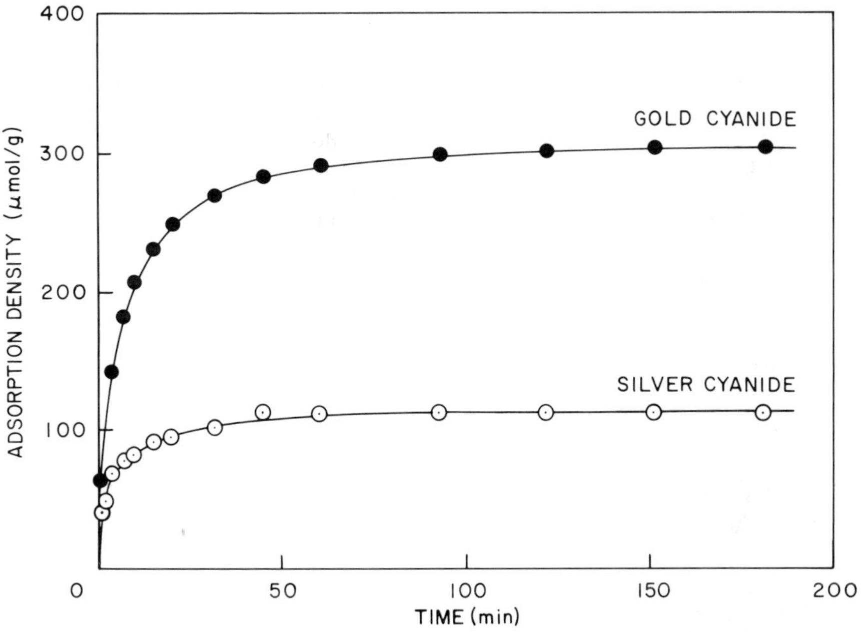

Figure 14. Gold and silver cyanide adsorption on charcoal at 25°C as a function of time. The initial gold and silver concentrations were 1.52 mmol/l, and the sodium cyanide concentration was 0.2%.

The adsorption differences may be explained at least partially by the ionic solvation theory (14,15) which states that larger ions are more favorably adsorbed than smaller ions when specifically adsorbed on metal surfaces. A summary of the proposed adsorption model is given as follows:

1. Silver or gold cyanide ion adsorbs on the active sites of the charcoal surface. Also, free cyanide ion can adsorb on the active sites but it is much less energetically adsorbed.

2. Since sodium and calcium ions are adsorbed only on charcoal in the presence of silver cyanide ions, it appears that these ions are adsorbed in the diffuse later above the silver or gold cyanide ion already adsorbed on the active sites.

3. Increasing the concentration of sodium and calcium ions increases the adsorption of silver cyanide. It seems probable that these cations adsorbed in the diffuse layer provide available sites for silver cyanide ions to be adsorbed in the diffuse layer.

4. Free cyanide ions cause a decrease in the adsorption of silver and gold cyanide on charcoal due to competitive adsorption. A sketch of the model is pictured in Figure 15 showing the large metal cyanide ions adsorbed in the Stern plane and the smaller sodium or calcium ion adsorbed in the Gouy plane with other cyanide ions adsorbed in the diffuse layer farther out.

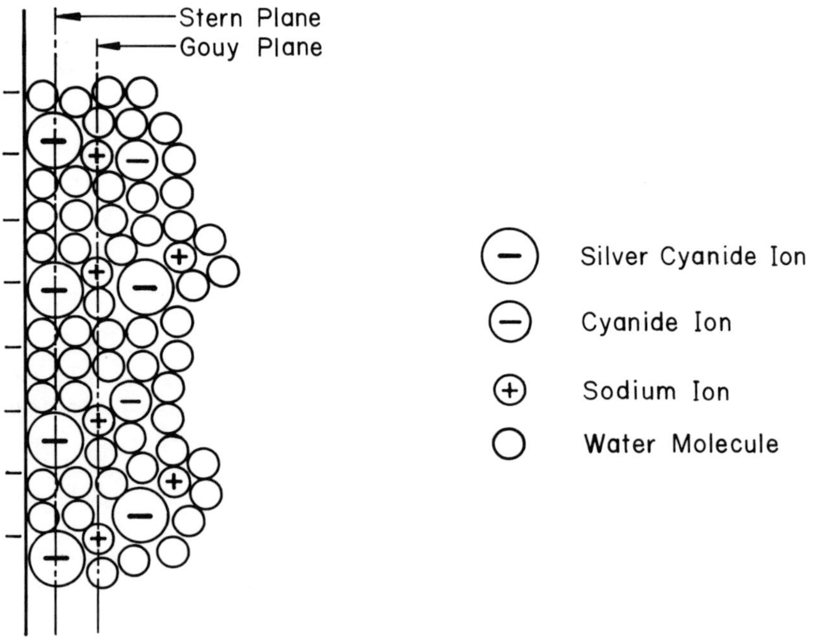

Figure 15. Model for the adsorption of gold cyanide and silver cyanide on charcoal.

Adsorption Rates

The amount of silver adsorbed as a function of time for various initial silver concentrations at 25°C is shown in Figure 16, and the amount of gold adsorbed as a function of time at various temperatures is shown in Figure 17. Initial rates were determined by drawing tangents to these curves at zero time. The adsorption of gold or silver cyanide into a porous charcoal particle appears to involve the simultaneous processes of pore diffusion and adsorption on active sites. In this kind of a model the adsorption rate can be characterized by a mass balance within a differential volume element utilizing the continuity equation:

$$\frac{\partial c}{\partial t} = \frac{1}{r^2} \frac{\partial}{\partial r} (r^2 \; De \frac{\partial c}{\partial r}) + G$$

where $\partial c/\partial t$ is the accumulation of metal cyanide in the charcoal, G is the adsorption rate, and the first term on the right hand side represents normal diffusion. D_e is the effective diffusivity. The details of the solution of the equation are beyond the scope of this paper but may be found in a paper by Cho, Dixon and Pitt (15). This model was developed by Crank (16) in 1948 and utilized later by Weber and Snoeyink (17). From the experimental data, the effective diffusivities of gold cyanide were found to be 3.19×10^{-5}, 4.14×10^{-5} and 5.30×10^{-5} cm^2/s for 25, 40 and 55°C, respectively. When these data are plotted on an Arrhenius-type plot, a straight line results from which the activation energy is determined to be 3.3 Kcal/mole. For silver cyanide a similar treatment yielded a value for De of 3.26×10^{-5} cm/s at 25° and 4.78×10^{-5} cm^2/s at 54°C. These values give 2.6 Kcal/mole for the activation energy. These low values of effective diffusion coefficient and activation energy are typical of a diffusion-controlled process and the values tend to confirm the proposed pore diffusion mechanism for the adsorption reaction. By using the effective diffusivity values, the diffusion model was used to compute adsorption rate curves, and the results were correlated with the corresponding experimental data. The correlations are shown for gold in Figures 18, 19 and 20 and for silver in Figure 21. The agreement between the experimental data and the computed adsorption rate curves is generally good which appears to confirm the initially assumed model of pore diffusion.

ACKNOWLEDGMENT

Gratitude is expressed to the U.S. Bureau of Mines, Department of the Interior, for financial assistance received under Contract G254027 which enabled this research to be performed.

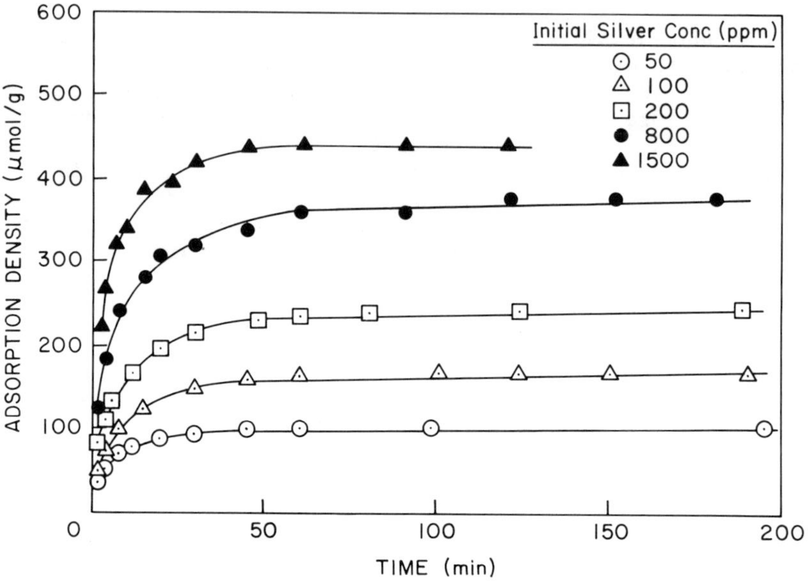

Figure 16. Adsorption of silver on charcoal as a function of time.

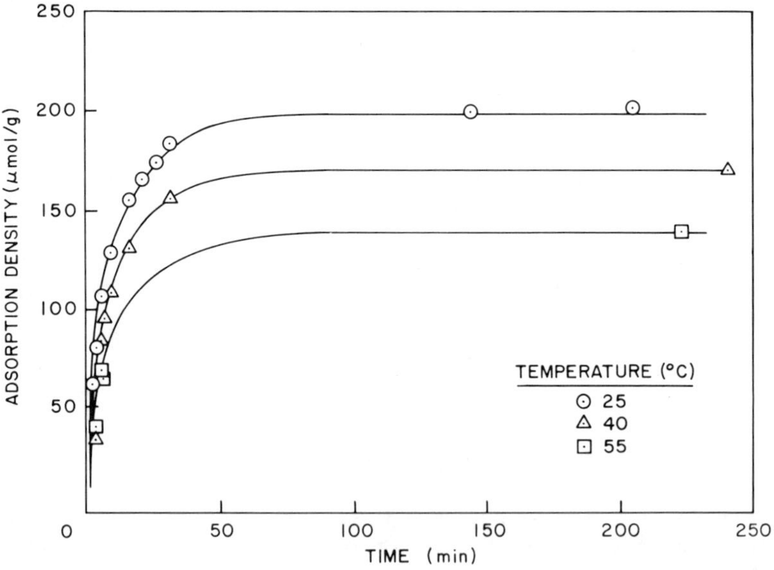

Figure 17. Adsorption of gold on charcoal as a function of time.

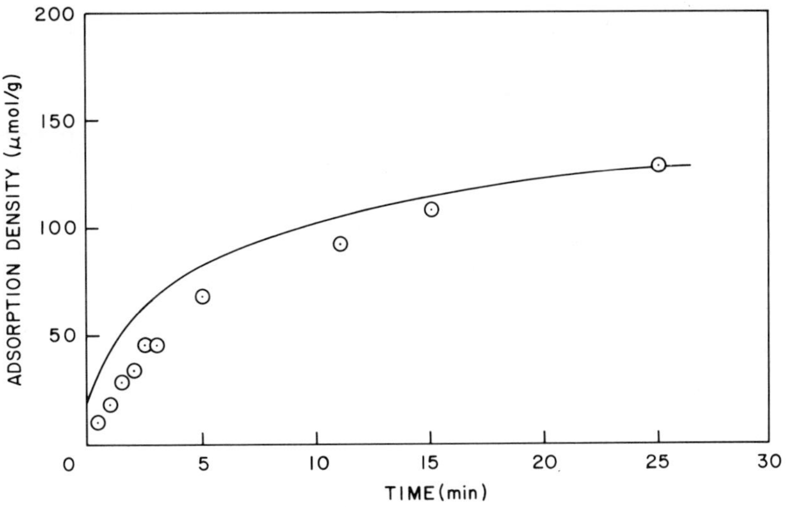

Figure 18. Correlation between the experimental adsorption density of gold and that predicted theoretically at 25°C. The initial gold concentration was 92.6 ppm, and the free cyanide concentration was 25 ppm. The amount of charcoal involved was 1.02 g.

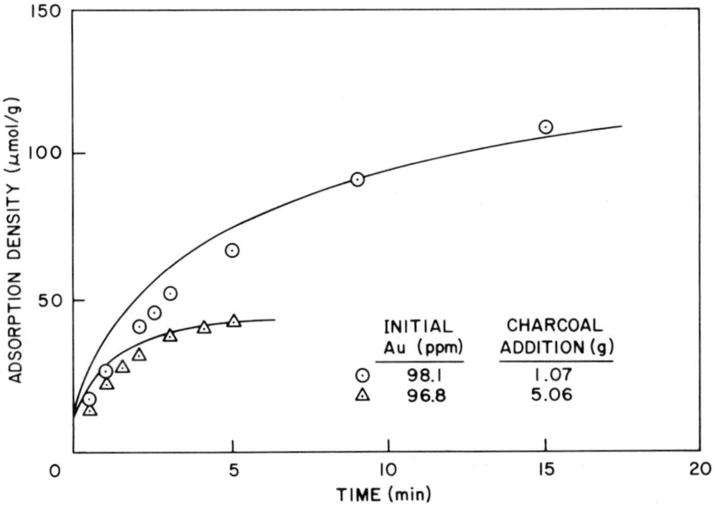

Figure 19. Correlation between the experimental adsorption density of gold and that predicted theoretically at 40°C. The free cyanide concentration was 25 ppm.

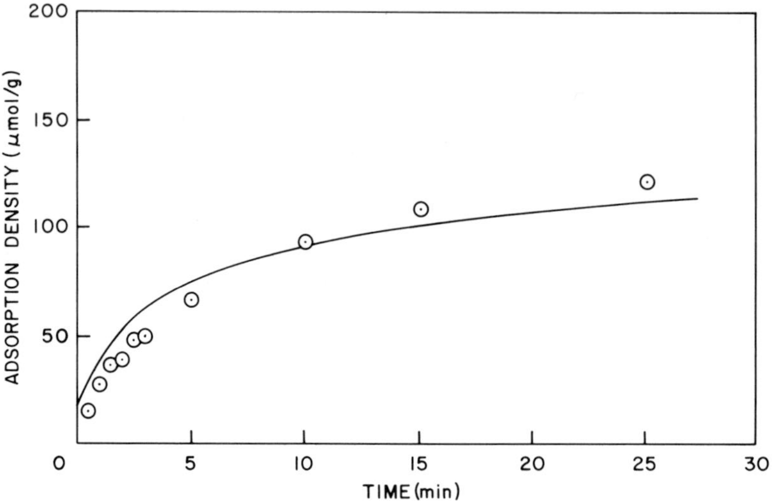

Figure 20. Correlation between the experimental adsorption density of gold and that predicted theoretically at 25°C. The initial gold concentration was 105.3 ppm, and the free cyanide concentration was 25 ppm. The amount of charcoal involved was 1.04 g.

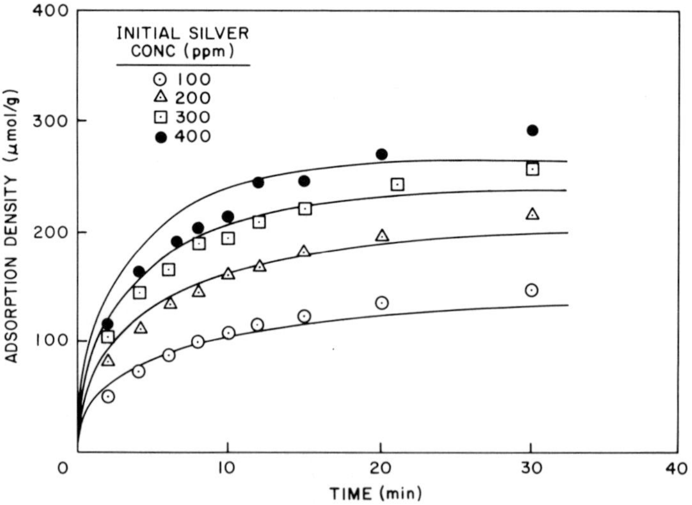

Figure 21. Correlation between the experimental adsorption density and that predicted theoretically for the silver-charcoal system.

REFERENCES

1. Lourtz, J. T., "Annales de Chemie," 8, pp. 131-34, 1788.
2. Davis, W. N., U.S. Pat. 227963, May 25, 1880.
3. Egleston, T. F., "The Metallurgy of Silver, Gold, and Mercury in the United States," J. Wiley & Sons, 2, New York, 1890.
4. Davis, H. B. K., U.S. Pat. 546113, Sept. 10, 1895.
5. Johnston, W. D., U.S. Pat. 522260, 1894.
6. Power, N. D., Australian Mining Std. No. 2305, 1897.
7. Scott, J. W. and Gross, J., U.S.B.M. Tech. Report 378, 1926.
8. Zadra, J. B., U.S.B.M. Rept. of Invest. 4843, 1952.
9. Ross, J. R., Salisbury, H. B. and Potter, G. M., Presentation at Annual AIME Meeting, Chicago, IL, 1973.
10. Dubinin, M. M., J. of Colloid and Interface Science, 23, p. 487, 1967.
11. Dixon, S. N., Master of Science Thesis, University of Utah, 1977.
12. Cho, E. H., Ph.D. Thesis, University of Utah, 1978.
13. Cho, E. H., and Pitt, C. H., Met. Trans. 10B, p. 159, 1979.
14. Bockris, J. O. M., Devanathan, M. A. V. and Muller, K., Proc. Roy. Soc., A274, p. 55, 1963.
15. Cho, E. H., Dixon, S. N. and Pitt, C. H., Met. Trans. 10B, p. 185, 1979.
16. Crank, J., Phil. Mag., 39, p. 140, 1948.
17. Snoeyink, V. L. and Weber, W. J., Environ. Sci. Technol., 34, p. 112, 1968.

Chapter 8

HOMESTAKE'S BULLDOG MOUNTAIN CARBON-IN-PULP SILVER PLANT

Steven Mitchell, Larry Trautman and Douglas Halbe*

*Steven Mitchell is Plant Metallurgist, Creede Operations
Larry Trautman, Assistant Metallurgical Superintendent, Lead Operations
Douglas Halbe, Manager of Metallurgy
Homestake Mining Company

BACKGROUND

Homestake Mining Company began milling operations at the Bulldog Mountain Mine near Creede, Colorado in 1969. The Bulldog Mill, rated at 350 tpd, produces a bulk flotation concentrate assaying about 180 oz/T Ag and 20% Pb, from heads of about 16 oz/T Ag and 2% Pb.

The circuit is straightforward with two-stage grinding to 1-5% plus 65 mesh, followed by rougher and cleaner flotation. Reagents used are Z-6, Z-11, Aerofloat 242 and Aero Promoter 404, with an Aerofroth 65-MIBC mixture for a frother. Additional information on the geology, mine and mill are available in references (1) and (2). A flow sheet for the mill and CIP plant together is attached as Figure 1.

Final flotation tailings normally vary from 1.5 - 2.0 oz/T Ag. These tailings are cycloned to produce sand for underground backfill. After cycloning, the sands are repulped and given a final scavenger float. This sand scavenger recovers a small additional amount of silver which recycles back to the main flotation circuit. The tailings are recycloned and pumped to the mine or, if the mine cannot accept sand, to tailings.

The majority of the silver in the flotation tailings exists in the fine fractions; cycloning upgrades the slimes and lowers the assay of the sand. The sand, after scavenging, typically assays 0.5 oz/T Ag, and the slimes assay about 3.0-3.5 oz/T Ag.

From 1969 to 1976 all of the slime tailings and the sand tailings not used underground were impounded in a tailings pond located about

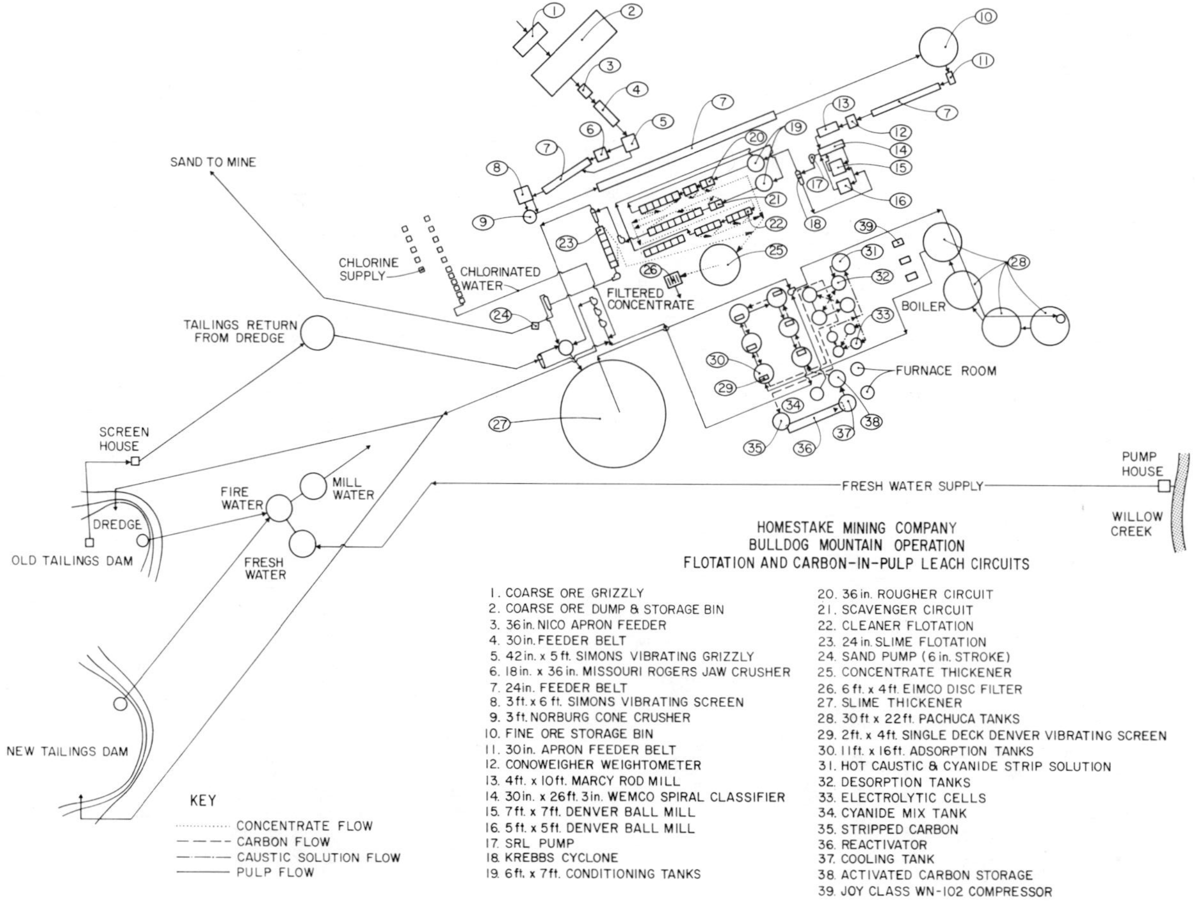

Figure 1. Bulldog Mountain flotation and carbon-in-pulp circuits.

two miles from the mill site.

CARBON-IN-PULP PROCESS DEVELOPMENT

In 1973, Homestake constructed a carbon-in-pulp (CIP) plant at the Homestake Gold Mine in Lead, South Dakota. This replaced the old slime filtration plant for cyanidation of the minus 200 mesh fraction of Homestake ore (3). The process, tested and piloted jointly by metallurgists from Homestake and the Bureau of Mines Salt Lake Metallurgy Research Center is metallurgically efficient and economical.

As described by the Staff Metallurgist at Lead at the time (4):

> From the resounding success of the carbon-in-pulp process on low-grade gold ores, it was decided to try it on the low-grade Bulldog Mountain slime tailings. Research was conducted in a joint effort by the Bureau of Mines and Homestake Mining Company (5). It was determined that 70-80% of the silver could be extracted from the slime by cyanide agitation leaching for 36 hours. Virtually all of the value could be adsorbed from solution on activated carbon in one hour. A lab pilot test and a field pilot test were run to confirm the results.

CIP PLANT

The Bulldog CIP Plant was designed to handle the total slime from the flotation plant (150 tpd) plus 250 tpd to be dredged from the old tailings dam during the summer session.

The plant was designed by the Galigher Company of Salt Lake City in close cooperation with Homestake engineers from the Lead operation. Engineered to share a common wall with the existing flotation mill, the dimensions of the CIP plant are 100 ft x 74 ft x 34 ft high, with the dissolution tanks outside. Figure 2 shows the layout of the CIP plant schematically.

The contractor for the construction was State, Incorporated, of Pueblo, Colorado, the contractor for the original plant. Construction was started in April 1976, and the plant was on-line in December of that year.

The plant has operated successfully since that time. With the tremendous current interest in gold and in the CIP process, both here and in South Africa, it seems appropriate to review the Bulldog plant and discuss some of the modifications and things we have learned since start-up three years ago.

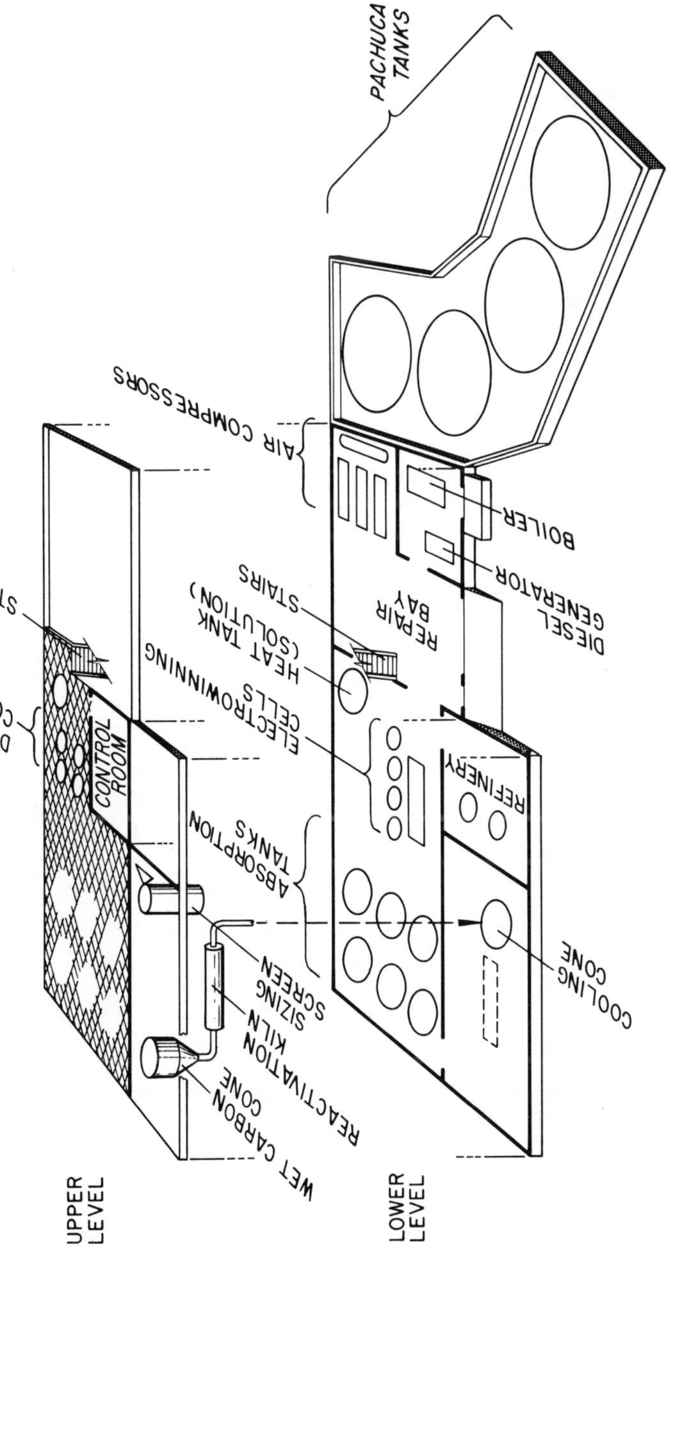

Figure 2. Schematic of plant layout.

TAILINGS RECLAMATION

Tailings from the old dam are reclaimed using a floating suction-type dredge. The dredge is operated from spring thaw until freezing over of the pond is imminent, generally from the end of May through the end of October.

The dredge operates eight hours a day and is maneuvered by means of electric winches and deadmen on the shore. The operator attempts to keep a relatively constant mixture of sand and slime. Slurry from the dredge is screened to remove any trash, then pumped by a four-inch Galigher horizontal pump to a 35 x 32-foot storage tank adjacent to the CIP plant. The dredge operates about 70-75% of available time. The major maintenance problem is wear of the seals in the cutter head cylinder and consequent leakage of water into the hydraulic system.

Solids in the surge tank are kept in suspension by means of three tangential jets located about four feet up from the flat bottom of the tank. A three-inch Galigher horizontal pump circulates slurry from the tank through these jets continuously.

Even though the amount of +200 mesh sand is small, it is larger than originally designed for. Everyone who has ever run a mill operating in conjunction with a sand backfill mining operation will know that as far as the mine is concerned, "they take all of the available sand, all of the time," and nothing will convince the mine staff otherwise. Unfortunately, when the CIP reclamation system was designed, we accepted that the amount of sand would be minimal and designed a handling system for essentially slimes; and we have been fighting this ever since. A number of modifications, however, have made the system fairly workable, and a few others are scheduled for additional improvement. The system is workable but not certainly at optimum.

Any overflow from the tank returns to the old tailings pond by gravity.

THICKENING

Overflow from the mill tailings cyclones (about 150 tpd) is pumped directly to a 50-foot outside thickener adjacent to the CIP plant. During the dredging season, slurry from the dredge storage tank also flows by gravity to the thickener at about 10 tph.

This thickener is the first critical step in the CIP process. Flocculant is added, and the slurry is thickened to 44-48% solids, then pumped by two four-inch ODS pumps to a 2 x 4-foot Denver-Dillon vibrating screen for trash removal.

A particular point of interest here, especially to those designing a CIP circuit, is the presence of this trash screen. Without such screening, a surprising amount of coarse trash--rocks, pieces of plastic from shot wire and powder bags, wood chips--begins to collect with the carbon.

The original plant design called for a curved punch plate trash screen with a revolving paddle scraper. This was not effective, and we have gone to a vibrating screen identical to those used in the adsorption circuit. Originally, we used 16 mesh screen here because finer cloth blinded too quickly with lime. Since we have started adding a phosphate for calcium and magnesium carbonate control, however, (see adsorption section), we have had little carbonate build-up on the screen and have been able to use 24 mesh screens--the same mesh as on the adsorption circuit carbon screens.

Some trash still builds up with the carbon and is usually most visible on the surface of the carbon in the desorption cones after desorption. Excessive amounts of trash plug up the eductors, but the limited amounts which are now present seem to pose no particular problem.

Bulldog Mountain ore is sometimes clayey, and at times difficulty is experienced in maintaining 48% solids in the thickener underflow. This high-clay ore is also difficult to screen through the trash screen, normally requiring more than the regular minimal amount of spray water. Fortunately, it is possible to operate satisfactorily at lower densities in the carbon adsorption tanks during these periods because the increased viscosity of the slurry allows the carbon to remain in suspension. The addition of small amounts of sand from the dredge circuit is noticeably effective in minimizing the occasional problem of these clays.

Table 1 lists a typical size and silver distribution of both typical dredged product and typical mill tailings. Note that the majority of the weight of the dredged material and the vast majority of the silver is in the minus 325 mesh fraction.

Lime and sodium cyanide are added to the slurry after screening, and the mixture is pumped to the Pachucas. Overflow from the fifty-foot thickener is recycled to the mill grinding circuit or, if too dirty, as is occasionally the case, back to tailings until the thickener problem can be cleared up.

Table 1. Size and Silver Distribution - Dredged Material and Slime Tailings

	Dredged Material			Slime Tailings		
Mesh	% Weight Distribution	Ag oz/T	% Ag Distribution	% Weight Distribution	Ag oz/T	Ag Distribution
+48	0.6	0.76	0.1			
48x65	3.5	0.56	0.6			
65x100	5.7	0.68	1.2			
100x150	6.5	0.68	1.4			
150x200	6.2	0.70	1.3			
200x325	8.0	0.92	2.3	0.9	1.72	0.7
-325	69.5	4.32	93.1	99.1	2.32	99.3
Total	100.0	3.23	100.0	100.0	2.31	100.0

DISSOLUTION CIRCUIT

The dissolution circuit was designed to give 36-hour contact time to 400 tpd of slimes at 48% solids. The four Pachuca tanks are 22 ft in diameter by 28 ft high, with one center air lift. The bottom is poured concrete with a 15° slope to the center. The pulp is transferred from tank to tank by a small air lift near the side.

Solution cyanide strength is maintained at 0.06% NaCN in the feed, with the pH between 10.5 and 11.0. Cyanide consumption in the dissolution circuit is 1.8 lb/ton. It has been determined that the upper pH limit is 11.0 for the reaction to proceed with reasonable speed and the lower limit is 10.2 at which point cyanide evolves to the atmosphere. Higher cyanide strengths, up to 0.10%, have been used in the plant, but silver recovery was not improved.

The most critical operating parameter is oxygen content of the slurry. The Pachuca tanks adapt themselves to giving excellent aeration as evidenced by the dissolved oxygen profile included in Table 2.

Table 2. Dissolved Oxygen Profile - Dissolution Tanks

Dissolution Tank	Oxygen Slurry Content
Pachuca Tank #1	91.5 %
#2	94
#3	95
#4	97

An interesting phenomenon has been noted both in this silver circuit at Creede and in Homestake's gold CIP circuit at Lead, South Dakota. Recovery in the Pachuca or dissolution tanks increases from tank to tank and reaches an apparent maximum, sometimes before the final dissolution tank is reached. On reaching the adsorption cones where gold and silver are removed from solution, additional dissolution takes place. This additional recovery is actually quite significant. Table 3 gives detailed typical data on this effect at the Lead CIP plant. The effect at Creede in the adsorption tanks is illustrated in Table 4.

ADSORPTION CIRCUIT

The discharge of the dissolution circuit contains from 0.80 to 1.0 oz Ag/ton in the solids and 2.2 oz Ag/ton in solution. At this point cyanide is about 0.04%, pH is 10.5-10.6, and the percent solids ranges from as low as 40% up to 48%.

Table 3. Plant Profiles - Lead

Sample Location	Temp °C	pH	NaCN %	Dissolved Oxygen-%	Solids Milliounces/T Au	Solution Milliounces/T Au
CIP Feed	18	8.7	–	14	76	0.5
#1 Dissol. O'flo	19	10.3	0.040	42	14	25
#2 " "	19	10.4	0.040	44	6	28
#3 " "	20	10.3	0.033	41	6	16
#4 " "	20	10.7	0.023	51	4	23
#5 " "	20	10.5	0.025	51	4	27
#6 " "	20	10.5	0.018	55	5	27
#7 " "	20	9.9	0.015	82	4	26
#1 Ads. O'flo	20	10.9	0.015	88	4.5	8
#2 " "	19	10.6	0.013	93	3	3
#3 " "	18	10.4	0.013	98	2	1
#4 " "	18	10.3	0.010	100	2	0.5

Table 4. Plant Profiles - Creede

Sample Location	Dissolved Oxygen-%	Solids oz/T Ag	Solutions oz/T Ag	Carbon oz/T Ag
November, 1979				
Ads Tank #1	71	0.92	2.05	340
" " 2	75	0.70	0.23	208
" " 3	70	0.60	0.17	159
" " 4	66	0.62	0.07	114
" " 5	68	0.62	0.08	77
" " 6	79	0.64	0.07	45
October, 1978				
Pachuca #1	-	2.20	1.91	-
" #2	-	1.58	2.16	-
" #3	-	1.30	2.18	-
" #4	-	1.28	2.14	-
Ads Tank #1	-	1.16	1.85	-
" " 2	-	1.02	1.42	-
" " 3	-	0.88	0.64	-
" " 4	-	0.94	0.11	-
" " 5	-	0.74	0.06	-
" " 6	-	0.62	0.03	-

The adsorption circuit consists of six stages, with one 11 ft. x 15 ft tank per stage. Each tank is agitated by a ship-type propeller agitator and three air lifts at the sides (Figure 3). Each slurry tank is equipped with a 2 ft x 4 ft Denver-Dillon vibrating screen above it. The screen cloth is 304 stainless steel, 24 mesh, 0.018 wire, 0.0237 opening. The slurry with carbon is air lifted by an external lift over the screen with the slurry discharging to the next tank, and the screened carbon returning to its original tank or for advancing to the preceding tank.

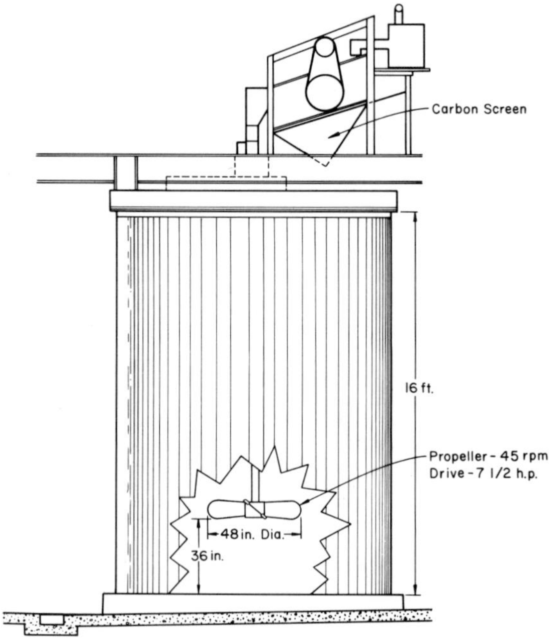

Figure 3. Adsorption tank and carbon screen.

Carbon, 6 x 16 mesh, adsorbs the silver from solution almost completely in the one and one-half hour residence times (at 400 tpd). Normally, 9,000 pounds of carbon are maintained in number one tank and 2,000-3,000 pounds in all others. Lead, on the other hand, keeps the largest load of carbon in the final adsorption tank. In Lead, normally Tank 1 has 4,500 lbs., Tank 2 has 4,500 lbs., Tank 3 has 5,000 lbs., and Tank 4 has 6,500 lbs. Carbon is advanced at Creede by batch on a three to four day cycle, depending on the solution strengths coming off the desorption cones. Seven thousand pounds are moved from the number one tank to a desorption cone. Carbon from tanks 2 and 3 are advanced to tank 1. Carbon is moved successively, counter-current to the pulp flow, ahead from tank to tank to maintain the desired carbon loading. Control of the advance is maintained by periodically grabbing slurry samples and checking carbon content. Fresh reactivated carbon is added to tank 6.

At Creede, it normally takes three to four hours to cut 7,000 pounds of carbon to the desorption cones, and then from 16 to 24 hours to complete the entire carbon transfer. Generally tank 3 is advanced from tank 2 to tank 1 at one time. Then tank 6 is moved to 5 to 4 to 3 in another batch run.

At Lead, the philosophy is somewhat different in that the carbon is moved no more than one tank at a time to maximize carbon loading gradient from tank to tank.

The question of why we do not use a finer carbon, such as 12 x 30 mesh, instead of the 6 x 16 mesh, has often been raised, since some researchers have reported higher loading on finer carbon. In our research and operation of carbon-in-pulp processes, both at Lead and Creede, we have found there is very little difference in the activities between the two sizes of carbon.

From an operating standpoint, we prefer the 6 x 16 mesh carbon because the larger mesh, less costly screens for screening the carbon can be used in the adsorption step. Also, with larger mesh screens there is less blinding of the openings with wood chips and fine carbon. Of concern, too, is the fact that it is possible to pass more tonnage of material through a large mesh screen.

One major problem in carbon-in-pulp circuits deserves particular mention at this time. This is carbonate build-up on the carbon. As with lime build-up in any milling circuit, the problem is acute where recycled water and fresh water are combined. We have found that the pores of the carbon apparently begin to fill with calcium carbonate. The change in appearance of the carbon, particularly under a microscope, is marked. The carbon takes on an almost glazed, sometimes whitish, appearance. This carbonate build-up has an immediate adverse effect on the adsorption of gold and silver on the carbon.

A particularly bad siege of carbonate build-up at the Bulldog occurred about a year ago. The ore to flotation was acid, requiring much larger than normal additions of lime for pH control. Shortly after this acid ore began arriving, the reactivation kiln was down for about two months for repairs. During this period the carbon became extremely fouled, and solution losses increased from our normal 0.05 oz/T or less to as high as 1 oz/T.

Once the calcium and/or magnesium carbonate is present on the carbon, the only way we know to remove it is an acid treatment, although heat reactivation may possibly have some beneficial effect until the fouling becomes too great.

Carbon was taken from the desorption cone and reactivated

through the kiln normally. From the reactivation discharge cone, it was jetted to the acid treatment cone located outside the mill building and treated with HCl in 1½ ton batches.

Although monitoring (with a MSA Drager tube monitor) has indicated that little HCN gas is given off during acid treatment, we prefer not to be inside a building with the process. Hydrochloric acid is added to the carbon-water mixture and allowed to react--a typical acid-carbonate effervescence. After the reaction is completed, the cone is drained, then flushed with mill water. After flushing, the carbon is educted back to the regular circuit to the sizing cone, sized over a 24-mesh screen for removal of fines, and re-introduced into the circuit.

Thermal reactivation both before and after acidification has been examined, and it has been found that the sequence makes little difference. We have been advised, however, that any residual HCl remaining in the carbon going from acid treatment to thermal reactivation would cause serious corrosion problems with the stainless steel lining of our kiln (6). For this reason we thermally reactivate before acidifying. Further, it has been found that addition of polyphosphate to the slurry will minimize scaling but will not remove carbonate already in place. Even with the polyphosphate, a small amount of lime build-up occurs, and we anticipate having to acid treat occasionally to keep the carbon at top efficiency. Our lowest barrens at Creede (.006-.007 oz/T) almost invariably occur shortly after a program of acid treatment.

At Lead about 3 ppm of Baroid S-35 is added to 600 gpm of the slurry feeding the CIP plant. This maintains about 1 ppm organic phosphate in the discharge solution from the adsorption circuit. The solution is analyzed for residual phosphate periodically using a Hach Kit colormetric determination.

At Creede S-35 is added to about 0.6-0.7 ppm organic phosphate (0.9 ppm total residual phosphate and 0.35 ppm inorganic phosphate). It is planned to increase this somewhat at Creede for a test evaluation.

Incidentally, the addition of phosphate has the distinct side benefit of reducing the rate at which the carbon screens blind. Before phosphate was added at Creede, it was necessary to change screens and acid treat them about every week or less. Now the screens do not blind with scale at all.

Table 5 shows results of activity tests on new, thermally reactivated, and acid-treated thermally reactivated carbon at both Lead and Creede. In both cases the acid treated carbon is more active than the fresh.

Table 5. Activity Test - Fresh, Reactivated and Acid-Treated Carbon

Lead

(milliounces Au/ton solution)

Time Minutes	Unreactivated Carbon	Reactivated Carbon	Acid-Treated Reactivated Carbon
0	22	30	30
10	11	15	12
20	5.5	10	6.0
30	5.0	6.0	3.0
40	3.5	4.5	2.5
50	3.5	3.5	1.0
60	2.5	2.5	1.0

Creede

(ounces Ag/ton solution)

Time Minutes	New Carbon	Reactivated Carbon	Reactivated Acid-Treated Carbon
0	1.96	1.96	1.96
5	0.66	0.33	0.17
10	0.29	0.16	0.06
15	0.20	0.10	0.04
30	0.11	0.06	0.02
45	0.08	0.04	0.01

It has also been found that when scaling is present, the production of fines in the reactivation furnace is significant. Table 6 shows some typical screen analyses of such carbon (in this case from Lead) both before and after reactivation. Note the amount of fines produced. When scaling of the carbon does not exist, the amount of fines produced during reactivation is almost insignificant.

Table 6. Carbon Degradation from Reactivation - Lead

	Unreactivated		Reactivated	
Mesh	% Weight	Cum. % Weight	% Weight	Cum. % Weight
+8	4.8	4.8	2.6	2.6
+12	33.0	37.8	17.7	20.3
+14	36.5	74.3	29.0	49.3
+16	11.2	85.5	19.0	68.3
+20	7.7	93.2	16.3	84.6
+24	5.3	98.5	11.3	95.9
-24	1.5	-	4.1	-

DESORPTION CIRCUIT

The 7,000 pounds of loaded carbon from the adsorption tanks is charged into one of four 6 ft x 7 ft stainless steel stripping vessels. The carbon is stripped of silver in three stages using hot caustic cyanide solution. Each charge is allowed four days to strip.

The strip solution is 172°F, 1% NaOH, and 0.3% NaCN, and flows at about 5-7 gpm. The solution is pumped through 316 stainless steel pipe and 316 stainless steel automatic ball valves in upward percolation through each cone (see Figure 4).

The stainless steel automatic ball valves are controlled pneumatically. The automatic valving system has worked very well; however, the most important single factor is a very clean, oil-free air supply. A Microalescer final air filter was installed sometime after start-up, and since that time the problems of sticking valves have been greatly reduced. Occasionally, a build-up on the valves does occur so that they will not seat, and it is then necessary to clean them in acid. At the bottom of the cones, there is a screen which keeps the carbon from getting into the valves. Initially, these screens were constructed of 304 stainless steel but due to corrosion problems, the screen material was changed to Type 316. These screens are spot welded to a frame, and they eventually pull away leaving small holes. These screens must be replaced every four months.

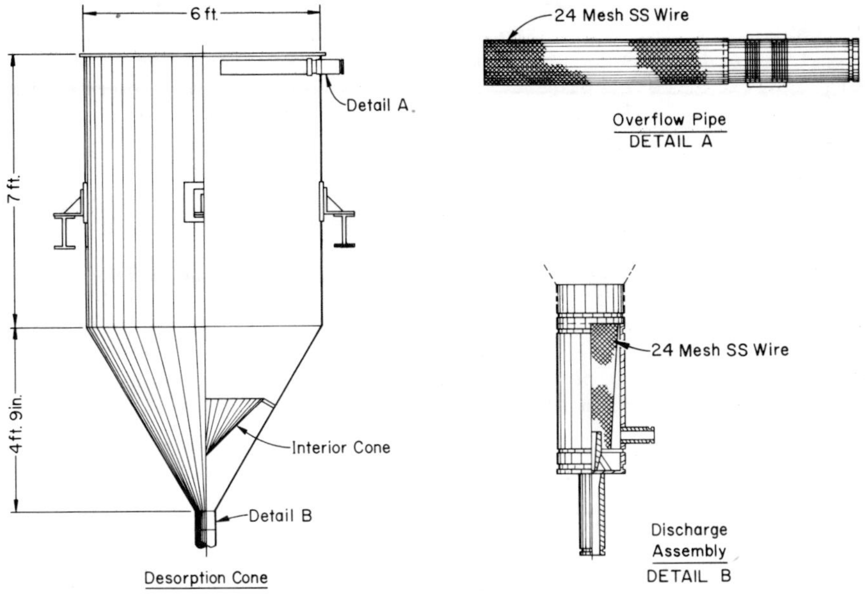

Figure 4. Desorption cone.

At full tonnage the discharge from the stripping cone is divided with one-half going to four Zadra-type cells in series and the other half going to a rectangular, eight compartment electrowinning cell. The barren solution is recycled through the circuit. During low tonnage periods, only the Zadra cells are used.

When a cycle is completed, the stripped carbon is hot flushed twice and cold flushed twice prior to being jetted to the reactivation feed cone. The carbon is jetted with 1½" cast iron, Type 264 SK water jet eductors, supplied by Ameteck-Schutte & Koerting Division. The stripped carbon normally assays 30-40 oz/T.

A few words about temperature of desorption is appropriate here. The desorption circuit was originally designed to operate at 190°F. Since start-up, however, we have been able to maintain the solutions no higher than 172°F. This temperature has provided adequate desorption for our circuit. We have discussed the possibility of some additional insulation and of additional heat input to reach 190°F (about the maximum obtainable at Creede's 9,360 feet elevation), but it has never been a high priority item for us, and thus has not yet been done.

In contrast with gold solutions at Lead, it is essential to operate at 195°F or higher. With temperatures more than a few degrees below this, desorption drops off rapidly. For example, at 185°, the gold comes off much more slowly, requiring twenty-four hours additional stripping time. In addition, instead of reaching a final level of gold in the carbon of 3-4 oz/T, stripping only to about 12-13 oz/T is possible.

ELECTROWINNING CELLS

The electrowinning cells originally installed were Zadra type, where the cathode is a center spool wrapped with steel wool and the anode is a surrounding stainless steel screen (Figure 5). Solution enters these cells through a slotted center feed tube and exits through an overflow pipe on the side of the tank. These cells have given good service from start-up.

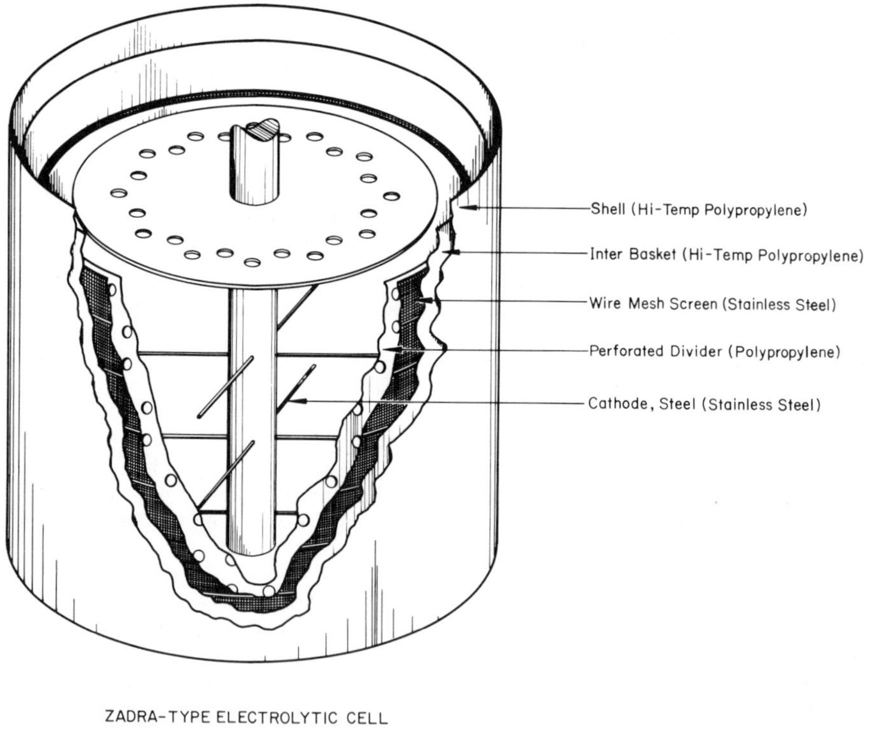

Figure 5. Zadra - type electrolytic cell.

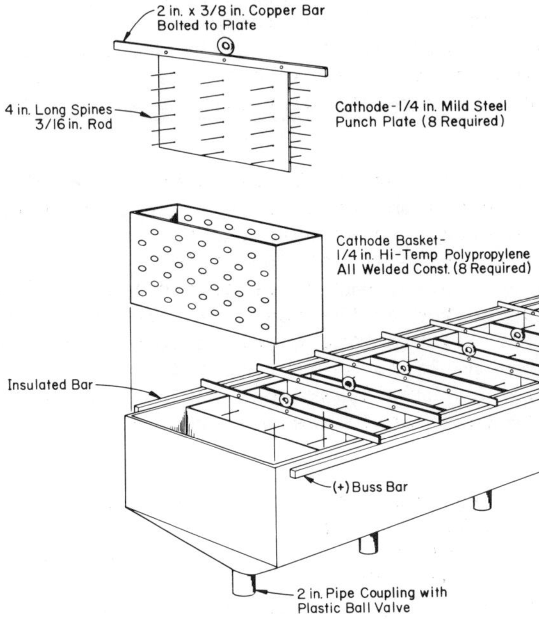

Figure 6. Rectangular electrolytic cell.

One modification to the fourth cell is currently under study. We are wrapping that cathode with a coarser (#3) steel wool and will evaluate whether this change will provide a longer steel wool life.

In the rectangular cell about two cathodes are normally pulled per day at high tonnage operation. At the end of the month, all cells are pulled for circuit cleanup.

Electrical connections in this cell are through a copper bus bar for the cathode (as shown in Figure 6) and individually bolted connections for each anode.

We experimented originally with using the entire mild steel tank of the cell as an anode. However, we fairly soon built up an oxide coating on the inside of the entire cell and developed a high resistance through the circuit. This lowered the cell amperage to lower than acceptable levels. Along with this, of course, corrosion of the tank itself was severe enough to make it unuseable after a number of months. The stainless steel anodes now in use seem to have solved these problems.

The original cathode design called for four sets of spokes on the spool to support the steel wool. No matter how carefully the steel wool was wrapped, however, as silver built-up, the cathode began to get square. The addition of extra spokes for a total of eight per row provide a much more circular cathode and posed no problem in wrapping.

Other modifications to the original design have been slight, namely, material in the anodes and contact bars and more holes in the cathode protecting basket. Holes near the bottom of the basket, though, have been minimized to eliminate possible short circuiting through the sludge and to prevent excessive sludge from falling through the holes to the bottom of the tank. We did experiment on wrapping the cathode with plastic window screen to eliminate what we felt might have been some possible short circuiting through loose strands of steel wool from a heavily-wrapped cathode to the anode. No positive benefit from this change was observed, and this practice has been discontinued.

The rectangular electrowinning cells are experimental. The tank is constructed of mild steel. The cathodes are rectangular steel plates wrapped with steel wool and the anodes are 1/8th-inch 316 stainless steel plates hung on each side of the cathode (Figure 6). High temperature, 1/4th-inch, polypropylene plastic covers fit between electrodes to provide a cell cover.

For the Zadra cells we are currently using twenty pounds per cathode of #2 grade steel wool, and for the rectangular cell two and one-half pounds per cathode.

In the Zadra cell circuit, solution flows from cells one through four in series, with cell four acting as a final scavenger. The solution through the rectangular cell is also scavenged through the final Zadra cell. Solution from the desorption cones is circulated until the strength falls to about 2 oz/T, corresponding to 30-40 oz/T left in the carbon.

Zadra cell cathodes are pulled after about four to five days of plating. After considerable experimentation, we now have adopted the following cell advance system. Cathodes for melting are always pulled from number one cell. A partially loaded cathode from cell two or three (the one which is more heavily loaded) is advanced to cell one, and a newly wrapped cathode is placed in the empty cell. Cell four, the scavenger, is pulled only once a month. By the end of this month, most of the steel wool is gone: dissolved or fallen to the bottom of the cell basket. The cathode in this cell contains a high percentage of impurities and requires careful fluxing to produce even a low-grade (800-fine) bar. This high impurity level is most obvious when we are operating at low-feed tonnage. At high tonnage, the quality of the dore' from this cell is not much lower than the other cells.

We are currently engaged in a small joint venture research project with St. Joe Minerals Corp. on electrolytic cell design. They are in the process of designing a mill and CIP plant for their El Indio, Chile, gold-silver-copper mine (7). Some of the first quantitative data to come out of this work deals with steel wool density. Tests were run with the Zadra cells using 20 pounds of #2 steel wool in the number one cell for a complete five day loading cycle, then 30 pounds in the cell for the next loading cycle. The silver extraction efficiency (Ag In - Ag Out)/(Ag In) was markedly higher with the greater weight of steel wool, as shown in Figure 7. Although this will have to be confirmed in further tests, it is certainly a point of extreme interest.

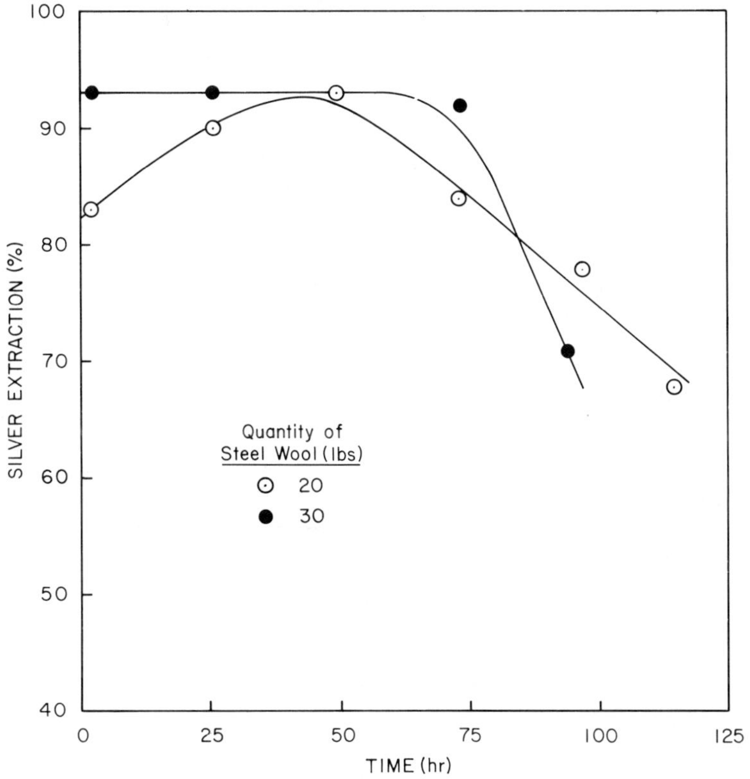

Figure 7. Reduction of silver on steel wool in the electrolytic cell.

The electrowon silver in either cell forms a loosely adhering plate on the steel wool. Approximately one-half of the silver sludge drops from the steel wool and is collected on the bottom of the cells. It must then be filtered and dried prior to refining. The silver sludge is approximately 85% silver.

The design of the Zadra cells (cathode, cathode basket, anode, anode basket, tank) is such that the vast majority of the sludge is collected in the baskets and pulled from the cell for treating. On occasion when there is significant sludge on the tank bottom, the most convenient method we have found is wiping it up with a sponge--not too convenient but not frequently necessary.

Most of the sludge in the rectangular cells is also caught in the cathode baskets. Enough falls to the bottom, however, to make it necessary to pump out the solution, then drain from the bottom of each two cells the sludge-electrolyte mixture. After settling, the clean electrolyte can be pumped off and the sludge removed and filtered.

One of the most important operations of the electrolytic circuit is the bleed-off of solution containing impurities on a periodic basis. In our Lead gold-silver operation, 90% of the strip solution is bled out on a weekly basis. If this is not done, copper plates on the cathode, which reduces the fineness of our dore' bullion.

At Creede about 10% of the electrolytic cell dishcarge is bled off to the tailings pond per week.

Typical impurity levels in the cell feed and discharge are shown below in Table 7. More detailed data on typical operating parameters is given in Table 8 below.

Table 7. Impurity Levels - Electrowinning Circuit

Element	Feed (ppm)	Discharge (ppm)
Fe	1,490	1,590
Zn	570	560
Cu	150	150
Pb	41	2
Ni	27	29
As	15	15
Sb	2	2
Mo	2	2
Al	<1	<1
Co	~0.8	~0.8
Bi	0.7	<0.5
Hg	<0.02	<0.02

Table 8. Steel Wool Loading - Electrowinning Cell

		Cell Data			Feed Solution		Discharge Solution			
Date	Hours	Volts	Amps	Liters/min	oz/T Ag	TSS(ppm)	oz/T Ag	TSS(ppm)	% Extraction	Current Efficiency
(Test 1)										
Aug 24	1	9.6	203	15.6	8.28	926	1.37	235	83	27
25	24	5.8	225	13.3	17.86	142	1.72	118	90	49
26	48	6.2	225	14.1	8.87	77	0.66	132	93	26
27	72	9.5	218	10.1	7.43	300	1.20	200	84	15
28	96	6.0	225	9.7	34.33	71	7.73	54	78	58
29	114*	5.2	225	8.9	25.40	43	8.03	81	68	35
(Test 2)										
Aug 29	1	9.1	225	11.9	7.41	534	0.54	70	93	19
30	24	6.0	225	12.2	33.73	70	2.24	92	93	87
31	48	5.9	225	13.3	-	-	-	-	-	-
1	72	5.8	225	12.6	26.68	66	2.20	56	92	70
2	93*	5.6	225	12.2	25.59	76	7.52	92	71	50

* Pulled

Test Data	Test I	Test II
Cell	Zadra	Zadra
Steel	21 lb - Grade 2	30 lb - Grade 2
Dry Weight Cathode	124.6 lb (18% moisture)	120.8 lb (15.5% moisture)
Oz Silver - bar	1545.5 oz	1589.2 oz
Fineness	923	926

1 $\% \text{ Extraction} = \frac{\text{oz Ag In} - \text{oz Ag Out}}{\text{oz Ag In}}$

2 $\text{Current Efficiency} = \frac{\text{gm Ag/hour actual}}{\text{gm Ag/hour theoretical}}$

Detailed data from the tests are attached as Table 8.

Visual observations of the loading of the silver in this and other tests have indicated that as the steel wool becomes loaded it sags badly, probably allowing loaded solution to bypass electrolysis. Since the solution flow in this cell is from the center out, most of the initial loading takes place on the steel wool nearest the feed pipe. We plan to experiment next with using coarser (and stronger) steel wool near the center at a fairly loose packing to plate out the initial silver, then a more dense packing of finer steel wool near the periphery of the electrode for better final stripping.

Later work on this project will involve testwork on-line with plant solutions in several small (4 liter) test cells to evaluate parameters such as voltage and resulting amperage, corrosion, effect of impurities (i.e. bleed-off rate), solution flow, and sludge removal.

REACTIVATION

The stripped carbon is reactivated for 20 minutes at 1100°F in an indirect-fired rotary kiln fired with fuel oil. Reactivated carbon is screened to remove fines, then recycled back to the adsorption circuit. About 2250 pounds of new carbon are added per month for make-up.

The tube through which the carbon flows is constructed of 321 stainless steel. Originally, the flights inside the tube were welded rigidly to the tube and due to constant heating and cooling, the tube pulled apart at the weld, allowing carbon to fall into the duct work. This plugged the burner box and duct work, which then had to be dismantled, cleaned and relined. Excessive cracking and spalling of the refractory were also noticed.

Due to the wearing out of the refractory and problems with the tube, the following modifications were made. A new 8 ft section of the stainless steel tube was installed with the flights spot welded, the duct work was moved to the side to keep the hot gases from impinging directly on the tube (Figure 8), and the refractory was changed from Kast-Set to Kruzite Castable. These modifications have greatly reduced the kiln downtime.

We also get a build-up on the flights which breaks off and used to constantly plug the carbon eductors. A 1/4th-inch trommel screen was installed on the carbon discharge end of tube, and this has eliminated the problem.

There is some degradation of carbon during reactivation, and this, along with abrasion from jetting, probably accounts for most of the total carbon loss referred to above.

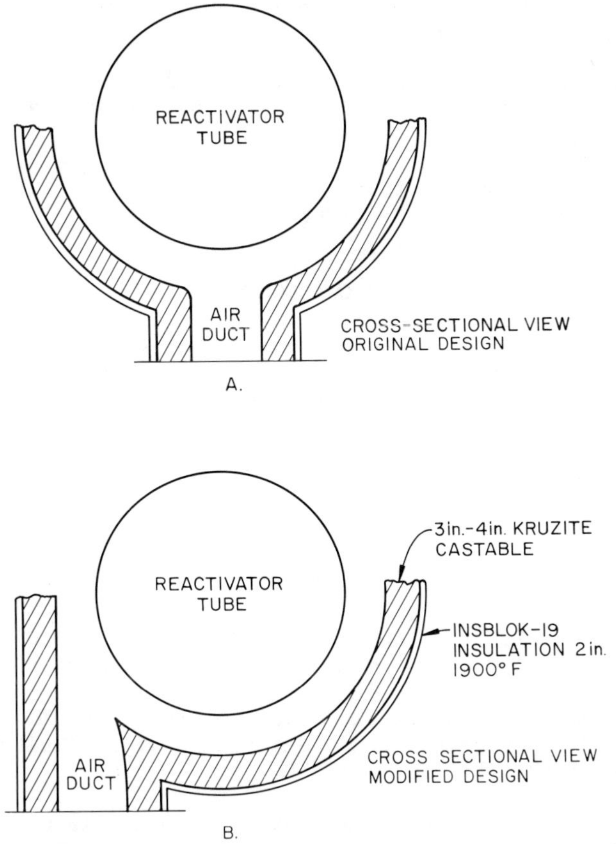

Figure 8. Reactor kiln modification.

REFINING

The cathodes are removed from the electrowinning circuits as noted above. The Zadra cells normally contain about 1500 ounces of silver when pulled--about four to five days of plating. A flux consisting of 30 pounds of sodium nitrate, 30 pounds of anhydrous borax, and 30 pounds of sand is mixed with the cathodes to slag off the steel wool in one melt.

The melt takes place in a Denver Fire Clay tilting furnace with a #150 starbide crucible. A rough bar is poured into a cast iron mold, and the slag is then removed.

The silver bullion at this point is 950 fine, and the slag contains approximately 10 oz Ag/ton.

ACKNOWLEDGMENTS

Special thanks are due to Ociel Plaza of Compania Mineral El Indio for his work at Creede on the electrowinning cell project, and also to George Potter for his very valuable comments and assistance during this research.

Thanks also to Ed Bolles, Sheri Ransom and Gary Young for the graphics work for this paper.

REFERENCES

1. Jackson, Daniel, Jr., "Homestake's Hard Work Pays Off at Bulldog Mountain Mine," Eng. and Min. J., pp. 65-70, May 1974.

2. Howell, Frank M., "Homestake's Bulldog Mountain Mill, Creede, Colorado," paper presented at SME Fall Meeting, Salt Lake City, UT, Sept. 1969.

3. Hall, Kenneth B., "Homestake Uses Carbon-In-Pulp to Recover Gold from Slimes," <u>World Mining</u>, <u>27</u>, No. 12, pp. 44-49, Nov. 1974.

4. Downing, Terry and Mitchell, Steven, "Homestake Mining Company, Bulldog Mountain Operation, Carbon-In-Pulp Plant," unpublished.

5. Hussey, S. M., Salisbury, H. B., and Potter, G. M., "Carbon-In-Pulp Silver Adsorption from Cyanide Leach Slurries of a Silver Ore," USBM RI 8268, 1978.

6. Private Communication, George M. Potter, June 1979.

7. Anon, "St. Joe Minerals Will Develop El Indio Au-Ag-Cu Deposit in Chile's High Andes," Eng. and Min. J., May 1979.

Chapter 9

NEW DEVELOPMENTS AT HOMESTAKE'S BULLDOG MOUNTAIN CARBON-IN-PULP SILVER PLANT

Richard Kunter and Douglas Halbe

Homestake Mining Company
Golden, Colorado

INTRODUCTION

Additional work has been done on the CIP circuit at Creede, and a brief description of this work is presented in this paper.

DREDGE

The original dredge for the Bulldog was built by Al Hack, Inc. of Denver. The dredge cutter head is mounted on a boom made of 6-inch diameter pipe which is raised and lowered by a cable-pulley system. Maximum cutting depth is 8 feet. Slurry from the dredge flows through a floating 6-inch diameter pipe supported on 55-gallon drums which runs from the dredge to the shore and then to a trash screen and pumping system. See Figures 1 and 2.

The cutter head, suction pump, winches for dredge movement, and all other operations are controlled from a small operations house located toward the rear of the dredge.

All of the basic dredge equipment is mounted on a chassis about 8 ft x 16 ft in size. Additional buoyancy was originally provided by sealed 55-gallon drums supported along the sides of the chassis by a small wooden framework. Additional framework and barrels on the front of the chassis on either side of the cutter boom provided additional stability, particularly when digging in hard material.

The dredge was quite stable, but two problems were apparent. For one, it could have been sunk by someone from the shore firing a rifle into all of the barrels. In prime deer-hunting and sportsman country

Figure 1. Dredge reclaiming flotation tailings from Upper Tails Dam.

Figure 2. Cutter head and hydraulic assembly on dredge.

such as Creede, this is not an idle fear. In addition, with the barrels it was extremely difficult to move the dredge into and out of the water without some damage to these supports.

The first problem could have been eliminated by filling the barrels with a spray-in type foam. The second problem was not as minor as it may initially appear. Winters at Creede are severe but, frequently, several days of sub-freezing weather in the fall may be followed by several weeks of Indian Summer. It is necessary to remove the dredge from the water for the winter to prevent ice damage either prior to or immediately after the first freeze. With the original design, this was a big job and usually resulted in minor damage to the barrels and framework. Then, even though the weather might then have moderated and made possible several more weeks of dredging, it was either impossible or not worth the effort to put the barge back into the water. This problem would continue to worsen as the level of the pond was lowered with continued dredging increasing the steepness of the banks.

The new design eliminated both of these problems. All barrels were replaced with pontoons running alongside and to the front of the dredge, again on either side of the cutter boom. These pontoons were made of steel plate and were compartmentalized. In addition, the pontoons were angled at each end so that the dredge can be pushed into or pulled out of the water readily. When the dredge was tested initially, there was a little more buoyancy in the front than the rear, resulting in a slight rear-ward cant. This was modified by the addition of several sets of barrels attached to a small structure off the stern.

One of the main original operating problems with the dredge was the low percent solids in the dredged product. Shortly after the system was commissioned, additional teeth were added to the cutter head to aid in digging the solids. This helped. In addition, the operators soon became more skilled in maintaining higher solids, and we moved from an initial 5-6% solids to 15-20% routinely.

Overflow for the 32 ft x 35 ft dredge storage tank flows by gravity back to the upper tailings pond. This overflow is normally very low in solids, and this does provide some upgrading in the dredge tank of the percent solids in the system.

Recently we have installed a larger hydraulic pump system which allows greater pressure on the dredge cutter head with a marked increase in tonnage output. Although we do not measure directly the tonnage from the dredge pond, tonnage to the CIP plant is measured. The tonnage of slimes from the mill remains relatively constant, so increases in throughput are due to tonnage from the dredge. In recent experiments we have operated at a sustained rate of 460 tpd over a 5 day period as compared with a maximum rate of 360 tpd the previous year.

PULP STORAGE TANK DESIGN

As noted in the original paper, slurry from the dredge is pumped to a 35 ft x 32 ft storage tank adjacent to the CIP plant. The original design called for keeping the slurry in suspension by means of three tangential jets located 9½ feet up from the flat bottom of the tank. A 6-in. x 4-in. Galigher horizontal pump circulates slurry from the tank through these jets continuously. A portion of the pumped slurry also went to a small constant head box on the top of the storage tank and flowed from there by gravity to the CIP plant thickener. Any overflow from the box dropped back into the storage tank.

This design would probably have been adequate for minus 200 mesh material, but once the dredge circuit started, it became evident that there was considerable plus 200 mesh material in the pond. The mine uses sand fill in the mining operation, and the design assumption was made that almost all of the available sand in the past had been pumped underground and that very little had gone to the pond.

The original pump suction was located on the side of the tank about one foot from the bottom. The sand present in the feed built up fairly quickly in the tank in a cone shape, with the top about 3 feet above the level of the tangential jets. The swirling action imparted by the jets kept the sand from building up at the very edge all the way to the bottom.

This big sand bar reduced the capacity of the tank by about 25%. More seriously, however, sloughing occurred periodically directly into the pump intake. It occasionally plugged the pump but, more frequently, just sent surges of sand through the pipeline to the thickener, sanding that line. Water flushing and constant attention to the line helped somewhat, but the system was not very workable.

After some discussion a proposal was accepted to install a withdrawal system similar to the systems used in the phosphate fields of Florida. These systems use a dual suction line with the suction placed in the center of the bins and are able to handle surprisingly coarse feed; namely, deslimed phosphate and silica sand sized with cyclones to a nominal 16 x 150 mesh.

Since the Bulldog Mountain dredge pumping system is of a relatively low percent solids (15-25%) as compared to the phosphate operations, some modifications were made. Feed from the phosphate storage systems is approximately 85% solids, and is diluted to 45-50% solids with the dilution water line placed near the suction line of the pump. Since the feed is much finer in the dredging operation than in the phosphate system, a cover was placed over the suction end of the lines to help prevent possible plugging in the event of an extended shutdown. The cover is a 4-ft x 8-ft sheet of ¼-in. plate raised off the bottom 18 in. A dilution water line was placed under the cover for use if required. The dilution line is a 2-in. line reduced to 1¼ in. at the outlet. The outlet is located 6-in. away from and aligned with the center of the 6-in. suction line.

An additional set of four tangential jets was placed two feet up from the bottom of the tank and located midway between the original jets. A second 6-in. x 4-in. Galigher pump was installed to provide sufficient volume of pulp flow for operation of both sets of jets when required as well as to provide a stand-by system for routine operation. The two pumps are manifolded to be used interchangeably or together.

We felt that, particularly after a shutdown, it might be preferable to start circulation with the upper set of jets, then cut in the lower set. The system has operated so well since start-up in May of this year that we have not had to shut down and have not had the opportunity to experiment with the system. We operate routinely, however, using only the lower jets.

The original feed line from the hill tank to the CIP thickener had been a two-inch gravity flow line laid down the hillside with an irregular slope and crossed one roadway on an elevated trestle. The line was relaid and the trestle raised to increase the minimum slope of the flatter section from 0 to 15° and to provide sufficient elevation for gravity feed to a DSM trash screen. In addition, as a safety factor, a 2-in. x 2-in. pump was installed at the base of the tank to supplement gravity flow of the slurry to the thickener. Because of the problems we had had previously with the pipeline plugging, the operators used this pump frequently when the system was first started up. Now, however, it is seldom, if ever, used.

All piping is supplied with both high pressure water and air for flushing, and a spare two-inch line was laid from the tank to the thickener in case of serious plugs.

It may sound as if there was a little engineering overkill in the design, but the dredging season is quite short at Creede, and we wanted to be able to minimize those start-up modifications which might have been required had we designed less liberally. In addition, we had fought the system for several seasons and had all decided, I think, that now we were going to make it work right - period! These "extras" if you wish to call them that, represented, in fact, minor additional cost, and not only provided an additional assurance of trouble-free operation but would also give some welcome flexibility in operation of the system.

The system started up with no significant problems and continues to operate in a completely satisfactory manner. The amount of sand in the tank when it is drawn down toward the end of graveyard shift is small. We've been pleased with the operation of the system; and the mill operators, after having expressed some major reservations about whether the design would work, are delighted with its operation.

Figure 3. Pulp storage tank and Bulldog Mountain flotation and CIP plant.

DSM WOOD CHIP REMOVAL SCREEN

One of the minor but occasionally troublesome problems in the Creede plant is related to the clayey nature of some of the ore from the mine. At times the thickener underflow is so clayey that it forms a rope which can almost be carried around without a bucket.

As described in the previous paper, ..."at times we have difficulty in maintaining 48% solids in the thickener underflow. This high-clay ore is also difficult to screen through the trash screen, normally requiring more than the regular minimal amount of spray water. Fortunately, we are able to operate satisfactorily at lower densities in the carbon adsorption tanks during these periods because the increased viscosity of the slurry allows the carbon to remain in suspension. The addition of small amounts of sand from the dredge circuit is noticeably effective in minimizing the occasional problem of these clays."

The addition of water as a spray to the trash screen did help somewhat as did additional sand when available, but the problem was not really solved. A simultaneous solution to this problem and the problem of trash removal from the CIP thickener overflow prior to recycle is under evaluation now.

We have installed a 4-ft., 2.3 mm opening DSM screen ahead of the CIP thickener. See Figure 4. At present either the dredge tank slurry or the tails from the flotation plant can go over this screen. Until maximum tonnages from the dredge circuit are definitely established, we will be unable to determine exactly the total area of screen required. However, the screen appears to operate very effectively in removing the wood chips, wire, coarse rock, etc., from the slurry. Since the slurry is fairly dilute at this point, the clay presents no problem.

Figure 4. DSM wood chip removal screen.

A side benefit is that the wood and other floatable material which would report to the thickener overflow is removed here also. We have periodically checked the wood for silver values but have found nothing, so it is sluiced into the tailings sump and goes to the lower tailings pond.

At this point we have experimented with both 0.75 mm and 2.3 mm opening screens, which would be expected to produce a +24 mesh or +16 mesh product. Both appeared to work satisfactorily in removing trash. The main criterion is that all +16 mesh material which would otherwise follow the carbon in the CIP circuit be removed.

CHANGES TO THE BULLDOG MOUNTAIN MILL WATER AND PIPING SYSTEM

A number of changes were made to the Bulldog Mountain Mill water and piping system. The intent of these changes was to make better use of existing fresh water consumption to decrease chemical contamination in the sand fraction returned underground, to decrease the volume of solution in the tails line thus releasing capacity, and to return water to the slime thickener for clean-up and re-use.

Fresh water was being used to cool the plant air compressors and the Nash vacuum pump for the concentrate filter. The total water volume from these two uses was estimated at 40 gpm and was being discharged to tails via floor sumps. It was decided that a further beneficial use could be made of this fresh water by collecting it in a common sump and pumping it to the sand cells. Repulping the sand fraction with fresh water instead of mill water in the sand flotation cells reduces considerably the amount of any dissolved chemicals in the sand tail slurry. These sand tails are cycloned and repulped again with mine water prior to being discharged underground as backfill. The use of fresh water in the sand cells reduces the possibility of any chemical contamination being introduced underground from the mill circuit and could be done without increasing fresh water consumption in the mill.

A backup fresh water line was also installed to ensure a sufficient supply of repulping water particularly during mill start-up. Existing mill water lines were left installed as additional backup.

In the past there has been a considerable volume of solution circulating in the mill piping system. In particular, return water from the tailings dam was overflowing the mill surge tank and returning to the tails sump. It was easier from an operator's viewpoint to allow the tailings return water to overflow the surge tank than take the chance of a low level in this tank. During the dredging season, however, the extra capacity in the tails line was needed for the increased amounts of solids being pumped to the tailings pond. It was either install a new tails line or take measures to reduce the volume of needlessly circulating water. It was felt that the percent solids in the tails being pumped (7 to 14%) was sufficiently low that an increase in pulp density could be readily accommodated and, in fact, might be beneficial in terms of less pipe wear.

Three sets of changes in mill piping and solution flows were proposed to implement the decision to reduce the solution volume in the tails line. Two of these have been completed, and the third is to be implemented at a later date.

The first and most significant change was the installation of float valves and a low level alarm on the mill surge tank. The high float was installed on the mill water return line, and a lower float was installed on the mine water return line. A low level alarm was installed in the mill control room. The lines with float valves had manual shut-off valves installed above the float valves.

A third float system is expected to be installed when the water treatment plant return water line is installed. The float on this line would be at the same level as the mine return float. When this is installed, it is expected that the mine return water valve would normally be closed and not used. There is an unfloated line to the surge tank from the slimes thickener overflow. The effect of these changes was to insure a satisfactory supply of water to the surge tank without the volume of overflow that was being returned to tails. The mill surge tank will now preferentially (assuming the valves are open) take make-up water from (1) the slimes thickener overflow, (2) the mill water return, and (3) the mine water return, in that order.

The second change made was to add additional piping and valves to allow the CIP sump pump to discharge to the flotation floor sump pump and, thus, to the slimes thickener for cleanup and re-use. The only options that were originally available were (1) to return this sump's discharge to the pachuca tanks or (2) to pump to the CIP plant tails sump and, thus, directly to mill tails sump.

The original options were left installed; however, the new piping allowed any drainage and spillage (and values) to be recovered immediately rather than be returned from the tails pond. This provided a faster return of any overflows in the CIP plant (and mitigates any value losses as the underflow of the slimes thickener is the feed to the CIP plant), as well as reducing solution volume in the tails line.

The third change that is yet to be implemented is the installation of high- and low-level alarms and mill-control room readouts on the sand surge tank. This tank receives the cycloned sand tails and mine water for repulping these tails prior to their being pumped underground for backfill. There is a 4-in. overflow on this surge tank which goes directly to the tails sump. The location of the tank does not allow ready observation of overflows which could allow a large volume of excess solution (and sand) flow to the tails line. It has been easier to allow an excess of mine water to flow to this tank than risk the possibility of bogging the pumps and sand line should the flow drop unexpectedly.

When the water treatment plant is completed, it is intended to install a line to this tank to replace the mine water currently being used for repulping. This again will use a better quality water thus reducing any contaminants being discharged underground. It may be considered prudent at that time to install a float valve system with a back-up water source to ensure a proper steady flow of water to the pumps. The alarms should allow a more closely controlled water flow avoiding overflows and obviating the potential of bogging the pump and sand lines.

ELECTROLYTIC CELL TEST PROGRAM

Currently, Homestake is engaged in a joint silver electrowinning development program with St. Joe Minerals Corporation. St. Joe's interest in this program is based on their current development of the El Indio gold-silver-copper mine in Chile (2). The El Indio plant will include a mill and carbon-in-pulp plant and is scheduled for start-up in April 1981.

In reviewing the literature, it does not appear that any comprehensive fundamental studies of silver-gold electrowinning from CIP desorption liquors has been made in the past. This fact is witnessed by current research and development studies being conducted in South Africa by Anglo-American and the National Institute of Metallurgy (NIM). Both Anglo-American and NIM have developed new gold electrowinning cells for CIP desorption liquors. Pilot studies on these cells are currently underway.

The Homestake-St. Joe development program has two basic objectives: (1) to determine the optimum operating parameters for "Zadra" type cells. Such parameters will provide for optimum operating economies and produce a product of the maximum possible purity, and (2) to determine changes in actual "Zadra" cell design which will improve the silver electrowinning operation. See Figures 5 and 6.

Testwork on this program is being conducted using a 4-liter "Zadra" cell which is a scaled-down version of the commercial cells used at Creede. The cathode spool is made from CPVC with numerous holes in the center column for solution flow. Steel wool is wrapped around the center spool and the spool assembly is inserted into a CPVC shell with holes for solution flow. Electrical contact is made by four stainless steel rods inserted into the steel wool from the top of the cell. The anode is stainless steel mesh and surrounds the cathode shell. There are six individual cells, each with its own rectifier. The cells are operated in parallel with a common head solution from the Creede plant.

Test variables which are currently being tested include:

(1) the effect of applied amperage,
(2) the effect of flowrate to cells,
(3) the effect of steel wool type and amount used, and
(4) the effect of cell operating temperature.

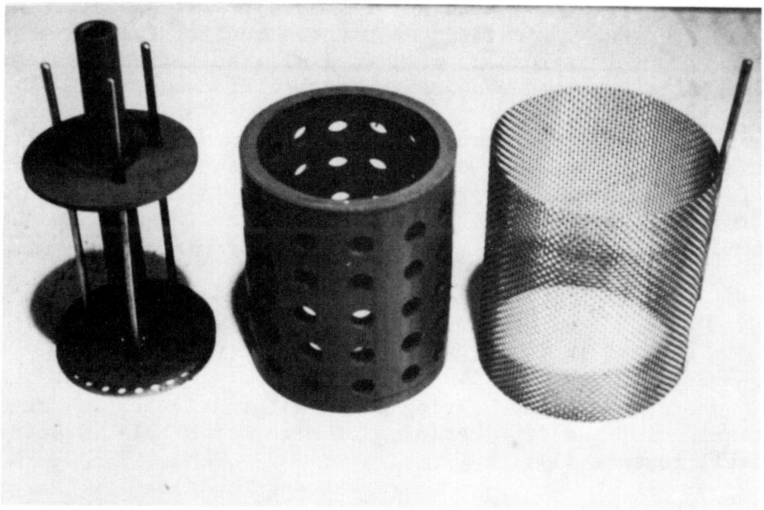

Figure 5. Zadra-type 4-liter electrolytic test cell cathode/anode assembly.

Figure 6. Zadra-type 4-liter electrolytic test cell.

In addition to these variables, later tests will measure the effects of various solution impurities and cell agitation on silver electrowinning. The foregoing variables will be correlated against individual cell current efficiency and cathode quality (nature of deposit). The silver-loaded steel wool will be refined to determine the purity of silver obtainable from each test condition.

Though these developmental tests are in progress, insufficient data have been generated to draw any significant conclusions at this time. It is hoped that information developed in this program will be the basis for improved silver electrowinning cell design and operation.

EVALUATION OF INDUCTION HEATING FOR USE IN THE FIRE REFINING OF SILVER AND GOLD

The use of induction heating as an alternative to gas or oil fired heating for the fire refining of precious metals is attractive for several reasons (1):

(1) compactness
(2) less waste heat, therefore cooler room temperatures
(3) cleaner operation
(4) no off-gas discharge
(5) energy efficient as less material heated, and
(6) less time to reach temperature.

The particular application of interest at the Bulldog operation involved the smelting of a steel wool/precious metal and flux change. Since this particular change is not an efficient susceptor, the crucible must be of a material that is a susceptor (plumbago or silicon carbide). It is also possible to use a furnace with an independent susceptor. The testwork was conducted to assess the technical potential of induction heating for the Bulldog charges.

The tests were conducted using steel-wool/silver and or silver sludge from the electrolytic cells to which fluxes were added. The objective was to fuse the charge so as to obtain a slag-metal melt which could be evaluated in terms of: dissolution of steel-wool, and silver reporting to the slag and fineness of silver (after a suitable contact period). Although silver was used in this evaluation, the results as they pertain to induction heating would be applicable to the refining of gold.

During a recent visit to the Rand Refinery in South Africa, the world's largest precious metal refinery, it was observed that they rely solely on induction heating for the refining, melting, and alloying of gold and silver bullion. The area in which several large induction furnaces were operating and containing several hundred kilograms of molten bullion was pleasantly cool and immaculately clean. The furnaces were all mounted with hydraulic rams so that they could be tilted for pouring with minimum difficulty and excellent control. This large scale bullion handling was very impressive and demonstrates the utility of induction heating in this application.

Two induction units were used in the testwork:

(1) Westinghouse Corp., 10 kva RF (400 kHz) generation
(2) Inductotherm Corp., 30 kva (4200 Hz) generation.

The Westinghouse unit was used in preliminary testwork with small uninsulated plumbago crucibles (1½ in. diameter x 3 in.). Later work used fire clay lined plumbago crucibles of the same size but smaller capacity. A single test was made with a cut down fire clay crucible to determine if adequate coupling could be achieved using a cylindrical pellet of steel wool/silver and fluxes as the susceptor.

The Inductotherm unit was used for larger charges with a Model T-30 furnace which contains a graphite susceptor. A plumbago crucible (3 in. diameter x 4 in.) and fire clay crucibles (2½ in. diameter x 3½ in.) were used with this furnace. The unit was also run with an induction coil and a fire clay lined silicon carbide crucible.

The first six tests were conducted to observe the behavior of a charge when subjected to induction heating. Small samples (12 gm of steel wool/silver sludge) were heated in a plumbago crucible until molten. The charge including fluxes occupied a volume of about 100 cm^3. The charge was kept molten from 15 to 30 minutes after which it was poured into a conical mold. The silver button was then separated from the slag. The silver button weight was typically 43 to 45% of the weight of the silver sludge. The flux composition used was: 1 to 5 parts borax, 1 to 5 parts nitre, and 1 part silica by weight to the weight of the steel wool.

The results of these tests indicated that the plumbago crucibles were effective susceptors but drastically reduced the oxygen content of the slag by formation of carbon monoxide. There was insufficient oxidation to completely oxidize the steel wool. In one test an iron button was formed on top of the silver button and the silver exhibited magnetic susceptibility. The temperature of these tests was estimated to be 1400°C.

The addition of fluorspar to the flux charge (0 to 3 parts by weight to weight of steel wool) and the use of a fire clay lined crucible increased the oxidizing conditions of the slag sufficiently to oxidize the steel wool completely and produce a refined silver button.

In one large scale test, a silicon carbide crucible, lined with fire clay, was used to smelt 2 kg of steel wool/silver material together with 1 kg of fluxes and produced a 950-gm refined silver ingot. The silicon carbide crucible acted as a susceptor, and the volume of the lined crucible was approximately 4000 ml. The total dry charge occupied about half of this volume and, when molten, was again reduced by half to ∿ 950 ml.

A single test on a pressed cylindrical pellet consisting of steel wool/silver plus fluxes indicated that satisfactory agglomeration, and thus contacting, of steel wool and flux components could be achieved. However, the pellet was a poor susceptor and direct heating of the charge was unsatisfactory. This may be a useful technique, nevertheless, to reduce the dry charge volume.

The results of these tests indicate that induction heating can be used successfully to fire refine silver/steel wool and flux composites. Fire clay lined silicon carbide crucibles appear to be the most satisfactory. While specific flux charges will have to be determined for differing steel wool/precious metal charges in order to provide the correct amount of oxidation, a guideline suitable flux charge appears to be: 1-5 parts borax, 1-5 parts nitre, 1 part silica, and 0-3 parts fluorspar by weight to 1 part steel wool. The required temperature and contact time between the slag and precious metal will need to be determined for a specific application. As a guide, temperatures in the range 1100 to 1250°C and contact times of 1 to 1½ hours (after the charge is molten) appear to be satisfactory.

An estimation of the volumetric requirements of a crucible to treat 1200 ozs of precious metals in a single batch would be 20,000 to 25,000 cm^3. The major portion of this volume is required for the fluxes and allows for a 20% feedboard. A crucible with internal dimensions of 11-in. diameter x 17 in. would be suitable. As an alternative to single large batch treatment, several smaller batches and a small crucible could be used which would reduce the size of furnace required but increase the amount of processing time. This would be an economic decision based on the amount of material to be refined.

ACKNOWLEDGMENTS

Particular thanks are due to Bruce Brogoitti, metallurgical consultant and formerly of Greyback Engineering, for his work, both conceptual and design, of the entire pulp handling system modifications and, in particular, his tireless efforts in convincing us that this dredge tank design would really work.

Don Gullickson, former Assistant Mill Superintendent at the Bulldog, and Steve Mitchell, former Homestake metallurgist at the

Bulldog, contributed many ideas and suggestions for the developments that have transpired, and it is with much appreciation that their work is acknowledged.

Curt Carey, a senior staff metallurgist for Homestake Mining Company, is doing the work on the electrolytic cell test program at Creede and contributed the section regarding this work.

BIBLIOGRAPHY

1. Martins, G. P., "Evaluation of Induction Heating for Use in the Fire Refining of Silver and Gold," Report #GC/GM-180, Unpublished.

2. Anon, "St. Joe Minerals Will Develop El Indio Au-Ag-Cu Deposit in Chile's High Andes," Eng. & Min. J., May 1979.

Chapter 10

CHLORINE EXTRACTION OF GOLD

Wendell E. Dunn, Jr.

Wendell E. Dunn, Inc.
Lead, South Dakota

INTRODUCTION

The early history of the chloride process has been recorded in a chapter of a Bureau of Mines bulletin (1) by one of the inventors of a chloride process, Stewart Croasdale, who is famous for the Pohle-Croasdale Process.

The earliest use of chlorine in metallurgy was in 1557 by Bartholome Medina. But apparently even earlier there was knowledge that salt could be used in gold and silver extraction because a treatise, de la Pyrotechnica, published in 1540 in Venice, asserted that silver could be obtained from its ores by a wet salt process.

In 1609 a priest, Alonzo Barba, invented a process using a copper vessel and boiling a solution of salt to extract silver chloride ores. The copper vessel was, of course, one of the reactants. This led to the Kronke process introduced in Chile in 1860 which substituted cuprous chloride for the kettle.

A major step was taken in Europe in the late 1700's to chloridize roasted silver ores in combination with amalgamation using iron to reduce the silver chloride.

By 1843 the Augustan process was introduced. Ores were roasted with salt and leached with a saturated salt solution. The silver chlorides passed into solution from which silver was precipitated by copper. In 1848 sodium or calcium thiosulfate solution of silver chloride was suggested by Dr. John Percy, and the idea was applied in 1856 by Von Patera in Bohemia.

Von Patera apparently realized there were significant losses by volatilization, and he added steam to the furnace during the chloride roasting. His process was adopted in Mexico in 1868 and in the United States in 1874 and reached its zenith in the next 25 years operating until the demonitization of silver by the U.S. government in 1893 closed every leaching plant treating silver ores in this country.

Platner in 1851 proposed converting gold to gold chloride by using chlorine gas and extracting the gold chloride with water. This process was introduced in Silesia. Here, the ore was either roasted dead or if it contained silver, it was first chloridized by roasting with salt. The silver was first leached out; then, chlorine gas was applied to the ore in vats, and the gold was recovered by subsequent leaching with water. Although the loss of metals by volatilization during chloride roasting was already known, Platner was the first to publish experiments and discussed at length the volatilization products of chloride roasting though he failed to record any loss of gold by volatilization.

In 1880 Charles Crosby at Nevada City, California, working with a pyritic ore where 75% of the value was gold and the remaining value silver, observed that roasting ore without salt showed no loss by volatilization, but when roasted with 3% salt, 30% of the gold and 50% of the silver were lost. He wrongfully attributed these losses to the tellurides of the ore, not to the salt.

C. H. Aaron was the first to recognize the volatilization of gold chloride and bring this into prominance in the metallurgical world. Apparently at his Melrose, California reverberatory furnace, he suffered a loss of $3,000* worth of gold by volatilization. This loss led to his investigation and many other investigations at western metallurgical plants of the cause of the loss. However, Aaron, after learning the cause of his loss, confined his efforts to preventing future losses.

With the understanding that gold chlorides were volatilizing came a long chain of experimental studies of both chloridization and more recently chlorination. The Pohle-Croasdale was the best known although it did not succeed in competition with cyanidation. Following this, the Bureau of Mines took up the study of gold chloridization processes and published a bulletin describing their experiments in 1923. Academic investigations of the several gold chloride species look back to the work of T. K. Rose in 1895 and the work of F. Meier 1901, F. Ephraim 1918, Fischer and Biltz 1928 followed by the work of Schafer in 1964 and many Russian investigators and culminating in the work of J. P. Hager and R. B. Hill (2) at the Colorado School of Mines in the 1970's.

*Worth $90,000 today

Two recent patents were filed by the Bureau of Mines as the result of practical laboratory chlorination of gold ores by Eisle, Heinan and Fischer (3,4).

The bibliography lists the more recent studies and patents. John Hager's papers (2,5) show the various chlorides of gold, their stability region and give the most recent thermodynamic values for their formation. He also studied the effect of ferric chloride complexing on gold chloride transport which had been reported earlier by numerous authors. The subject of vapor phase complexing is discussed extensively by Hastie. Much of the pioneer work on gas phase complexes was done by E. W. Dewing who discovered the gas phase $NaAlCl_6$. Its analogue $NaFeCl_4$ was discovered later by this author.

From quite another direction in chlorination processes comes technology which is applicable to the problem at hand. With the pioneering success of the chloride pigment and titanium metal efforts of the duPont Company in the early 50's came the elements of high temperature chlorinator design and chloride condensation trains. Although the following decade revealed the difficulties of high temperature chlorination to the many companies which followed the duPont efforts, nearly all have successfully and profitably surmounted their problems.

Having personally been engaged for three decades in this latter technology, this paper brings together the two streams of experiment and thought and attempts to show that there are viable gold extraction chlorination process concepts which are operable and economic and which have strong environmental advantages over the reigning cyanide processes.

We now know from the literature: (1) a low temperature gold complex with ferric chloride exists, (2) the thermodynamics of the gold chloride reaction to Au_2Cl_6 and the decomposition of Au_2Cl_6 to AuCl and $AuCl_2$.

We also know from chloride pigment technology and ilmenite beneficiation studies of Chlorine Technology Limited: (1) how to design and operate fluidized chlorinator beds for minus 20 mesh particles, (2) how to cool chlorine vapors arising from chlorinators, (3) how to extract ferric chloride from gas streams by a salt or salt melt adsorption reaction, and (4) how rapidly gas diffuses into particles.

From the early mining experiences described in the introduction, we know that it is possible to chlorinate and volatilize gold values even with chloridizing agents such as salt.

To bring together these technologies, an example of a possible flowsheet with rough heat and mass balances is given for an ore with

Homestake ore characteristics.

Before going into detailed descriptions of a flowsheet, the basic concepts will be discussed. First, the work of Hager and Hill has established by mass spectrographic methods the gas phase species which are encountered in gold chlorination. Their studies pointed out a 25-fold increase in gold vapor transport in the presence of ferric chloride, but they did not study the thermodynamics of the complex.

Others (7,8) have indicated factors even higher than 25 as the transport increase by iron chloride complexing. A rough calculation of the bond strength of the mixed complex gives about a 4 Kcal/mol, a bond strength similar to other such complexes.

The breakdown of the Au_2Cl_2 complex and the adverse equilibrium effect of temperature on the reaction of Au to $AuCl_3$ is very marked, and it is only the appearance of AuCl or Au_2Cl_2 (9,10) at much higher temperatures that makes high temperature transport feasible. However, kinetic studies now indicate that it is quite feasible to operate at low temperatures to form the tri-chloride dimer or tri-chloride gold-iron complex, and it is obviously desirable to take advantage of this reaction.

It is apparent from the equilibrium constant that the pressure of chlorine is important, appearing as the third power, and it is also clear from the free energy equation

$$\Delta G^\circ_R = 24{,}600(\pm 4400) + 59.8(\pm 8.2)T$$

for the reaction

$$2Au_{(c)} + 3Cl_{2\,(g)} = Au_2Cl_{6\,(g)}$$

that reaction falls off rapidly with temperature. So, it is a contest between adequate kinetics and adequate equilibrium value as to whether gold can be effectively transported at low temperatures. The formation of an iron chloride complex is worth about 65°C temperature increase in this situation which brings the kinetic rates to an even more favorable value.

It is apparent from the minute amount of gold which is to be volatilized that the chlorinator design will not strive to utilize the entire chlorine input as would be the case for a rutile chlorinator or an ilmenite beneficiator. Rather, the chlorinator will operate with a high chlorine pressure even in the exit gases.

The chlorinator temperature will not be significantly influenced by any chlorination reaction, again differing from standard chlorination practice in the titanium pigment industry where the reactions are extremely exothermic.

It is obvious, then, that chlorine must be recycled frequently and the chlorine partial pressure maintained at a high value probably between one and two atmospheres absolute.

This means that the chlorine must be stripped of its gold chloride values, cleaned and partially cooled so that it can be purged and repressurized. A small stream bleed will be necessary to remove inert gases.

Although in the past, carbon adsorption has been practiced both in cyanide and chloride technology, it seems better to absorb this gold complex in a much cheaper and effective sequesterant salt. Either sodium or potassium chloride would be economic, and the amounts needed would be much less than 5 pounds per hour.

The ore preparation will probably only require crushing to minus 20 mesh before roasting to convert the sulfides to oxides with evolution of sulfur dioxide. This sulfur dioxide would be sent to a contact sulfuric acid plant, or if desired, converted to sulfur for sale. It is interesting to note that the price of Texasgulf sulfur has just been raised to $127 per ton (6).

The conversion of pyrite and arsenopyrite is expected to allow easy diffusion of chlorine into the gold contained within the oxidized pyrite phase.

The roasting should be done at the lowest possible temperature, with oxygen or an oxygen-air mixture in order to enrich the gases to the contact sulfuric acid plant.

Following the roasting of the crushed ore, heat can be removed or recovered to bring the roasted ore temperature down to a temperature about 350°C. Part of the heat removal can be accomplished in reheating the recycled chlorine.

The chlorine gas, having been stripped of its gold complexes by absorption with NaCl to form a liquid chloride melt, is cooled to about 80°C, scrubbed with sulfuric acid, and repressurized with an axial flow compressor to about 15 psig. A bleed stream of the chlorine with its inerts is further cooled and compressed to liquify the chlorine. This is distilled to separate chlorine from the inerts.

The chlorinator will require solids feed with gas isolation between the roasting bed and the chlorination bed gases. This can be accomplished very effectively by solids transfer techniques developed by CTL.

In order to get a high yield of gold from the fluidized bed chlorinator, it may be divided into stages either internally or externally. Final design of the chlorinator would be dependent upon laboratory studies of the reaction kinetics for a particular ore (11).

The chloride complex is a low melting, approximately 150°C, salt melt which can be dissolved in water or electrolyzed directly. This paper does not deal with recovery of gold from its gold chloride complex.

FLOWSHEET

Figure 1 is a block diagram showing the main stages of the process. Crushed ore is fed to a fluidized roaster, and the resulting sulfur dioxide is fed to a contact sulfuric acid plant. The roasted ore goes to a two-stage chlorinator which is fed recycled chlorine. Gases from these stages go to a condenser where the gases are contacted and reacted with sodium chloride to form a salt-gold chloride melt extracting the gold values from the gas stream for further processing to recover the gold. The chlorine and its inerts are cooled, scrubbed with sulfuric acid and sent to the compressor for repressurizing. A portion of the chlorine is withdrawn from the recycle stream and liquified so that the chlorine may be distilled and the inerts rejected.

Tables 1 and 2 give mass and heat balances for the roaster and chlorinator based on a flow rate of 200 tons per hour of ore.

DISCUSSION

Flowsheet

Roaster

A simplified block flow diagram has been drawn with operating temperatures and flows in order to present the basic aspects of a chlorination process with a siliceous gold ore containing modest amounts of pyrite. At an iron level of 5% it is possible to generate sufficient heat to roast ore without additional carbon. It has been assumed that the ore roasting temperature is held below the combustion temperature of graphite.

The question of whether to use oxygen or air or a combination thereof for the roast has been resolved in favor of oxygen because although there is an increased power and investment cost, the reduced roaster volume and the higher SO_2 concentration to the contact sulfuric acid process should outweigh the additional cost of oxygen. Detailed cost analysis for local considerations might vary this decision. Fortunately for either case, there is sufficient heat

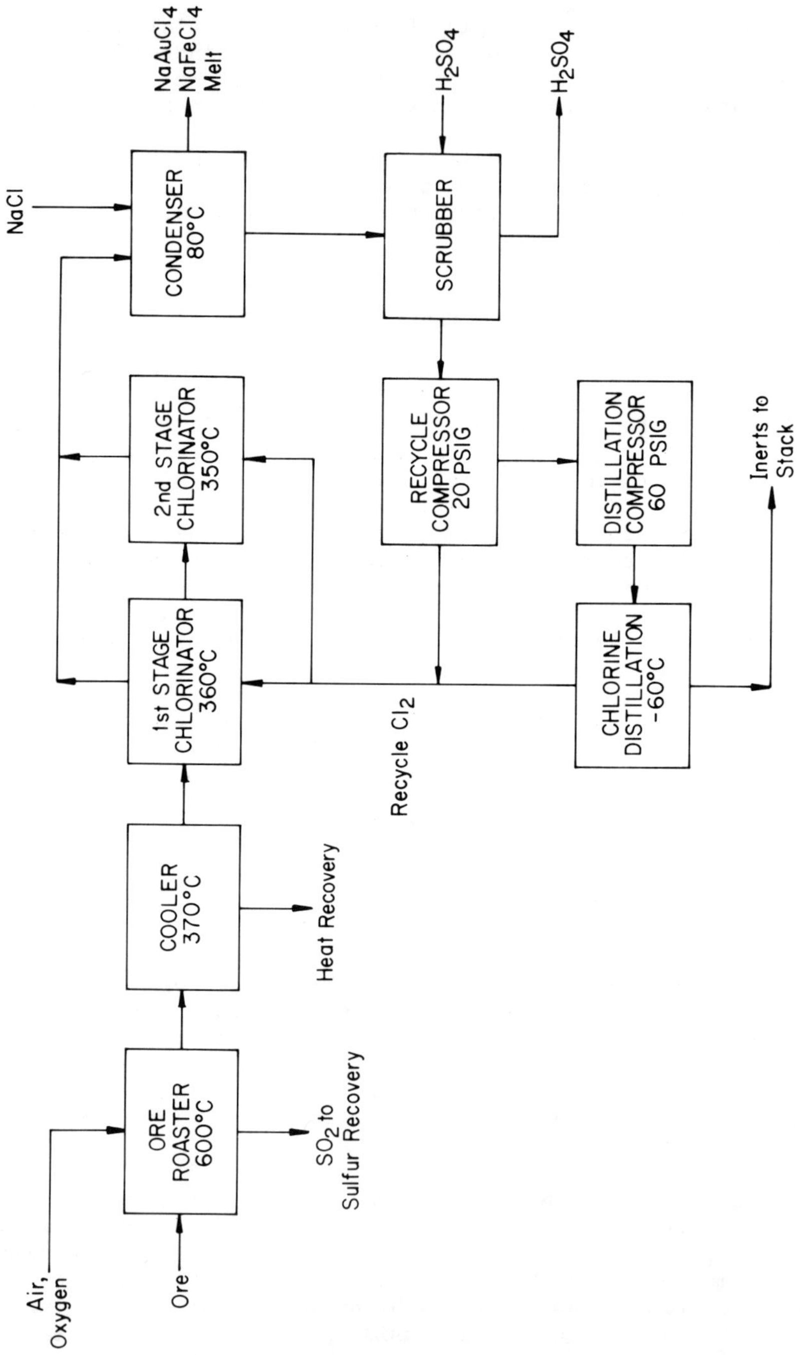

Figure 1. Flowsheet for the chlorination of gold ore.

Table 1. Roaster Heat and Mass Balance at 900°K.

Material	Flowrate (mol/hr)		$H_{900}-H_{298}$	Sensible Heat (MPCU/hr)		$H_{R_{298}}$	Heat Input
	In	Out	(PCU/mol)	In	Out	(PCU/mol)	(MPCU/hr)
SiO_2	5,918.8	5,918.8	9,204	0	54,477		
Fe	358.2						
S	661.2					98,825	65,343
As	53.4						
Graphite	33.4	33.4	2,318	0	77		
O_2	996.6			0			
SO_2		661.2	6,937		4,587		
Fe_2O_3		179.1	20,051		3,591		
As_2O_3		26.7	23,400		646		
					63,378		65,343
	(troy ounces/hr)						
Au	34.4	34.4					
Ag	6.8	6.8					

Notes

1. Reactor Heat Loss:

Bare Shell	(900°K)	11,238 MPCU
	(850°K)	5,250 MPCU
Refractory Liner		374 MPCU

2. $2\ FeS_{2(s)} + 5\frac{1}{2}\ O_{2(g)} \rightarrow Fe_2O_3 + 4\ SO_{2(g)}$ $\Delta H_{R_{298}} = -395.3$ kcal/mol

Table 2. Chlorinator Heat and Mass Balance

Material	Flowrate (mol/hr)		H_T-H_{298}	$H_{623}-H_{298}$	Sensible Heat (MPCU/hr)		WC_p(700°K)
	In	Out	(PCU/mol)		In	Out	(MPCU/°K)
SiO_2	5,918.8	5,918.8	4,400	4,400	26,042.7	26,042.7	95.64
Fe_2O_3	179.1	179.1	9,801	9,801	1,755.4	1,755.4	6.41
Graphite	33.4	33.4	1,956	1,956	65.3	65.3	0.15
Cl_2	586.0	586.0	440	2,769	257.8	1,622.7	
Au	0.012						
Ag	0.004						
$AuFeCl_6$		0.012					
$AgFeCl_6$		0.004					
					28,121	29,486	102.2

Note:

H_T = Enthalpy at Inlet Temperature (Cl_2 at 80°C, Ore at 350°C + ΔT)

Roaster Temperature:

Heat Deficit = (Heat Out - Heat In) + Reactor Heat Loss

= (29486 - 28121) + 640 = 2005 MPCU/hr

Increased Roaster Temperature for Heat Balance, ΔT = 2005/102.2 = 19.6°K

Roaster Feed Temperature = 623 + 19.6 = 643°K = 370°C

available without carbon addition as a fuel.

The roaster design would depend upon the decision whether to use oxygen or air. Also to be considered is the optimum roasting temperature. For convenience, a temperature of 900°K was assumed, but it is highly likely that a lower temperature, 50° lower or more, is desirable and may be possible. At 100° lower temperature, i.e. an 800°K reactor, it could have cold feed and a completely bare shell and still be in heat balance.

The roaster pressure has been set above the chlorinator operating pressure which is elevated to improve the reactivity of the chlorine with gold. The increased pressure is also desirable for the contact sulfuric acid SO_2 feed gas.

Chlorinator

An overall heat and mass balance have been prepared for the chlorinator. The chlorinator will undoubtedly be designed as a two- or three- stage reactor with these stages possibly incorporated within a single shell. The most desirable operating temperature is believed to be about 350°C. Although chlorine flow will cool the reactor and although there is no significant reaction heat, nevertheless it will also be necessary to cool the roasted ore feed.

It would be possible to operate a bare shell chlorinator were it not for the fact that an accumulation of the gold-iron chloride complex would form (10) on the walls and represent a vast inventory of gold which could be recovered only at reactor clean-out annually or bi-annually. Hence, the chlorination reactor would probably be designed with a refractory lining to prevent accumulation of gold complex on its walls. The reaction conditions are so mild that there will be no refractory corrosion problem.

A chlorine flow of 0.2 foot per second was selected for the heat and mass balances. This should give a minimum fluidization rate to allow a good mixing of the solids and yet represent the lowest steady state gas flow likely. Alternatively, a pulsed flow of chlorine might be used to reduce further the amount of chlorine flowing into the reactor and through the condensation stages. As mentioned earlier, there is no attempt to completely utilize chlorine, rather the chlorine pressure is kept high by the chlorine recirculation in order to insure good reaction rates for the gold complex formation.

Chlorinator Kinetics

Kinetic studies made by Landsberg and Hoatson (12) of direct gold chlorination in the absence of any promotor indicate that in 760 torr chlorine, a 200-mesh particle would chlorinate at a maximum rate at 3.45% per minute. A 20-mesh particle would chlorinate at one-tenth this rate. Since we are operating in the presence of iron chloride which should increase the rate by a factor of at least 25 and at a chlorine pressure which should increase the rate between 18 and 13.5 fold depending upon the point in the reactor bed, it would appear that the kinetics, unless limited by diffusion, would be very adequate.

It is expected that the mineral structure will be sufficiently disrupted in roasting to make the gold particles readily accessible to gas diffusion. Since liquid diffusion is successful reaching particles at and below 200 mesh, it would seem likely that larger size particles could be accepted by a gas diffusion process to equally good effect. A Bureau of Mines report on chlorination of various ores indicated improved yields as particle sizes were reduced but high conversions of gold to gold chloride were possible even at 40 mesh (7).

It is expected that Homestake ore at one hour retention time should give complete conversions for ore crushed to minus 20 mesh, but in order to assure a high degree of conversion in a single stage, the bed must operate with extremely low average bed concentration. The use of two stages will allow the first stage to operate at a high gold concentration with a final polishing stage operating at very low gold concentrations giving overall, a high gold conversion.

The retention time chosen for the bed would appear to be greatly in excess of need, but because the cost of added retention time for solids in a fluidized bed is small and because there is no penalty for excessive retention time and some benefits in reactor operating stability, a one-hour retention time has been selected. Laboratory experiments, of course, can establish yield vs reasonable retention time.

The flowsheet indicates a small amount of iron powder to be added to each chlorinator stage. This is thought necessary because equilibrium calculations indicate that at the chlorinator temperature, iron oxide and chlorine do not react sufficiently to give the necessary amount of $FeCl_3$ complex.

Condensation Train

Following the formation of the gold complex in the chlorinator, the vapor complex exits from the reactor as a very low concentration of complex in chlorine. The dust is cycloned from the gas stream at a temperature close to the chlorinator temperature before the gold complex can condense. The gases are then contacted with a sodium chloride containing melt so that the gold-iron complex which is volatile can be converted into the gold equivalent of sodium tetrachloroferrate, $NaAuCl_4$.

The bond strength of the expected salt complex, $NaAuCl_4$, should be sufficient to remove the gold chloride from the vapor complex $AuFeCl_6$ and contain it as a liquid melt below 150°C.

This reaction and the gas cooling would take place in a high velocity flue 18" in diameter and about 100 feet in length either horizontally or vertically disposed. The flue can have 180° bends if desired.

The liquid salt complex would be collected at the flue end by a cyclone and the gas, then at about 150°C, further cooled to 80°C by contact with sulfuric acid which scrubs the chlorine.

The chlorine is then compressed in an axial flow compressor and recycled to the chlorinator stages at 80°C. A bleed stream is taken off for high compression and distillation to prevent inert gas buildup.

The gold complex reaction with salt, although probably quite exothermic, is too small in amount to affect heat balance so that heat transfer is the only important thermal factor.

Economics

While it is not the purpose of this paper to make a detailed economic study of this chlorination process, it is of great interest and essential that the economics of the process outlined here be comparable with cyanide process economics. Hence, very rough but conservative values have been estimated for ingredients, by-products, labor and capital cost.

Because the ore analysis taken as representing a typical Homestake ore has enough sulfur to make roasting with oxygen or air self-sufficient in heat and the amount of chlorine and salt associated with the volatilitization and condensation of the complex is essentially of negligible cost, the ingredient estimate is primarily one of energy cost.

The two major energy requirements, apart from the present ore handling electrical requirement, are the energy to separate air for the oxygen feed and the energy to recycle the chlorine and to purify the plant chlorine stream.

The added by-product value of sulfur sold either as sulfur or sulfuric acid is far higher than the ingredient labor and capital costs of the process.

It is estimated from recent prices that the by-product sulfur may be sold for a value of \$40-\$52 per ounce of gold recovered. The energy cost to supply oxygen to the process would be about \$4.45 for

95% oxygen at a power cost of $0.03 per kilowatt hour. With heat recovery equipment (not shown) this energy cost can be reduced as air is substituted for oxygen. However, this balance involves the efficiency of a contact sulfuric acid plant, and a detailed economic study would be necessary to determine the optimum oxygen-air ratio.

Actually, energy costs at $0.03 per kilowatt hour for an oxygen separation system would vary somewhere between $2.90 and $7.42 depending upon the efficiency of the air separation plant; a mid-value efficiency has been assumed.

The cost of repressurizing the chlorine is only 21¢ per ounce of gold and may be less for lower operating pressures, the operating pressure being set by very conservative kinetics estimates. The cost of purifying the chlorine stream would appear to be of the same order of magnitude or less depending upon the size of the Cl_2 bleed stream.

The capital cost of the process considering only battery limits* equipment roughly scaled from ilmenite beneficiation economics would very probably be less than $2.5 million. For a capital cost of 50% per year which would include return on investment, maintenance costs and depreciation one would calculate a $5.20 cost per ounce of gold.

Labor costs estimated from the ilmenite beneficiation studies would be about $0.87/ounce at an average labor cost of $25,000 per man direct and indirect labor.

The total chlorination cost is estimated to be approximately $10.32 per ounce of gold which would not include recovery of the gold from gold chloride by electrolysis or other means but which would include costs of extraction from minus 20-mesh ore with a return on investment of 17% after taxes. No cost has been assessed for cooling water and general plant utilities.

Similarly, the added by-product value of $40-$52 per ounce would have to bear the expenses of roasting and sulfur recovery or the contact sulfuric acid plant. Its ROI has not been estimated.

CONCLUSIONS

A study of the literature, particularly recent literature, indicates that the likelihood of extraction of gold and silver values from ores using a chlorine low-temperature process to volatilize these values is technically feasible.

*Does not include ore bins, crushing, grinding, materials handling, air separation and the sulfuric acid plant.

Fluidized bed chlorinators and condensation techniques developed in titanium ore chlorination and beneficiation are directly applicable to the process which has much less difficult conditions than existing titanium tetrachloride processes which have operated for three decades at tonnages about one-half an order of magnitude less ore input of the Homestake tonnage but with a total chlorination of the titanium ore.

It is believed that salt complexing of the gold chloride vapor complex can be accomplished and would provide a more efficient recovery from the chlorinator gas stream than adsorption on active carbon (13).

The environmental hazards of the process are insignificant compared with hazards which were acceptable to the EPA in the case of the recent large duPont Chloride Pigment installation near the suburban area of Wilmington, Delaware.

BIBLIOGRAPHY

1. Varley, T., Barnett, E. P., Stennson, C. D. and Bradford, Robert, U.S. Bureau of Mines Bulletin No. 211, 1923.

2. Hager, John P. and Hill, Robert B., "Thermodynamic Properties of the Vapor Transport Reactions in the Au-Cl System by a Transpiration-Mass Spectrometric Technique," Met. Trans. B, 1, 1970.

3. Heinen, Harold J., Eisele, Judith A. and Fischer, Dennis D., "Recovery of Gold from Ores," U.S. 3,825,651.

4. Eisele, Judith A. and Heinen, Harold J., "Recovery of Gold," U.S. 3,834,896.

5. James, Stephen E. and Hager, John P., "High Temperature Vaporization Chemistry in the Gold-Chlorine System Including Formation of Vapor Complex Species of Gold and Silver with Copper and Iron," Met. Trans. B, 9B, 501-508, 1978.

6. Texasgulf Quarterly Report, June 30, 1981.

7. Engel, A. L. and Heinen, H. J., "Chloride Volatilization of Oxidized Lead Ore from Eureka, Nev.," Bureau of Mines, Report of Investigation 5409.

8. Eisele, Judith A., Fischer, D. D., Heinen, H. J. and Kesterke, D. G., "Gold Transport by Complex Metal Chloride Vapors," Bureau of Mines, Report of Investigation 7490.

9. Janssen, E. M. W., Fohmer, J. C. W. and Wiegers, G. A., "The Preparation and Crystal Structure of Gold Monochloride, AuCl," J. Less-Common Metals, 38, 71-76, 1974.

10. Janssen, E. M. W., Pohlmann, F. and Wiegers, G. A., "The Phase Diagram of the Gold-Chlorine System," J. Less-Common Metals, 45, 261-273, 1976.

11. Engel, A. L. and Heinen, H. J., "Chloride Volatilization and Other Tests on a Gold-Copper Ore," Bureau of Mines, Report of Investigations 4612.

12. Landsberg, A. and Hoatson, C. L., "The Kinetics and Equilibria of the Gold-Chlorine System," J. Less-Common Metals, 22, 327-339, 1970.

13. Cho, Eung Ha, Dixon, Steven N. and Pitt, Charles H., "The Kinetics of Gold Cyanide Adsorption on Activated Charcoal," Met. Trans. B, 10B, 1979.

14. Kellog, Herbert H., "Thermodynamic Relationships in Chlorine Metallurgy," Journal of Metals, June 1950; Transactions AIME, Vol. 188.

15. Jensen, J. W., Doerr, R. M. and Visnapuu, A., "Experimental Extraction of Gold by Chemical Vapor Transport," Annual Meeting, AIME, Denver, CO, February, 1970.

Uranium I

Session Chairman

Bruce R. Palmer
South Dakota School of Mines
and Technology
Rapid City, South Dakota

Chapter 11

ALKALINE CARBONATE LEACHING AT UNITED NUCLEAR-HOMESTAKE PARTNERS

Kenneth E. Skiff and John P. Turner

United Nuclear-Homestake Partners
Grants, New Mexico

INTRODUCTION

United Nuclear Corporation and Homestake Mining Company are partners in a major uranium mining and milling venture near Grants, New Mexico. The partnership was formerly called Homestake-Sapin Partners and was recently renamed United Nuclear-Homestake Partners (UN-HP).

The mill, originally built in 1958, had a capacity of 1,650 tons per day. With subsequent improvements and the addition of the adjacent Homestake - New Mexico Partners mill, the UN-HP mill is currently rated at about 3,500 tons per day.

An alkaline type leach process is utilized by the mill. Milling consists of five basic steps:

A. Ore handling and preparation

B. Extraction

C. Solid-liquid separation

D. Precipitation and purification

E. Product preparation

MILLING

Ore Handling and Preparation

The primary source of ore for the UN-HP mill is from its four underground mines located in the Ambrosia Lake Bed area. Lesser amounts of ore are received from the United Nuclear Corporation, Homestake Mining Company, and Cobb Nuclear Corporation. These mines are all within thirty miles of the mill site.

Two basic types of ore are processed by the mill. Sandstone accounts for approximately 80% to 85% of the mill feed with the balance being limestone. The principal mineralization of these ores includes coffinite $[U(SiO_4)_{1-x}(OH)_{4x}]$, uraninite $[(U^{+4}{}_{1-x},U^{+6})O_{2+x}$, ideally $UO_2]$, tyuyamunite $[Ca(UO_2)_2(VO_4)_2 \cdot 5\text{-}8H_2O]$, and carnotite $[K_2(UO_2)_2(VO_4)_2 \cdot 3H_2O]$. This mineralization generally occurs as an impregnation, a pore filling, or as a cement between sand grains. In the ore there is often the presence of carbonaceous materials in association with the uranium mineral. In addition to uranium, there are trace amounts of molybdenum, selenium, titanium, gold and silver.

Ore Receiving

Trucks hauling 22 to 28 tons of ore deliver their loads to the mill from the mines. When a loaded truck arrives at the mill, it is weighed, and the ore is sampled for moisture. The moisture determined from the ore in the trucks is used as the basis to calculate the number of dry tons of ore for metallurgical accounting. Each mine's ore is kept separate in lots for control purposes through crushing and sampling. Lots vary from 200 to 2,000 tons each.

The ore grade ranges between 0.04% and 0.30% U_3O_8. To minimize fluctuations in the grade of the feed to the mill, the ore lots are crushed selectively.

Crushing

An 18-inch grizzly covers a sub-grade hopper on the ore pad. Ore is discharged from this hopper on to an apron feeder. This ore then discharges through an anchor chain curtain on to the crusher feed belt. As the ore discharges from the crusher feed belt on to a vibrating grizzly, the minus 2-inch fraction is allowed to by-pass the crusher. The crusher discharge and vibrating grizzly undersize are combined and sized to plus or minus 3/4 inch by the use of vibrating screens. The 3/4-inch screen oversize is recycled to the crusher, and the undersize is transferred to the sample plant.

Drying

Ores containing more than 8 to 9% moisture generally require drying. When required, a 10-foot diameter by 80-foot co-current fired rotary dryer is used. The burner has a capacity of 70 million BTU per hour. Firing can be with either natural gas or light fuel

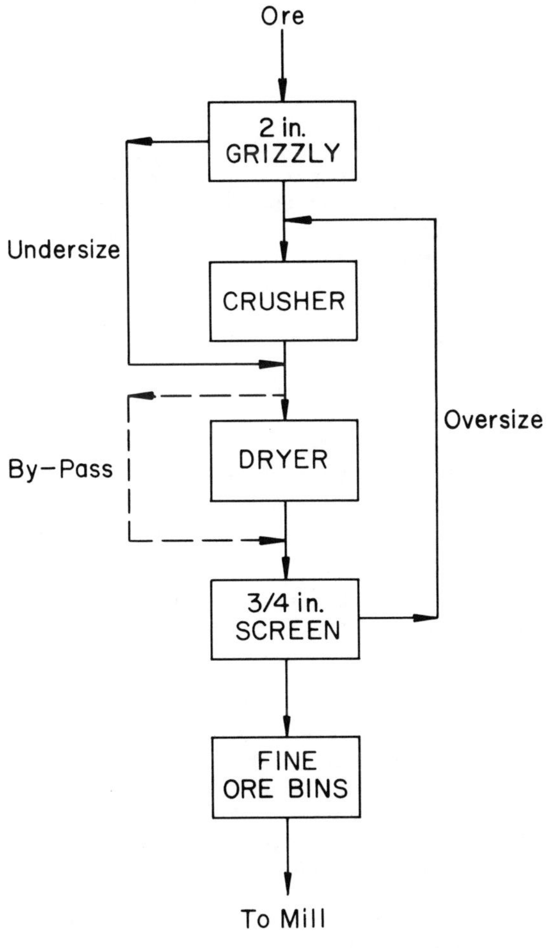

Figure 1. Flowsheet for the crushing and drying circuit.

oil.

The feed to the dryer is the vibrating grizzly undersize (minus 2-inch) and the crusher discharge. Ore discharges from the dryer at about 5% moisture. A conveyor belt returns the ore to the vibrating screens in the crusher plant.

Sampling

A representative sample of each ore lot is obtained by using four stages of samplers. The first stage operates on the entire crushing plant output and discharges the sample into a small surge bin. A load sensing device is mounted in the surge bin to control the speed of the belts feeding the remaining samplers. This insures a steady stream of falling ore for each sampler. A roll crusher is in the sample stream between the second and third sampler to reduce the particle size to minus 1/4-inch. The fourth and final stage of sampling cuts a quantity equivalent to one pound of sample for each four tons of ore crushed. After the final sample is cut, the ore is dried continuously and crushed to minus 10 mesh. With further sample preparation, chemical and metallurgical testing data are obtained from this sample for metallurgical accounting.

Fine Ore Bins

Four concrete silos comprise the fine ore bins with each being 35 feet in diameter by 50 feet in height. At full capacity the ore bins contain approximately 6,000 tons of ore. Since each lot of ore must retain its identity until the ore reaches the ore bins, there is very little blending of the ore prior to milling. A conveyor belt transports the ore from the sample plant to the top of the ore bins. A belt tripper is used to discharge the ore into anyone of the four ore bins or into a truck hopper for return to the ore pad.

The ore bins are used to segregate different types of ores in order to process the refractory ores such as limestone ore that require a finer grind and a longer retention time during leaching to maximize the uranium recovery. The mill utilizes two parallel circuits in grinding, thickening, and leaching. These circuits are referred to as the north and south circuits. Most of the sandstone ores are processed in the north circuit, while the south circuit uses a secondary grind and a longer atmospheric leaching time to process the limestone and other refractory type ores. These ores are directed to the two ore bins that feed the mill's south circuit.

The feed to the grinding circuit is withdrawn from the ore bins by means of two belt feeders under each ore bin. A collecting belt transfers the discharge from the feeder belts to the ball mill belt feed.

Grinding

Each primary grinding circuit consists of a ball mill which is operated in closed circuit with a spiral classifier.

The grinding circuits utilize 10-foot by 66-inch Hardinge Conical ball mills. The ball charge in each mill is approximately 22 tons. Equal weights of 2 and 2-1/2 inch forged steel balls are used for charge make up. Each ball mill is driven by a 400-hp, 4160-volt, motor geared through a pinion shaft to the ball mill. Operation of the ball mills is maintained at 20 rpm or 83% of critical speed. The ball mill discharge density is maintained at 65% solids in a sodium carbonate and sodium bicarbonate solution containing 37 grams Na_2CO_3 per liter and 7 grams $NaHCO_3$ per liter.

A 72-inch Wemco spiral classifier is operated in closed circuit with the ball mill. Overflow from the classifier is controlled at 20% solids. The overflow product of the classifier is about 10% plus 48 mesh and 35% minus 200 mesh.

The north circuit spiral classifier overflow goes directly to the north thickening circuit. As compared to the south circuit containing limestone and other refractory ores, the overflow is fed to a 20-inch cyclone which is in closed circuit with a regrind ball mill; this mill is 6 feet in diameter and 6 feet long. A 200-hp, 480-volt, motor is connected to a conventional belt drive system that provides the drive to the mill. The mill contains approximately 4 tons of 1-inch forged steel balls.

The regrind cyclone overflow product is about 5% plus 65 mesh and 50% minus 200 mesh. This product then goes to the south thickening circuit.

Preleach Thickening

The UN-HP mill has two thickening circuits. Overflow from each grinding circuit is pumped to its respective 20-inch cyclone in the thickening circuit. The cyclone overflow advances to the thickener where a polyacrylamide flocculant is used to aid in settling and clarification. Both thickeners overflow by gravity to a common mill solution storage tank for recycle to the grinding circuit. The thickened slurry from the thickener is removed at about 40% solids. The thickener underflow and the cyclone underflow are recombined in their respective preheat tank of the leaching circuit.

EXTRACTION

With the mineralization contained in the ores, oxidation of tetravalent uranium to the hexavalent state is required. Oxygen

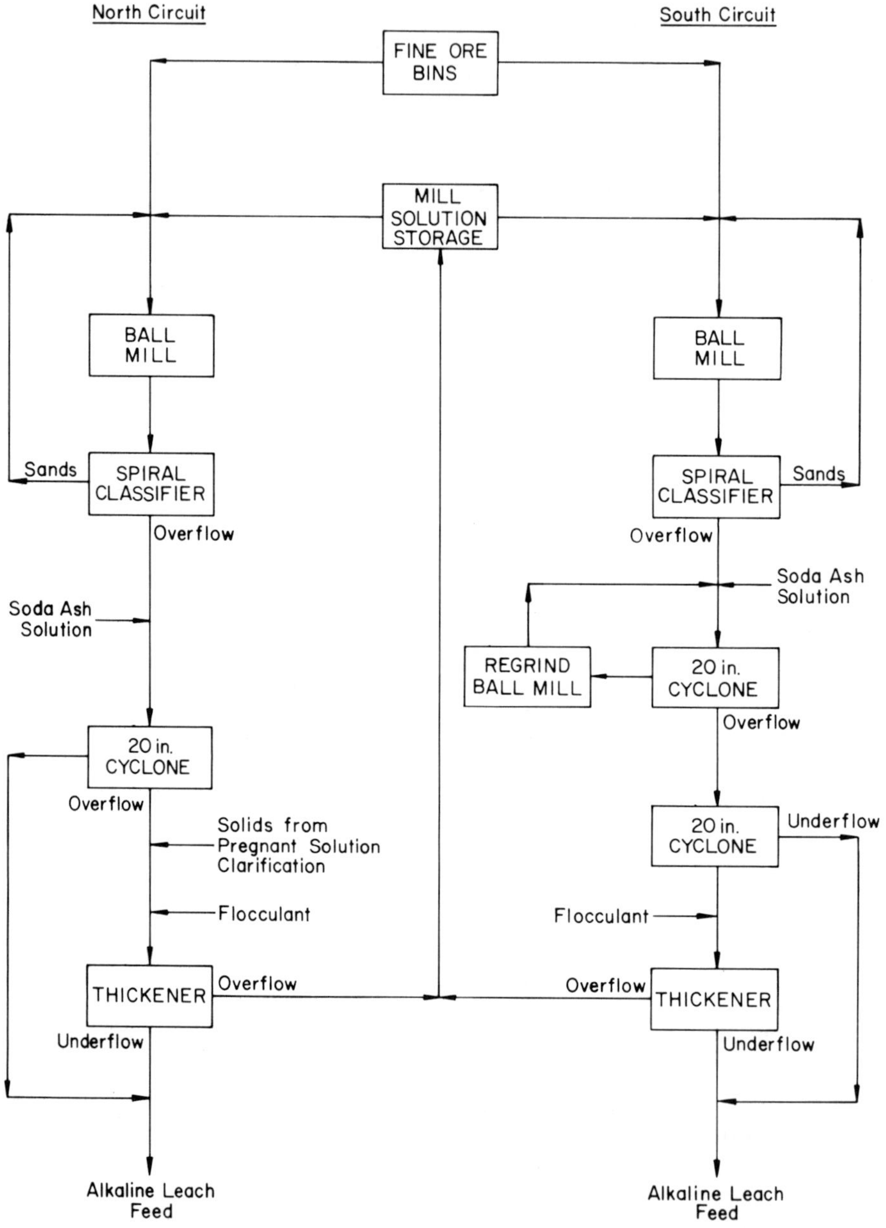

Figure 2. Flowsheet for the grinding and classification circuit.

available in the air is the most commonly used oxidant. Hexavalent uranium dissolves in the presence of carbonate to form the complex ion, uranyl carbonate, according to the following reaction:

$$2\ UO_2 + 6\ CO_3^{=} + O_2 + 2\ H_2O \rightarrow 2\ UO_2(CO_3)_3^{-4} + 4\ OH^- \qquad (1)$$

The uranium cannot be dissolved in a solution containing only sodium carbonate because the high hydroxyl ion activity prevents Reaction (1) from occurring. In a solution containing sodium bicarbonate, however, the hydroxyl ion activity is lower and Reaction (1) occurs readily.

Leaching

The leaching of the uranium ore is accomplished in a two stage circuit. The first stage consists of a pressure leach where the pressure is maintained at 60 psig, and the temperature is maintained at 200°F. Retention time of the slurry in the pressure leach is 4.5 hours.

The second stage consists of an air agitated atmospheric leach at a temperature of 170°F. The atmospheric leach slurry has a retention time of 12 hours for the slurry coming from the north leach circuit and 24 hours for the slurry coming from the south leach circuit.

Pressure Leach

Pressure leaching is accomplished in two separate circuits, each containing eight autoclaves. The autoclaves are 12 feet in diameter and 16 feet in height. They are domed pressure tanks equipped with a top mounted 'Lightnin' mixer. Each tank has two, turbine-type, 42-inch diameter impellers mounted on a 14 foot, 4-inch diameter shaft. Air for oxidation is fed into the bottom of the autoclaves through a 30-inch bubble-cap diffuser. Pressure is maintained by an automatic bleed-off valve on the pipe header connected to each autoclave. Heat for leaching is supplied by steam coils. Temperature and pressure are maintained by automatic controls.

Each of the two circuits is operated as a series of eight autoclaves; retention time of ore in each circuit is 4.5 hours. The first unit in each leaching circuit is a preheat tank in which the thickener underflow and the cyclone underflow are recombined and heated to 150°F. This preheated slurry at about 55% solids is pumped to the first autoclave leaching tank of its circuit. The flow through the autoclaves is by gravity, and the feed to each autoclave enters the slurry surface and discharges from the bottom through an

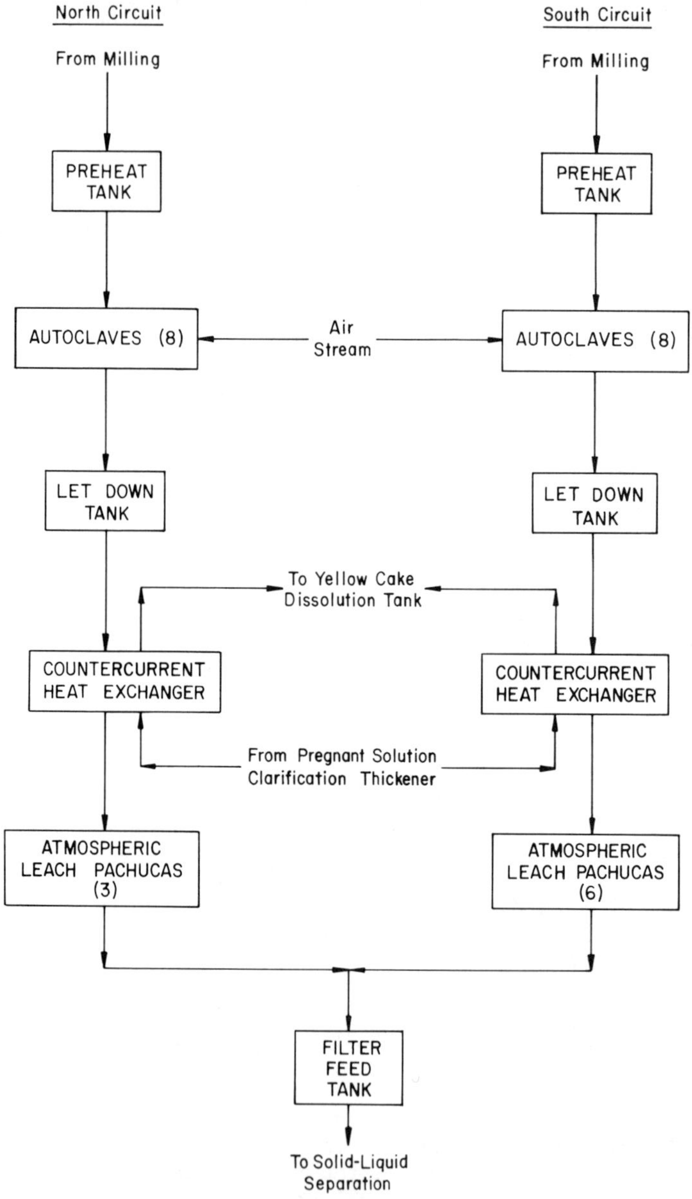

Figure 3. Flowsheet for the extraction circuit.

internal riser. A six-inch drop between autoclaves provides the head for flow through the piping in the circuit. The operating volume of each autoclave is approximately 11,000 gallons. The discharge from the last autoclave in each circuit flows into a letdown tank. From the letdown tank, the leached slurry discharges through a concentric tube heat exchanger that cools the slurry from 200 to 170°F. The pressure of the system provides the energy to push the partially leached slurry through the heat exchanger to the atmospheric leach. The pressure of the system provides the energy to push the partially leached slurry through the heat exchanger to the atmospheric leach.

Atmospheric Leach

Atmospheric leaching also consists of two separate circuits. The north circuit pressure leach discharges to the first of three pachuca tanks with a retention time of 12 hours. The south circuit pressure leach discharge is split where one-half of the flow goes to the first of another three pachuca tanks and the other half of the flow goes to the first of another three pachuca tanks. Retention time for the south circuit is 24 hours. Each pachuca tank is 19 feet in diameter by 38 feet tall. Four 10-inch air lifts provide agitation for each pachuca. Heat for the circuit is provided through steam jackets on the air lift pipes. Additional air for oxidation and agitation is supplied through five pipes which are suspended from the top of each tank. The air enters the slurry approximately three feet from the bottom of the tank.

The slurry from the pressure leach circuit is added to the top of the first pachuca tank of its circuit and flows by gravity through the tanks and discharges into a sump where the slurry is pumped to the solid-liquid separation circuit.

SOLID-LIQUID SEPARATION

Filtration

The slurries from both atmospheric leach circuits are pumped to the filter feed tank in the solid-liquid separation circuit. Slurry is then pumped to each of the first stage of filters where the soluble uranium values are removed by three modified stages of counter current filtration. A dilute solution of flocculant is pumped into the filter feed line where the filter feed and flocculant are mixed. A thorough mixing takes place as the filter feed slurry and flocculant flow through the pipes, the filter feed valves, and a small mixing chamber on each filter tub.

Each filter stage contains five 650-square foot and two 570-square foot rotary drum vacuum filters. The filters are 11-1/2 feet

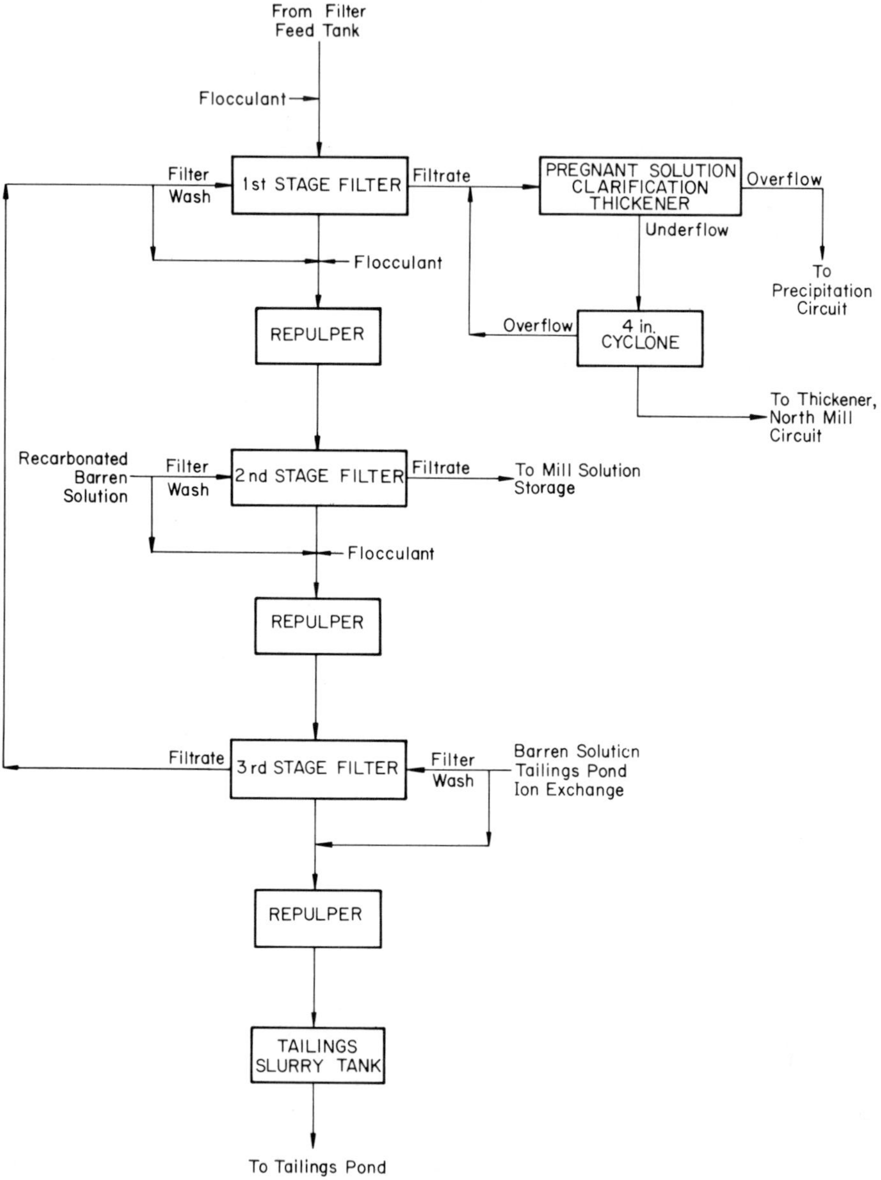

Figure 4. Flowsheet for the solid-liquid separation circuit.

in diameter and are equipped with polypropylene grids. A heavy duty nylon filter cloth is used to cover the filters. A 14-gauge stainless steel wire is wound on the filters to retain the filter cloth.

The first stage filter cake is washed with a hot filtrate solution from the third stage of filters. The filtrate is sent to clarification, and the first stage filter cake discharges into repulpers for repulping with third stage filtrate. The repulped slurry flows by gravity through an agitated sump and into the second stage of filters. Flocculant is added to the repulper solution to aid in the second stage of filtration. The second stage of filters is operated in much the same manner as the first stage except that recarbonated barren solution from the precipitation circuit is used as a filter wash and in the repulpers. Tailings pond solution is used for wash and in the repulpers on the third stage. The wash solution for each stage of filters is maintained at 100°F. The second stage filtrate is sent to the mill solution circuit, and the third stage filtrate is used as a wash and repulper solution on the first stage of filters. The filter cake from the third stage of filters is repulped with recycled tailings pond solution and slurried for tailings disposal.

Clarification

The filtrate produced by the first stage of filters is the pregnant uranium solution for the precipitation circuit. The solution must be clarified before precipitation to remove slimes that have penetrated the filter cloths. Clarification occurs in a thickener where a major portion of the silica and slime contaminants settle out. The pregnant solution for precipitation is then pumped from the thickener through heat exchangers to heat the solution from 125 to 180°F before flowing into the precipitation circuit. The heat is obtained by cooling the pressure leach discharge slurry from 200 to 170°F.

Tailings Disposal

The filter cake from the third stage filters is repulped with recycle solution from the tailings pond ion exchange system and transferred through launders to a tailings slurry tank. Disposal of the tailings slurry is handled by three Ash pumps in series and a cyclone truck. The pumping capacity of these pumps is rated at 1,500 gallons per minute.

Tailings disposal encompasses an area of 110 acres. Construction of the pond is done by pumping the tailings slurry at about 40% solids through two, truck mounted, 20-inch cyclones. Underflow from the cyclone is deposited on the dike that surrounds the pond, and the

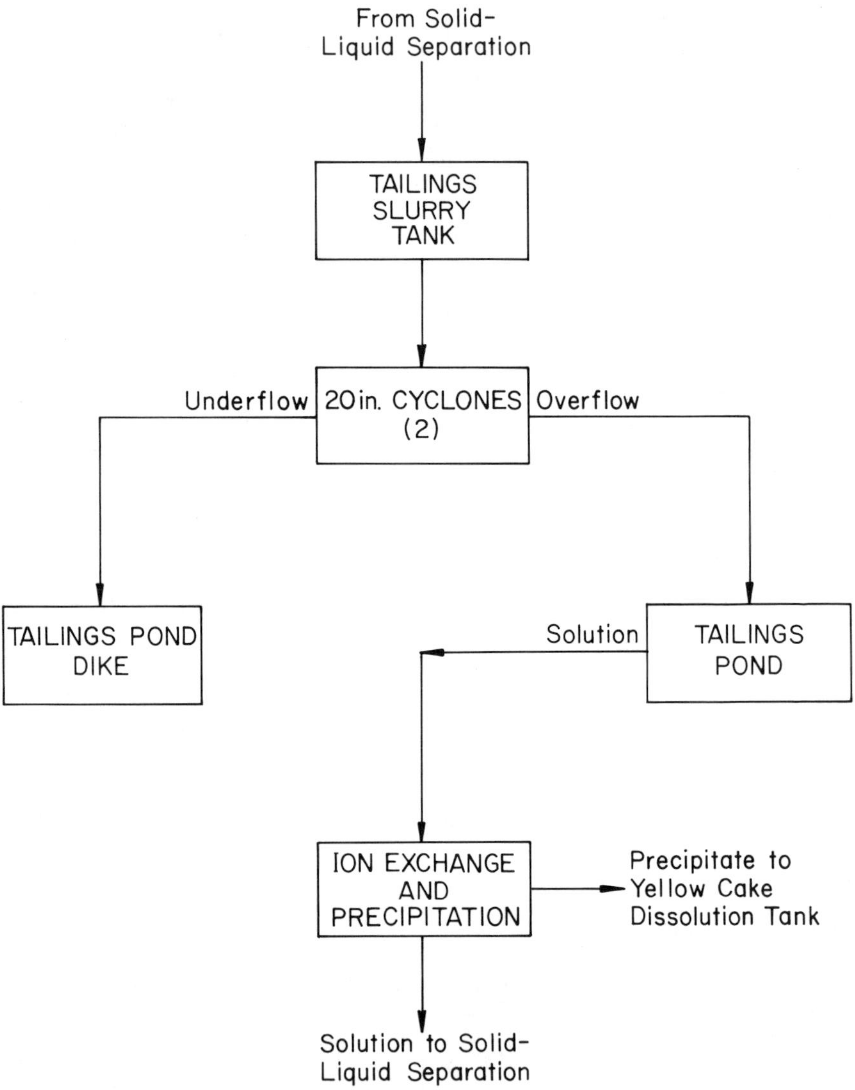

Figure 5. Flowsheet for the tailings disposal circuit.

cyclone overflow is directed into the pond where the slimes settle out of the tailings solution. Recovery of the tailings solution is through two centrally located decant towers. The reclaimed tailings pond solution flows underground to a pump basin and is returned to the mill where it is processed in an ion exchange circuit for uranium removal.

Tailings Pond Ion Exchange

Prior to the reclaimed tailings pond solution being returned to the third stage filters, the solution is treated in two upflow ion exchange columns for removal of soluble uranium.

UN-HP is installing a NIMCIX system to replace the existing inadequate facilities. The new plant will treat 1,200 gallons per minute of solution yielding a tail of less than 1 ppm U_3O_8.

PRECIPITATION AND PURIFICATION

Precipitation

After being heated to 180°F, the pregnant solution from clarification is pumped to the precipitation circuit. Precipitation is conducted in two stages. First, the pregnant solution is mixed in the dissolution tank with recycled yellow cake to increase the soluble uranium content. Second, caustic soda is added to precipitate the uranium. When the pH is raised above 12.0 with caustic soda, uranyl carbonate decomposes to form carbonate and sodium diuranate. The latter salt is a yellow precipitate commonly called yellow cake.

$$2\ Na_4UO_2(CO_3)_3 + 6\ NaOH \rightarrow Na_2U_2O_7 + 6\ Na_2CO_3 + 3\ H_2O \qquad (2)$$

The technique of recycling yellow cake was developed to achieve a more complete precipitation of the uranium in the original pregnant solution. The dissolution tank operates with a yellow cake recycle equivalent to 500% to 700% of the uranium in the incoming pregnant solution. In five hours of contact with the pregnant solution, a portion of the yellow cake dissolves, and the soluble grade of uranium in the solution increases. The precipitation efficiency of the circuit will vary with the level of soluble uranium in the feed. The undissolved yellow cake in circulation does not appreciably affect the precipitation.

The precipitation circuit consists of nine agitation tanks in a series. The solution has a retention time of about 12 hours. A solution of 50% caustic soda is metered into the first tank of the precipitation circuit in sufficient quantity to neutralize the

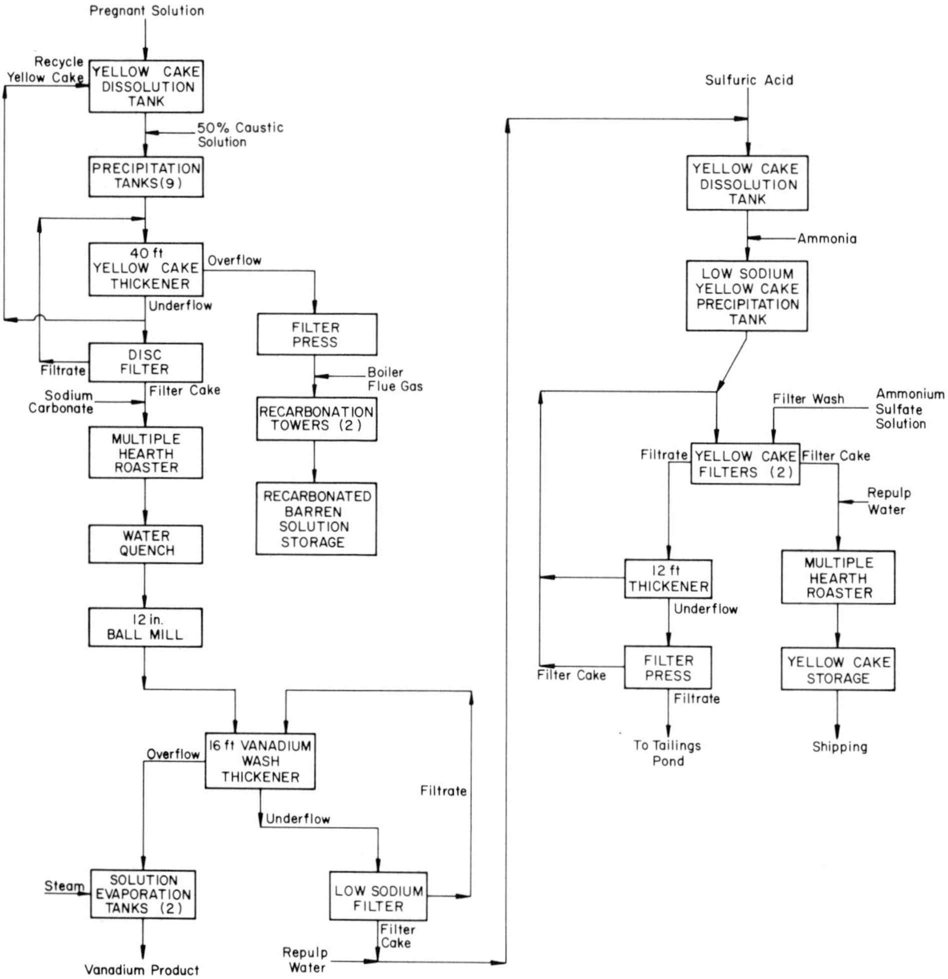

Figure 6. Flowsheet for precipitation, purification, and product preparation.

sodium bicarbonate and also to maintain an excess of 5 grams per liter caustic soda in the barren solution. The caustic soda may also be added to the pipeline feeding the tank to change the characteristics of the precipitated yellow cake and the efficiency of precipitation which are affected by the point of addition.

The yellow cake slurry from the precipitation circuit flows into a 40-foot diameter by 12-foot deep thickener. The thickener is insulated with a floating two-inch styrofoam lid and two inches of fiberglass insulation on the sides. Because the thickener operates at 170 to 180°F, cover insulation is necessary to minimize evaporation and heat loss which create thermal currents which hinder settling. The thickener underflow is pumped at about 35-40% solids to the vanadium removal section or to the yellow cake dissolving tank for recycle. The thickener overflow is pumped through three plate and frame filter presses for final clarification and then to caustic barren storage.

Recarbonation

The caustic barren solution produced in the precipitation circuit contains sodium carbonate and a small quantity of sodium hydroxide. To reuse the barren solution, the caustic must be converted to sodium carbonate and sodium bicarbonate. This conversion is accomplished in two packed towers in which the caustic barren solution is contacted with boiler flue gas. The CO_2 in the flue gas neutralizes the sodium hydroxide and converts some of the carbonate to bicarbonate. The recarbonated barren solution is pumped to the solid-liquid separation circuit for use as a wash and repulping solution on the second stage of filters.

Purification

The primary precipitation of yellow cake produces a product which, when washed and dried, will assay around 75 to 77% U_3O_8, 5 to 6% V_2O_5, 2 to 2.5% CO_3 and 7.5% Na. Although the uranium content is satisfactory, the vanadium, carbonate, and sodium content of the precipitate exceeds contract specifications. Removal of these contaminants is required before the yellow cake is acceptable to the upgrading process plants.

Vanadium Removal

The removal of the vanadium and, quite incidentally carbonate, from the yellow cake is accomplished by roasting followed by water leaching. The yellow cake thickener underflow is pumped to a disc filter to dewater the slurry. The filter cake is agitated with a small quantity of sodium carbonate before being fed to the yellow cake roaster.

Yellow cake roasting is accomplished in an 8 foot 6 inch diameter, six-hearth Pacific furnace. The yellow cake is dried at around 1600°F. The calcined yellow cake discharges into a water cooled screw conveyor which cools the yellow cake to approximately 200°F before discharging into a ball mill conveyor. Water is used to dissolve the vanadium and carbonate contaminants in the yellow cake. The leached yellow cake slurry is collected in a 16-foot diameter by 10-foot deep thickener. The thickener overflow containing the vanadium is filtered and sent to vanadium storage. The vanadium solution is concentrated and sold to a nearby vanadium producer.

Sodium Removal

Vanadium solution contained in the yellow cake slurry in the 16-foot yellow cake thickener underflow is removed on a four and one-half foot diameter by six-foot long vacuum drum filter. Then, the yellow cake is washed and repulped with water. This yellow cake containing approximately 7.5% sodium as sodium diuranate is dissolved with sulfuric acid at a pH of around 2.2. Next, ammonia is added to maintain a pH of around 7.4. The uranium reprecipitates as yellow cake, ammonium diuranate and basic sulfate, generally containing less than 0.5% sodium. This product is filtered and washed with ammonium sulfate solution on two six-foot diameter by eight-foot long vacuum drum filters in series before being dried in a four-hearth Pacific roaster and packaged as described under the section on Product Preparation.

PRODUCT PREPARATION

The filter cake from the second yellow cake filter discharges into an agitated sump from where it is pumped to the yellow cake dryer. Yellow cake drying is accomplished in an 8 foot 6 inch diameter, four-hearth Pacific furnace which is fired to about 1000°F. By the time the yellow cake reaches the bottom of the yellow cake dryer, the moisture and ammonia have been removed. The yellow cake is discharged from the dryer through a pulverizer into a hopper where it is held until packaging.

Yellow cake is packaged in 55 gallon open-head drums. A vibrator under the drum loading station is used to settle the contents of each drum. After the drums are sealed, sampled, and weighed, the drum is cleaned and placed in storage until it is shipped.

Each drum contains a maximum of 1,000 pounds of yellow cake and about 45 drums comprise a shipping lot.

ACKNOWLEDGMENTS

The authors wish to express their thanks to the UN-HP management and Robert D. Lear, Hydrometallurgical Superintendent, for their support and assistance in the preparation of this paper.

Chapter 12

HEAP LEACH IN THE EDGEMONT AREA

R. G. Woolery, S. Ramachandran and M. R. Woolery

Metals Division
Union Carbide Corporation
Niagara Falls, New York

INTRODUCTION

Since 1975, Union Carbide Corporation has been conducting uranium exploration in southwestern South Dakota. This program has resulted in the discovery of an ore body about eight miles north of Edgemont in Fall River County. The site which was previously mined for uranium in the 1950's and 1960's consists of 950 lode claims (approximately 20,000 acres) and is within the Black Hills National Forest. Preliminary development plans indicate that fewer than 200 surface acres will be involved in the mining and heap leaching of the ore.

The ore occurs in both the Fall River and Lakota formations. The exploration to date has developed something in excess of two million tons of ore averaging ~0.1% U_3O_8. About 75% of the ore is expected to be mined from underground with 25% coming from five to seven small pits. The surface mining will be extensions of previously mined areas with a disturbance of 35-40 new acres near 35-40 previously disturbed acres requiring a total reclamation of 70-80 acres.

Several options for processing the ore including conventional milling by UCC, conventional milling in joint venture with another producer, and heap leaching are under consideration. This paper is concerned with the preliminary assessment of the heap leaching option based on the limited test work conducted on surface samples and on available cores from the deeper ones. The final design for a commercial operation will not be available until an underground bulk sample has been obtained and evaluated.

Union Carbide has been operating heap leach facilities commercially since 1973 at Gas Hills, Wyoming and Maybell, Colorado. During this period, we have made considerable progress in heap construction and operation. This method of treatment has proven to be highly successful on both low-grade and/or small reserve properties. Heap leaching, as practiced by Union Carbide, employs the concept of self-containment in that no liquid or solid wastes are discharged and, consequently, has been recognized as an environmentally acceptable practice. This feature, along with its inherent economic advantages, are significant factors in the selection of a process for this project.

METALLURGICAL DEVELOPMENT - SURFACE ORE

Bench Scale

The initial ore sample from the Edgemont Project was obtained from the surface where the uranium mineralization was readily identified as oxidized (reportedly carnotite and tyuyamunite). Standard agitation slurry leach studies on a crushed (2.5 mm x D) sample indicated the uranium to be readily extractable at an acid consumption of ~12 kg H_2SO_4 per ton. In this respect the Edgemont ore responds to leaching in a manner similar to our operating heaps at Gas Hills, Wyoming and Maybell, Colorado. However, in both of these operating heaps, the ore is an unconsolidated sand and crushing is not required prior to heap construction whereas the Edgemont ore is hard rock. This necessitated an evaluation of the maximum size amenable to leaching. It was found that even the coarse rock pieces available (up to 90 mm) had appreciable permeability to dilute acid solutions. This led to a further investigation into the feasibility of heap leaching a coarsely-crushed ore.

Exploratory tests were performed by three different techniques: (1) Acid soaking (or static leaching) on samples at sizes ranging from 0.15 mm x D through 80 mm rocks, (2) small-column percolation on 0.15 mm x D through 7 mm samples, and (3) bench-scale acid spraying of 80 mm and 100 mm rock pieces. The latter two series were included to determine the effect of dynamic conditions where the ore is continuously contacted by free-flowing acid solution. The ore grades for these various size ranges varied randomly from a low 0.02% U_3O_8 to a high 0.11% U_3O_8 (V_2O_5, 0.05 to 0.19%). The results of this test program showed:

(1) As the particle size increases, the V_2O_5 extraction decreases whereas the U_3O_8 leaching is not affected significantly. U_3O_8 extractions of +90% were obtained in almost all of the cases.

(2) Percolation leaching has a higher rate of uranium extraction than the acid-soaking technique.

(3) Crushing the 80 mm x D material further does not necessarily improve the U_3O_8 extraction.

In summary, the bench-scale program indicated that a heap constructed of primary crushed ore should respond well to dilute acid leaching.

Semipilot Scale

Following the laboratory studies, a semipilot-scale percolation test was conducted to confirm the laboratory bench-scale data as well as to obtain reasonable design data. A schematic diagram of the experimental column setup is shown in Figure 1. An FRP column of 0.46 m diameter and 3.8 m height was mounted on a steel structure that acted as a support as well as a gravity drainage system. A perforated "dump" plate at the bottom provided liquor drainage during the test as well as a means for removal of the leached tailings from distinct zones at the end of the test. Approximately 635 kg of 65 mm x D ore was skip loaded into the column on top of a sized, cleaned gravel bed overlain by a stainless steel screen. The leach liquor was sprayed on top of the ore bed at a rate controlled by the nozzle size and the air pressure. Changing the rate during the test was accomplished by the timer on the pressure regulator. The conditions of the pilot-column test are provided in Table 1.

Table 1. Chord Uranium - Pilot Column Leach Test

1. <u>Ore Characteristics</u>		
	1. Ore I.D.	Edgemont Oxidized Ore
	2. Particle Size	6.5 cm x D
	3. Heads	0.059% U_3O_8
	4. Weight	647 kg Wet 642 kg Dry
	5. Moisture	0.79%
	6. Total Contained Values	379 gm U_3O_8
2. <u>Column Characteristics</u>		
	1. Diameter	0.46 m
	2. Height	3.7 m
	3. Area	0.165 m^2
	4. Ore Bed Height	3.2 m
	5. Ore Bed Volume	0.53 m^3
	6. Ore Bulk Density	1215 kg/m^3
	7. Sweep Volume @ 200 l/t	148 liters

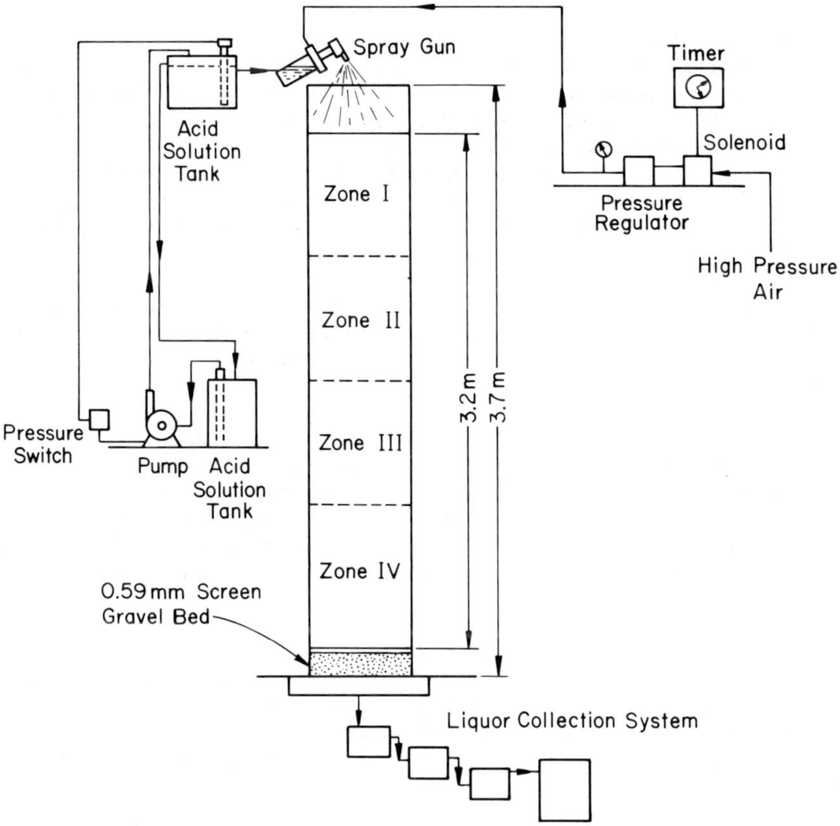

Figure 1. Arrangement of equipment in the pilot column leach test.

The test consisted of spraying dilute acid solution onto the ore bed until the effluent pH was ~1.0. After leaching, the ore bed was washed with 1.5 pH wash water until 500 liters/ton of product liquor were collected. Additional washing was continued through recycle until the rate of increase of the uranium liquor grade was minimal. After thorough drainage of the bed, the bottom plate was lowered and the leached residue collected in four zones, dried, sized, sampled and assayed.

The pilot-column test was run for 27 days during which time the effluent stream was periodically sampled and analyzed. The results of this sampling program are reported in Table 2. As shown in Figure 2, the pilot test produced a high rate of extraction with good overall uranium recovery. Over the entire test, 95% of the uranium was recovered in 516 liters of product liquor for an average grade of 740 ppm U_3O_8. However, 90% of the uranium was extracted at an average product grade of ~2.0 g/l. Maintaining a flow rate through the column was trouble free indicating good permeability of the ore.

The residue analysis of the four zones was uniform with respect to U_3O_8 grade (av. = 0.0030% U_3O_8) indicating that neither channeling nor ore segregation was a problem. The details of the residue analyses are shown in Table 3.

Because our present plans do not include the recovery of a vanadium by-product, this element must be removed from the process stream if a closed cycle is employed. For this reason, the fate of the vanadium in the leach operation was determined and reported in Table 2. At a head grade of 0.066% V_2O_5, about 60% of the vanadium is solubilized along with the uranium under the conditions of the pilot test.

This test, thus, confirmed the excellent amenability of the surface ore at Edgemont to the heap leach approach. Considering the anticipated higher ore grade of the commercial heap of 0.1% U_3O_8 and using scaleup experiences from our other heap leach operations, the overall uranium recovery for a commercial heap has been estimated at 87 ± 2%.

METALLURGICAL DEVELOPMENT - UNDERGROUND ORE

Bench Scale

The second phase of the metallurgical development program was directed toward the heap leach amenability of the underground ore. Core samples from the Lakota formation, sufficient for only a limited bench-scale evaluation, were available for this purpose. The underground samples were observed to be higher in calcite and to contain appreciable quantities of carbonaceous matter. The uranium occurrence was also found to be in the reduced state (probably as coffinite

Table 2. Edgemont (Chord) Uranium - Pilot Leach Data.

Sample No.	Total Hours	Liquor Volume (1)	pH/emf	Sweep Volume	Flow Rate $l/m^2/d$	Liq. Grade gpl U_3O_8	Liq. Grade gpl V_2O_5	Cum. Grade, gpl U_3O_8	Cum. Grade, gpl V_2O_5	Est. Dissoln. Cum. % U_3O_8	Est. Dissoln. Cum. % V_2O_5
	Start	----	------	----	------	----	----	----	----	----	----
1.	69	13.4	1.5/618	0.09	71	3.27	1.68	3.27	1.68	11.5	5.3
2.	141	33.7	1.28/645	0.35	68	3.18	1.80	3.20	1.77	39.6	19.5
3.	165	11.4	1.6/664	0.44	69	2.97	1.82	3.16	1.78	48.6	24.4
4.	189	9.8	1.06/667	0.51	59	2.62	1.75	3.08	1.77	55.3	28.4
5.	213	10.3	1.08/650	0.59	62	2.41	1.61	2.99	1.75	61.8	32.3
6.	237	11.8	1.17/675	0.68	72	2.07	1.48	2.87	1.72	68.2	36.4(1)
7.	309	32.1	1.14/636	0.92	65	1.48	1.19	2.51	1.58	80.7	45.3
8.	333	21.2	1.16/651	1.07	128	1.05	0.95	2.29	1.49	86.6	50.0
9.	357	28.8	1.22/652	1.29	174	0.55	0.58	2.00	1.33	90.7	53.9
10.	381	27.9	1.35/632	1.50	168	0.21	0.28	1.75	1.19	92.2	55.8
11.	405	28.3	1.52/614	1.71	164	0.10	0.16	1.55	1.06	93.0	56.8
12.	429	34.2	1.69/584	1.96	207	0.059	0.094	1.35	0.94	93.5	57.6
13.	453	23.5	1.71/552	2.14	142	0.043	0.072	1.25	0.86	93.8	58.0
14.	477	25.4	1.60/559	2.33	147	0.039	0.067	1.15	0.80	94.0	58.4
15.	501	23.8	1.47/583	2.50	150	0.071	0.087	1.07	0.75	----	----(2)
16.	525	23.3	1.56/576	2.68	144	0.027	0.054	1.00	0.70	----	----
17.	549	26.8	1.56/581	2.88	162	0.025	0.053	0.94	0.66	----	----
18.	573	25.8	1.49/587	3.07	156	0.031	0.057	0.88	0.62	----	----
19.	645	84.1	1.50/589	3.70	168	0.044	0.062	0.74	0.53	94.9	59.4
20.	669	14.3			86						(3)
21.	693	5.5			33						(4)
	693	515.5		3.84	111						

Table 3. Summary of Residue Characteristics.

Zone	Wet Weight (kg.)	Dry Weight (kg.)	Moisture % (Wet Basis)	Average Assay, %		Distribution, %		
				U_3O_8	V_2O_5	Wt.	U_3O_8	V_2O_5
1	187	166	11.2	0.0031	0.0211	26.0	26.5	22.4
11	166	147	11.2	0.0010	0.0318	23.1	7.6	30.0
111	175	156	10.9	0.0041	0.0186	24.3	32.7	18.5
1V	189	170	10.1	0.0038	0.0268	26.6	33.2	29.1
Weighted Average	717	639	10.8	0.0030	0.0245	100.0	100.0	100.0

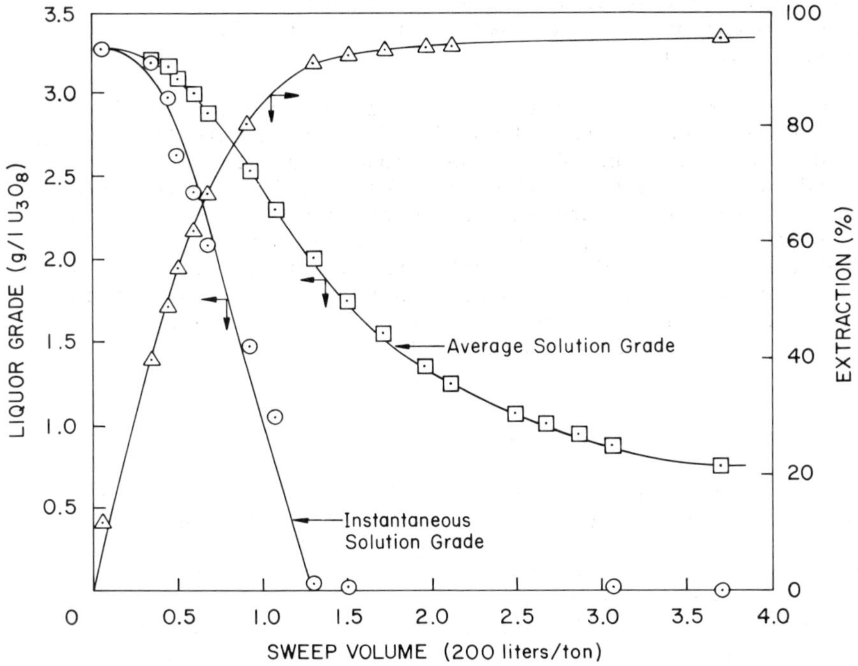

Figure 2. Grade of the leach liquor and uranium recovery as a function of the volume of solution passed through the leaching column.

or extremely fine-grained uraninite). Mineralogical examinations indicated that the uranium was finely disseminated through the sandstone layers parallel to the thin beds of carbonaceous matter.

Based on these ore characteristics, it was anticipated that the more reduced underground ore would not be as amenable as the original surface ore. A test, comparable in conditions to the pilot-scale test, conducted in a small 0.3 m x 0.9 m column resulted in only 50-60% uranium extraction. These results confirmed our assumption that good metallurgy on this deeper ore will require more severe leaching conditions.

In the second small column test, the ore was reduced in size to 15 mm, the acid was increased to 37 kg/t, and an oxidant ($NaClO_3$) was added at 1.25 kg/t. Under these conditions the uranium extraction improved to 87%. These results show the need for a more extensive laboratory program to determine the optimum processing techniques with respect to particle size, acid concentration, contact time, and oxidant requirement. Unfortunately, the supply of available underground core samples was exhausted. Therefore, this part of the program, as well as the pilot-scale test, has been delayed until an underground bulk sample becomes available.

HEAP DESIGN

The proposed heap design for the Edgemont Project is similar to those employed successfully at our other heap leach facilities. See Figure 3. The heap site is currently planned for the Robinson Flats area and will take advantage of the existing topography plus some cut and fill in base construction, so that upon completion of the project, the heap will be buried in accordance with the reclamation requirements in effect at the time.

In preparation for heap construction, the area where the heap is to be constructed is graded to a 1% slope dipping toward the front of the heap. This grading will be accomplished almost exclusively by excavation using fill only for a few low spots. The end result will be a partially buried heap with the uphill section of the base rising 45° to the natural contour of the land as shown in Figure 6. Where fill is required, it will be placed in lifts not to exceed 20 cm in its loose form and compacted according to the American Association of State Highway Official's Code T 180, "Method of Compaction." The surplus fill from the base preparation will be stored uphill from the heap for use in restoration. Any top soil recovered from this operation will be stored separately to be utilized for revegetating the restored heaps.

A 60-meter monitor drain will be placed under the portion of the heap that will be leached the first year. The design of this drain, as shown in Figure 8, consists of digging and lining a trench ~1.0 m

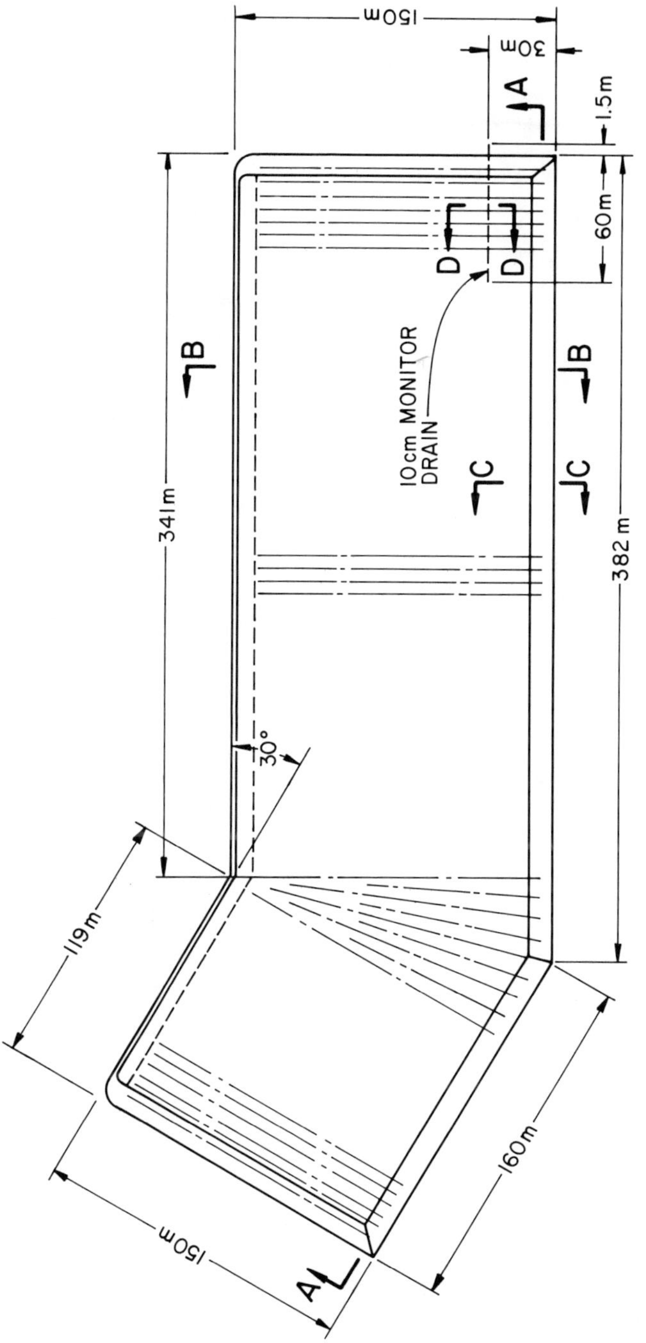

Figure 3. Plan view of the ore heap.

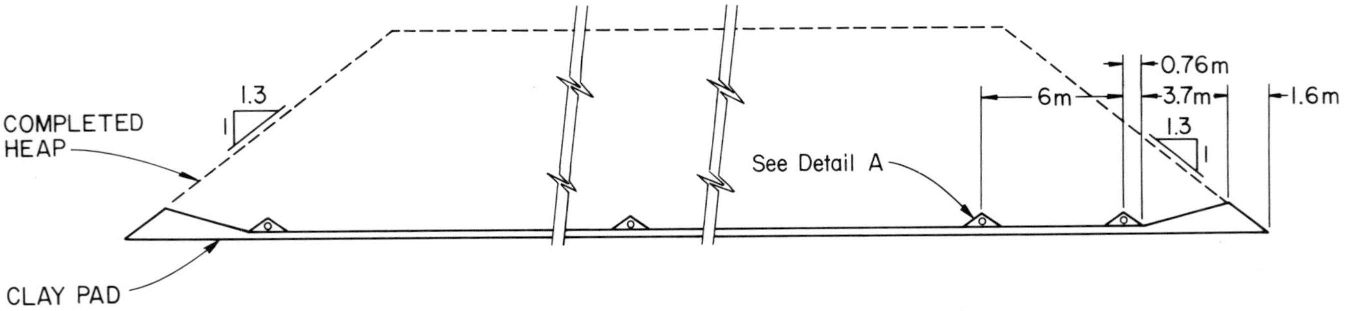

Figure 4. Elevation Section A-A (Figure 3).

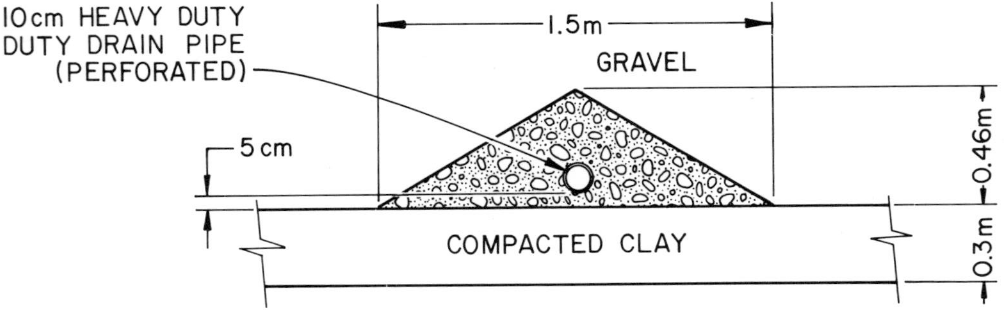

Figure 5. Heap drainage system, Detail A of Section A-A (Figure 4).

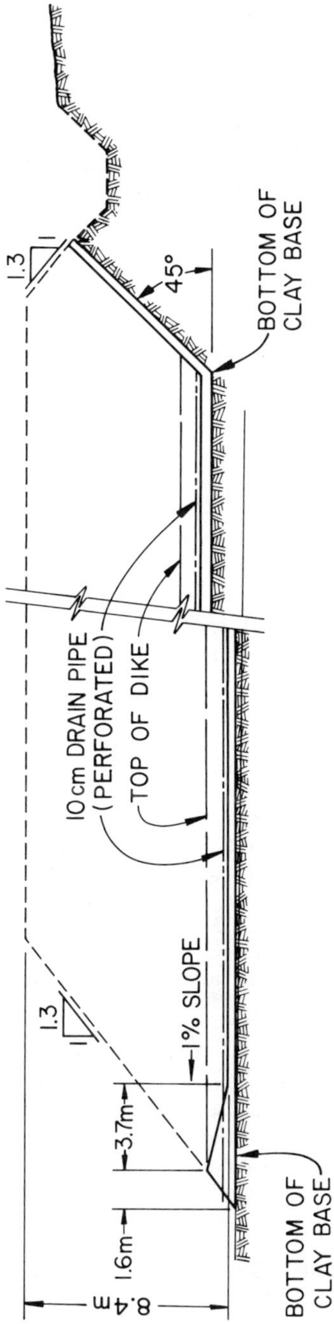

Figure 6. Elevation Section B–B (Figure 3).

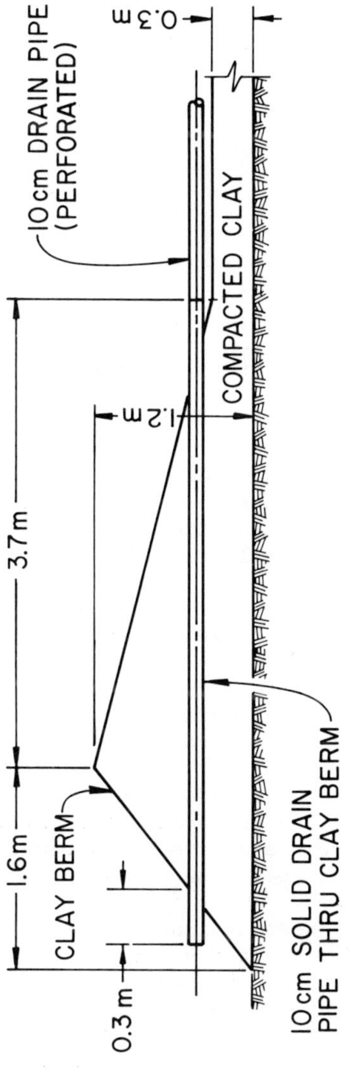

Figure 7. Clay berm, Section C–C (Figure 3).

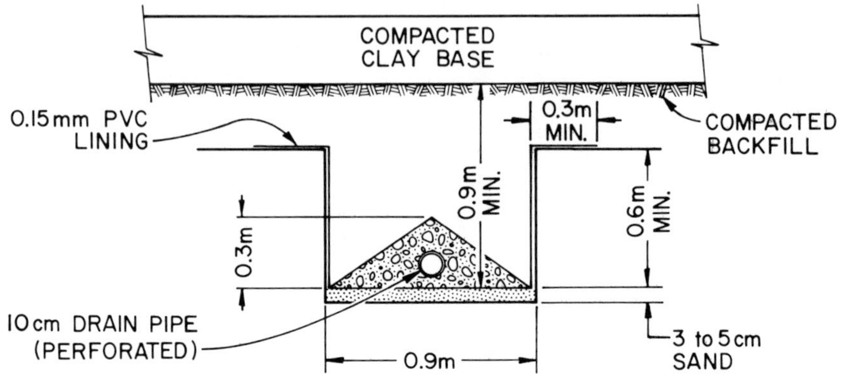

Figure 8. Monitor drain, Section D-D (Figure 3).

wide and ~1.0 m deep. The trench will be lined with 0.15 mm PVC, and a 10-cm perforated drain pipe will be placed on the liner. The pipe will be covered with ~0.3 m of sized gravel, backfilled and compacted to the original grade.

Upon completion of the base preparation, a clay liner will be installed to prevent losses of process liquor and to inhibit seepage. The liner is constructed in two lifts. About 20 cm of base clay is spread over the base after moisture adjustment, and it is compacted to ~90% of its maximum dry density. This procedure is repeated for the second lift providing a compacted clay base of at least 30 cm in total thickness. The base extends under the entire heap and up the excavated incline on the uphill side of the heap. A 0.9 m high clay dike will extend around the entire periphery of the heap base excluding the excavated incline. These clay dikes are prepared and compacted to the same standards as the clay base.

The heap drainage system will consist of 10-cm diameter perforated drain pipes spaced ~6 m apart extending from the inclined base to the front dike. Each drain is laid on ~5 cm of sized aggregate, then covered with additional aggregate to a total height of ~45 cm. (See Figures 5 and 7). Solid pipe is substituted for the perforated drain pipe for extension through the clay dike and into the common collection header.

Placement of the ore on the base requires that it be constructed to its planned height in a single lift. This is accomplished by constructing a truck ramp on the edge of the clay liner to the desired elevation of the heap. The ore is dumped at the top of the ramp and pushed over the edge by a dozer. The ore will assume its natural angle of repose (~37°) and will extend to the top of the clay dikes and the excavated incline around the perimeter as shown in Figures 4 and 6. Lower grade ore will be placed selectively around the perimeter as this section is not leached (as effectively) as the center. The top perimeter of the heap is graded slightly toward the center of the heap for ~2 m to ensure that any runoff will be contained within the heap.

During the life of the project, all heap operations are continued simultaneously. Base construction, mining, and heap construction schedules are designed to precede leaching far enough in advance to ensure that the planned production rates are met. The heap is estimated to contain 660,000 tons, and only the area necessary to achieve the desired production rate will receive leaching solution.

PROPOSED OPERATING PRACTICE AND SCHEDULE

The operating practice will be similar to the procedure employed in our operating facilities which has been described previously (1). Sections of the heap length are leached in stages with the first 1,200 l/t directed to the recovery circuit and the final 600 l/t of wash solution advanced to the sections under leach by direct recycle.

Uranium recovery from the heap product liquor will be by solvent extraction with subsequent removal of the vanadium from a bleed stream on the SX raffinate. The precipitated vanadium will be collected in sludge ponds and the vanadium depleted solution will rejoin the uranium raffinate for recycle to the heaps. The rich organic will be stripped with a sodium carbonate solution, and the uranium product liquor will be shipped by truck to our Gas Hills facility. A schematic flowsheet of the proposed processing circuit is shown in Figure 9.

Predictability of heap operation has yet to become an accurate science. The heaps vary considerably along their length, and for any one section, our estimates of operating schedules can have significant error. However, these variables have a way of averaging out over the course of production, and our ability to determine average production data is within ± 5%.

In determining the production schedule for this feasibility study, the following conditions have been assumed:

1. Weather conditions would permit a heap operation for a total of 7 months/year or 6 months equivalent of full operation.

2. A production target of 182,000 kg U_3O_8/year.

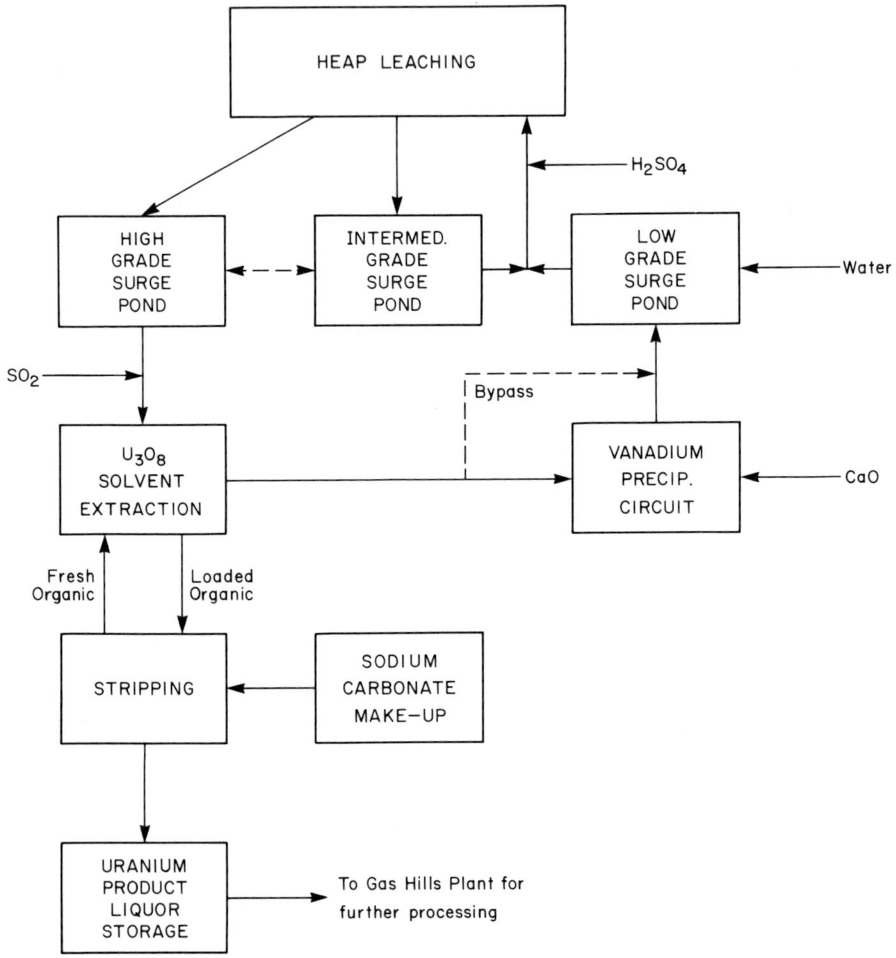

Figure 9. Flowsheet for leach liquor processing.

3. An average ore grade of 0.10% U_3O_8.

4. A final tailings assay of 0.01% U_3O_8 after leaching and washing.

5. A total of 6 sweep volumes (1,200 l/t) for leaching and 3 sweep volumes (600 l/t) for washing are required.

6. A processing plant capacity of 990 l/min.

7. 1,300 tons of ore per meter of heap length (150 m x 8.4 m).

8. Percolation rate = 100 $l/m^2/day$

On the bases of these assumptions, we have proposed an operating schedule as shown in Table 4. This schedule covers the first 3+ years and is shown in terms of sweep volumes. A sweep volume (S.V.) is an arbitrary unit of 200 l/t and is a convenient term applicable to all heaps regardless of heap dimensions. In all cases, the first 6.0 S.V. are considered leaching, and the final 3.0 S.V. are considered wash and are recycled without SX treatment to the heap. Because of different stages encountered during the operating seasons, a small annual variance in production rate and product liquor grade is anticipated.

The "leach section" is also an arbitrary and calculated value and is determined on the basis of uranium production level, S.V. requirement for extraction, percolation rate, and ore grade. For the production rate employed and the assumptions used, a leach section of 108 m was calculated. On the basis of this proposed operating schedule, annual uranium production is estimated at 205,000 kg for the first year, 157,000 kg the second year, and 181,000 kg for the third year.

Upon completion of production, the earth removed during base preparation and stored above the heap is used to cover the heap in accordance with the restoration requirements. A typical cross-section through the restored heap is illustrated in Figure 10. The trench on the uphill side of the heap will be excavated prior to heap operation to divert any natural runoff water around the heap area.

FUTURE PROGRAMS

The option of heap leaching Edgemont ore presented here has been prepared for comparative evaluation with other processing alternatives. It is based on the results obtained from the ore samples available. However, before a final design can be determined, it will be necessary to establish the character and amenability of the underground ore. This can only be determined from a bulk sample of the

Table 4. Proposed Operating Schedule

Notation: 1.0 Sweep Volume (S.V.) = 200 1/t R = Recycle to next section
Leach Section (Length = 108 m D = Drain Heaps (Cease Spraying)

Each Section	I	II	II	III	III	IV	V	V
Days	Year #1		Year #2		Year #3			Year #4
0	Start		Start		Start	Start		Start
10	-		-		-	-		-
20	0.5 S.V. Out		4.0 S.V. Out		7.0 (R)	0.5 S.V. Out		4.5
30	1.0		4.5		7.5	1.0		5.0
40	1.5		5.0		8.0	1.5		5.5
50	2.0		5.5	Start	8.5 (D)	2.0		6.0
60	2.5		6.0		9.0	2.5		6.5
70	3.0		6.5 (R)	0.5 S.V. Out		3.0		7.0
80	3.5		7.0	1.0		3.5		7.5
90	4.0		7.5	1.5		4.0		8.0
100	4.5		8.0	2.0		4.5		8.5 (D)
110	5.0		8.5 (D)	2.5		5.0		
120	5.5	Start	9.0	3.0		5.5	Start	9.0
130	6.0			3.5		6.0		
140	6.5 (R)	0.5 S. V. Out		4.0		6.5 (R)	0.5 S.V. Out	
150	7.0	1.0		4.5		7.0	1.0	
160	7.5	1.5		5.0		7.5	1.5	
170	8.0	2.0		5.5		8.0	2.0	
180	8.5 (D)	2.5		6.0 (D)		8.5 (D)	2.5	
190		3.0 (D)					3.0	
200	9.0			6.5		9.0	3.5 (D)	
210		3.5					4.0	
g U_3O_8	123	84.8	36.1	120.5		120.5	60.3	
Grade		720 ppm		620 ppm			710 ppm	

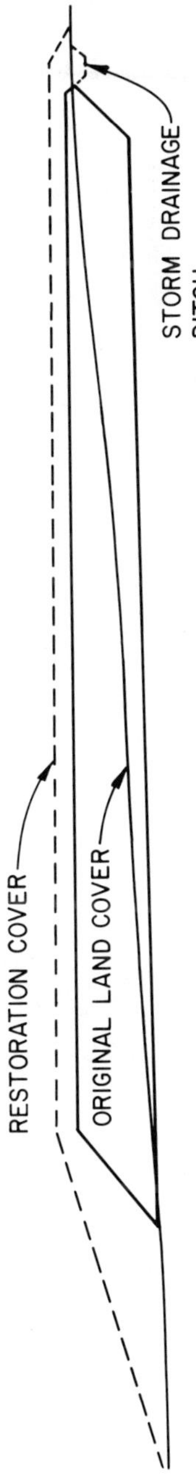

Figure 10. Typical restored cross-section for Section B-B (Figure 3).

underground ore sufficiently large to complete the bench-scale evaluation and a pilot-scale test on a representative sample. When the bulk sample becomes available, our remaining program will consist of:

1. Bench-scale testing of leaching variables associated with underground ore.

2. Pilot-scale testing for confirmation of the bench-scale results and to obtain heap design data.

3. Determination of the vanadium removal rate necessary for optimum operation of the uranium recovery circuit.

4. A re-evaluation of the proposed operating practice and the economic impact due to the more severe requirements of leaching unoxidized underground ore.

Unless the underground ore proves to be more refractory than anticipated, the heap leach option for processing Edgemont ore must be considered favorably. This method of processing uranium ores has been highly successful within Union Carbide, and our past experience should prove invaluable in designing a successful and an environmentally acceptable operation.

REFERENCE

1. Woolery, R. G., Hansen, D. J., Weber, J. A. and Ramachandran, S., "Heap Leaching of Uranium - A Case History," Mining Engineering, March, 1978.

Chapter 13

CARBONATE LEACHING KINETICS OF LOW GRADE SANDSTONE ORES

S. Y. Han, C. A. Petrache and M. E. Wadsworth

Department of Metallurgy
University of Utah
Salt Lake City, Utah

INTRODUCTION

More than 97% of the total uranium production of the United States comes from sedimentary deposits (1). These sandstone deposits were formed by reduction of migrating uranium-bearing solutions from volcanic silicate rock sources. Such deposits occur in two major forms, either as relatively flat, lens-like deposits or as crescent-shaped roll front deposits. The lens-like deposits are found as discrete masses enclosed in reduced sandstone which is generally surrounded by oxidized sandstone. The roll front deposits are found along crescent-shaped interfaces between reduced and oxidized sandstone (2).

In-situ mining of selected sedimentary deposits has proven to be a viable method of extraction and competitive with conventional mining for relatively deep low-grade ores (2). In in-situ leaching, uranium is recovered by drilling injection and collection wells in the ore body, pumping a suitable lixiviant to dissolve the mineral, extracting uranium from the pregnant solution by ion exchange, and finally regenerating and recycling the lixiviant. Even though there is increasing interest and practice in developing the in-situ technique for obtaining uranium values from deeper, lower-grade reserves, there is a lack of knowledge of the basic leaching kinetics and combined transport processes.

In-situ leaching operations, in general, appear attractive for the following three general types of deposits (3):

(1) low-grade surface dumps or deposits which lie above the natural water table,

(2) low-grade deposits located below the natural water table but accessible by conventional mining techniques for purpose of rubblization, and

(3) deposits too deep for economic conventional mining.

Type (2) refers to the leaching of inundated deposits which exist at relatively shallow depth (less than 1,000 ft) and is the subject of this study. In-situ leaching of porous ore deposits inundated by water has the benefit that the hydrostatic head will increase the oxygen concentration and, thus, the kinetics of dissolution.

The dissolution of primary uranium minerals occurs by electrochemical surface reactions. Nicol et al. (4) obtained electrode kinetic measurements to prove the electrochemical model for the carbonate leaching of UO_2 using oxygen as the oxidant. The following two overall half-cell reactions describe the leaching behavior of UO_2:

$$\text{anodic dissolution;} \quad UO_2 + 3CO_3^{-2} = UO_2(CO_3)^{-4} + 2e^- \qquad (1)$$

$$\text{cathodic oxygen discharge;} \quad \tfrac{1}{2}O_2 + H_2O + 2e^- = 2OH^- \qquad (2)$$

The above reactions constitute a corrosion couple in which the solid behaves as an internally short-circuited system with anodic and cathodic regions. The rates of reaction are subject to analysis using mixed potential corrosion theory.

Hiskey (5) recently carried out an autoclave analysis of the leaching of pure UO_2 under oxygen pressure and has effectively explained the results from electrochemical theory. The purpose of this study is to see if these basic concepts may be operative for very low grade uranium ores. The mixed potential is the potential which results when the anodic and cathodic currents are balanced such that $i_a = -i_c$ where i_a is the anodic current density, and i_c is the cathodic current density. The rate of leaching may be expressed in terms of the anodic current density,

$$-\frac{dn_{UO_2}}{dt} = \frac{i_a}{z_a F} = -\frac{i_c}{z_c F} \qquad (3)$$

where z_a and z_c are the total number of electrons transferred for the anodic and cathodic processes, and F is the Faraday constant. The net current density for the reaction couple above may be expressed in a simplified form by the Butler-Volmer equation using Hiskey's findings,

$$i = i_a + i_c \qquad (4)$$

$$i = zFk_a[HCO_3^-] \exp(\beta_a FE_a/RT) -$$

$$zFk_c[\frac{KP_{O_2}}{1+KP_{O_2}}] \exp[1-(1-\beta_c) Fe_b/RT] \quad (5)$$

where i_a and i_c are the partial anodic and cathodic current densities, k_a and k_c are rate constants, K is an equilibrium constant for oxygen adsorption, β_a and β_c are the transfer coefficients for the anodic and cathodic processes. Equation (5) assumes the partial anodic and cathodic branches of the coupled half-cell reactions are each in their own respective Tafel regions. The basic concepts of mixed potential theory will be applied here to explain observed kinetic data obtained from the batch leaching tests. The general purpose of this study was to determine the intrinsic particle kinetics for application to conditions anticipated in in-situ extraction systems.

When the system reaches steady state, E becomes the mixed potential, such that $E_a = E_b = E$, and the net current is zero. The transfer coefficients for purposes of analysis are assumed to be one-half, $\beta_a = \beta_c = 0.5$. Equation (5), for i = 0, may be used to evaluate the exponential voltage term giving,

$$e^{FE/RT} = \frac{k_c}{k_2[HCO_3^-]} [\frac{KP_{O_2}}{1+KP_{O_2}}] \quad (6)$$

The rate of leaching may now be expressed by

$$-\frac{dn_{UO_2}}{dt} = \frac{i_a}{zF} = (k_a k_c)^{1/2} [HCO_3^-]^{1/2} [\frac{KP_{O_2}}{1+KP_{O_2}}]^{1/2} \quad (7)$$

Several research studies have noted the leveling off of rate with increasing oxygen pressure (5,7). This has been explained in terms of a single-site adsorption mechanism. Also, the rate increases with hydroxyl ion concentration to a maximum and then decreases. Reported kinetics on pure minerals are consistent with reactions:

$$UO_2 + HCO_3^- = UO_2 \cdot HCO_3 + e^- \quad (8)$$

$$UO_2 \cdot HCO_3 + OH^- = UO_2 \cdot CO_3 + H_2O + e^- \quad (9)$$

$$UO_2 \cdot CO_3 + 2CO_3^{-2} = UO_2(CO_3)_4^{-4} \quad (10)$$

The oxygen adsorption reaction is,

$$UO_2 + O_{2(g)} \rightleftarrows UO_2 \cdot O_2 \tag{11}$$

$$[O_2]_s = \frac{R_o K P_{O_2}}{1+KP_{O_2}} \tag{12}$$

where $[O_2]_s$ is the surface concentration of adsorbed oxygen, K is the equilibrium constant for adsorption, and R_o is the total concentration of surface sites. The electrochemical discharge of oxygen may be expressed as,

$$O_{2(s)} + H^+ + e^- \rightleftarrows HO_{2(s)} \tag{13}$$

$$(HO_2)_{(s)} + H^+ + e^- \rightleftarrows H_2O_{2(s)} \tag{14}$$

$$H_2O_{2(s)} + 2e^- \rightleftarrows 2OH^- \tag{15}$$

Reactions (13) and (14) retard the discharge of oxygen at high pH values due to the low H^+ concentration (7). Reaction (15) is much faster in alkaline solution, explaining the enhanced oxidation of UO_2 by hydrogen peroxide in basic solutions.

In this study pH was held constant by maintaining a constant molar ratio, $[CO_3^=]/[HCO_3^-] = 2.5$. All tests were carried out at ambient temperature, and the oxygen pressure range studied varied from 3 to 250 psi.

EXPERIMENTAL PROCEDURE

The low-grade sandstone uranium ore used in this study was obtained from the Reserve Oil and Gas Mine near Grants, New Mexico. The as mined ore was sealed with doubly covered plastic lining inside a steel drum to minimize oxidation by air. About 5 g of ore were split from each sample to determine the average grade of the sample.

The leaching tests were carried out in a one-gallon autoclave manufactured by Autocalve Engineers. For each run, 300 g of a random ore sample were added to 1 liter of reagent-grade ammonium carbonate-ammonium bicarbonate solution. The following $[CO_3^=]/[HCO_3^-]$ molar ratios were used for the tests: 1.5/0.6, 1.0/0.4, 0.5/0.2, 0.3/0.12, 0.2/0.08 and 0.1/0.04. Different levels of oxygen pressure were applied during each run except for five runs which were purged with pure deoxygenated nitrogen. For all tests, the stirring was maintained at 800 rpm which was discerned to be adequate to eliminate solution boundary diffusion as rate controlling. All tests were carried out at room temperature. Sample aliquots were collected at selected time intervals through a filtered sample line.

Size Distribution, Grade and Ore Characteristics

To determine whether there was any specific size of the ore particle that has a high uranium concentration, dried ore, as received, was carefully sized by standard analysis. The cumulative weight fraction finer is shown in Figure 1. The 50 percent passing size is approximately 0.35 mm. Each size fraction was analyzed for the uranium content. Figure 2 illustrates the results obtained. The average overall grade was 0.04% uranium.

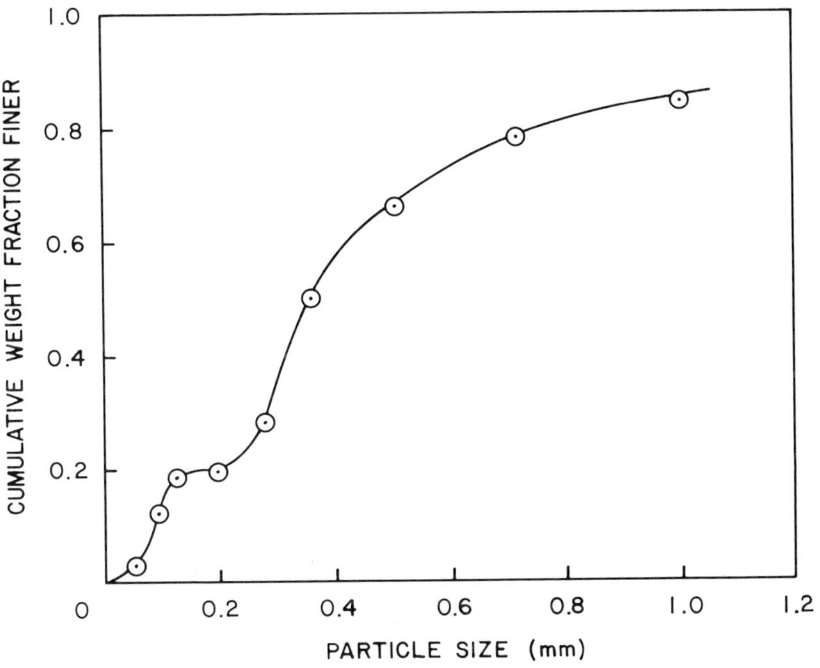

Figure 1. Particle size distribution for the as-received ore.

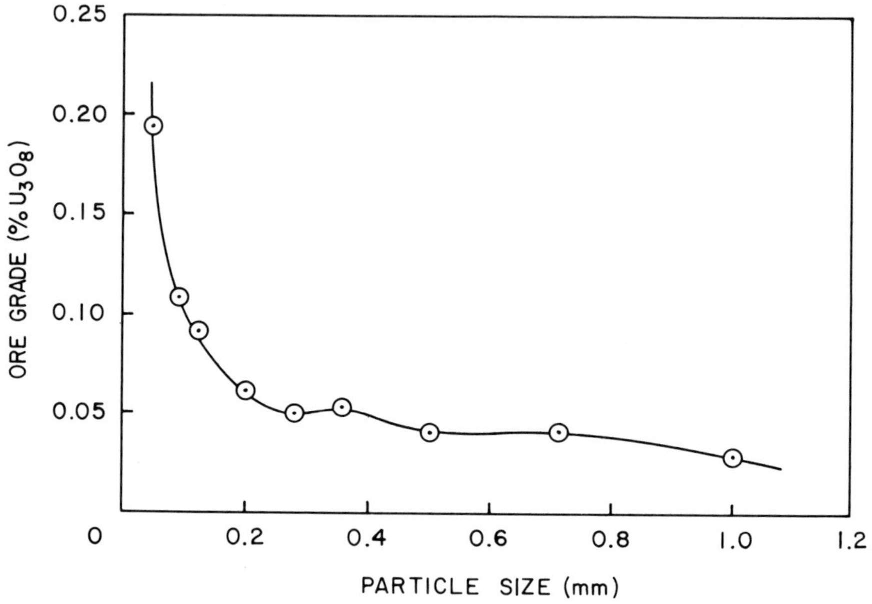

Figure 2. Ore grade (% U_3O_8) after sizing as a function of particle size.

A petrographic analysis[a] showed that the sandstone ore used in these tests is rather impure and consists of quartz, microcline, plagioclase, shale, granite, and other sandstone constituents found in arkose or subarkose rock.

Method of Analysis

Solutions were analyzed by the spectrophotometric dibenzoylmethane method which is suitable for the determination of uranium concentrations in the range of 0.001 to 5 g per liter. Dissolved uranium was oxidized to the +6 state, then separated from most interfering elements by tributyl phosphate extraction. The yellow color formed with dibenzoylmethane was measured in the organic phase at a wavelength of 410 nm.

For the determination of the uranium content of the ore head sample, the ore was first decomposed by heating with aqua regia and

[a]Hazelton Geologic Services, Salt Lake City, Utah.

evaporated to dryness. The cooled residue was wetted with hydrofluoric acid, then diluted with water to a pre-selected volume. The resulting solution was analyzed by the method described above.

EXPERIMENTAL DATA

Effect of Carbonate Concentration on the Rate of Leaching of the Oxidized (U^{+6}) Fraction

Samples were leached in a solution which had been boiled and purged with deoxygenated nitrogen. Leaching was carried out under the same nitrogen atmosphere. Oxygen analysis with an oxygen specific electrode indicated the oxygen content of air-equilibrated water had been reduced from 7.1 ppm O_2 to approximately 1.6 ppm by nitrogen purging. The latter content is equivalent to an equilibrium oxygen pressure of 3 psi at the ambient barometric pressure of 0.86 atmospheres. Experimental results are shown in Figure 3. These data illustrate the rate of leaching at four separate carbonate concentrations from 0.28 to 1.4 M combined carbonate plus bicarbonate. Subsequent analysis, correcting for the 1.6 ppm oxygen present, indicated that the ore was approximately 40% oxidized or the fraction oxidized, f_o, is 0.4.

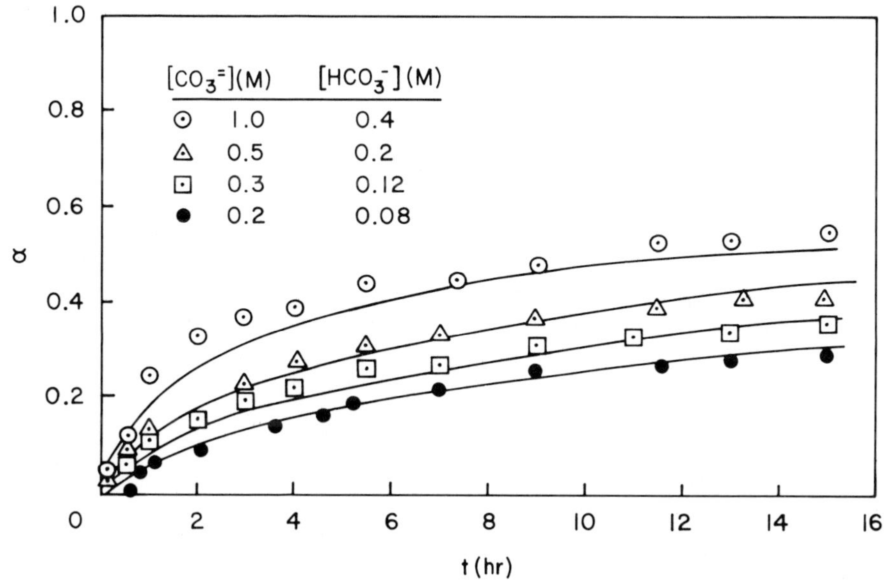

Figure 3. Release of uranium in the absence of O_2 for several concentrations of $CO_3^{=}$ and HCO_3^{-}. The lines show the dependence of the fraction reacted on time as predicted by the rate model.

The observed data were well correlated by rate expressions for cylindrical geometry involving diffusion through a matrix of low porosity and high tortuosity. The cylindrical geometry may be rationalized if the uranium crystallites represent deposits between particles of organic or inorganic material. Considering cylindrical geometry and diffusion of carbonate through a porous matrix, the rate of reaction may be expressed by the equation

$$\frac{dn}{dt} = -\frac{2\pi rhD'}{\sigma}\frac{dC}{dr} \tag{16}$$

where n is the moles of uranium remaining in a single uranium domain, r is the domain radius at any time t, h is the thickness of the domain, D' is the apparent diffusivity, C is the total carbonate concentration and σ is the stoichiometry factor [σ = 3 for the formation of $UO_3(CO_3)_3^{-4}$]. A domain in this instance may be a partially oxidized uranium particle or a mixture of totally oxidized uranium and an organic matrix. The kinetics are characteristic of results expected for diffusion through pores with diffusion the rate-controlling process. Accordingly, the apparent diffusivity is given by,

$$D' = \frac{D\Omega}{\tau} \tag{17}$$

where D is the diffusivity for carbonate ion diffusion, Ω is the porosity and τ is the tortuosity. If α_o is the fraction of the oxide portion which has been leached, then

$$\frac{d\alpha_o}{dt} = -\frac{M}{r_o^2 h\rho G}\frac{dn}{dt} \tag{18}$$

where M is the molecular weight of uranium, r_o is the original radius of the domain, and ρ and G are the density and grade (percent uranium), respectively, of the domain. For cylinder geometry,

$$r = r_o\ (1-\alpha_o)^{1/2} \tag{19}$$

Equation (16) may be integrated under steady-state conditions resulting in the expression,

$$\alpha_o + (1-\alpha_o)\ \ln\ (1-\alpha_o) = k_oCt \tag{20}$$

where k_o is the rate constant for the oxidized fraction and

$$k_o = \frac{4D'M}{3r^2 \rho G} \tag{21}$$

Figure 4 illustrates the data plotted according to Equation (20) using $\alpha_o = \alpha/f_o$ where α is the fraction of the total uranium content which has been leached in time t.

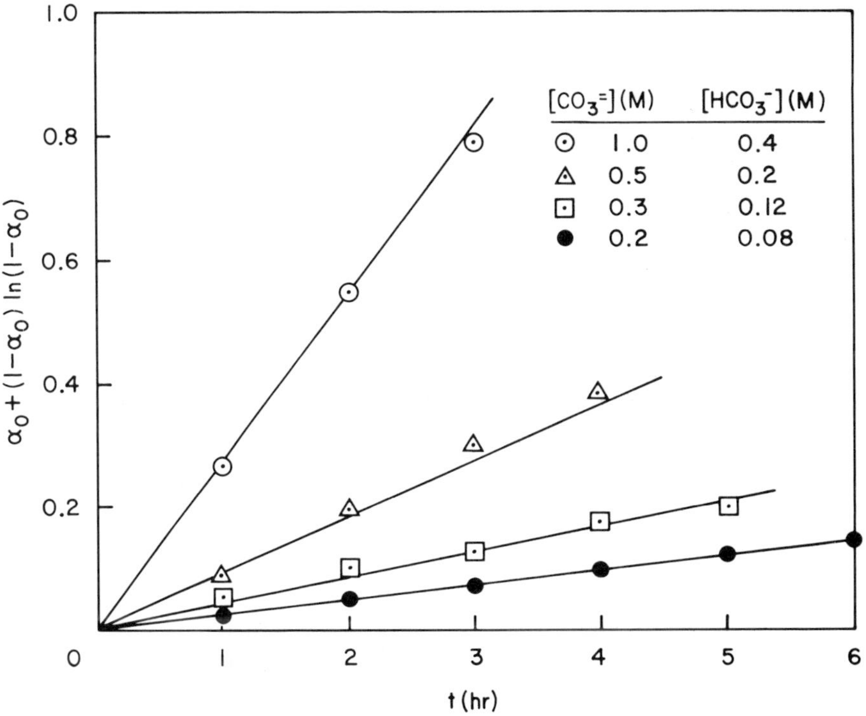

Figure 4. Correlation of the rate data with the leaching model for the dissolution of the oxide ore fraction. Leaching was conducted in an oxygen-free atmosphere.

Effect of Carbonate Concentration and Oxygen Pressure on the Leaching of the Total Uranium Content

Figures 5 through 8 illustrate results obtained for the oxygen pressure range 50 to 250 psi and for four levels of carbonate concentration. The results illustrate the general observation that at high carbonate concentration, the kinetics are much more oxygen pressure sensitive than at lower carbonate concentration. At low carbonate concentration, the oxygen pressure dependence diminished appreciably.

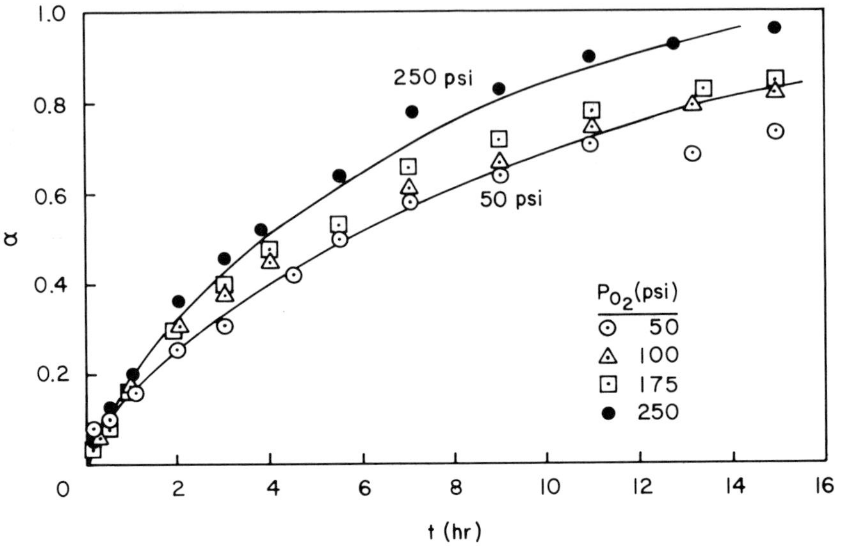

Figure 5. Effect of oxygen pressure on the leaching of uranium ore for $CO_3^=$ and HCO_3^- concentrations of 0.5 and 0.2 M, respectively. The lines show the dependence of the fraction reacted on time as predicted by the rate model.

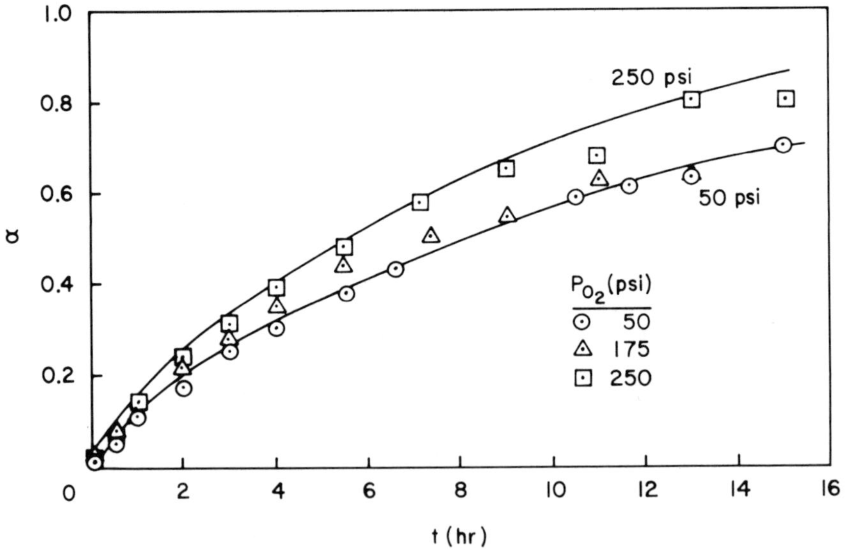

Figure 6. Effect of oxygen pressure on the leaching of uranium ore for $CO_3^=$ and HCO_3^- concentrations of 0.3 and 0.12 M, respectively. The lines show the dependence of the fraction reacted on time as predicted by the rate model.

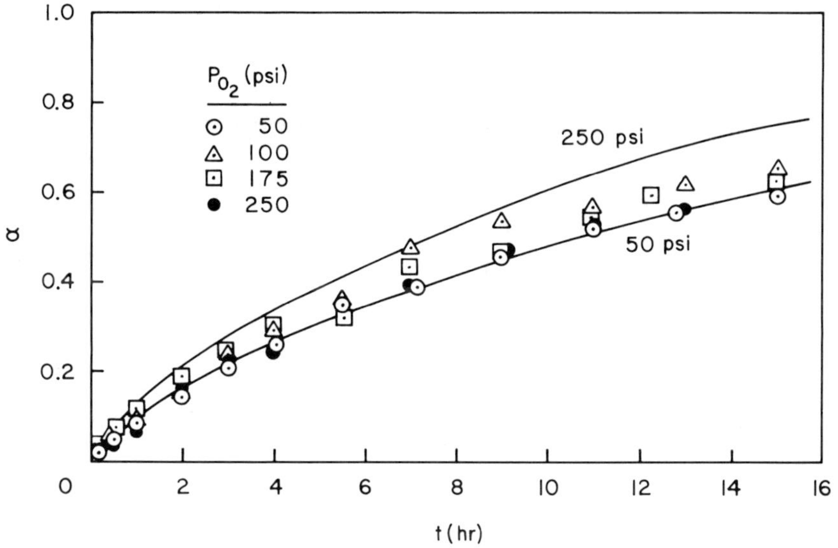

Figure 7. Effect of oxygen pressure on the leaching of uranium ore for $CO_3^=$ and HCO_3^- concentrations of 0.2 and 0.8 M, respectively. The lines show the dependence of the fraction reacted on time as predicted by the rate model.

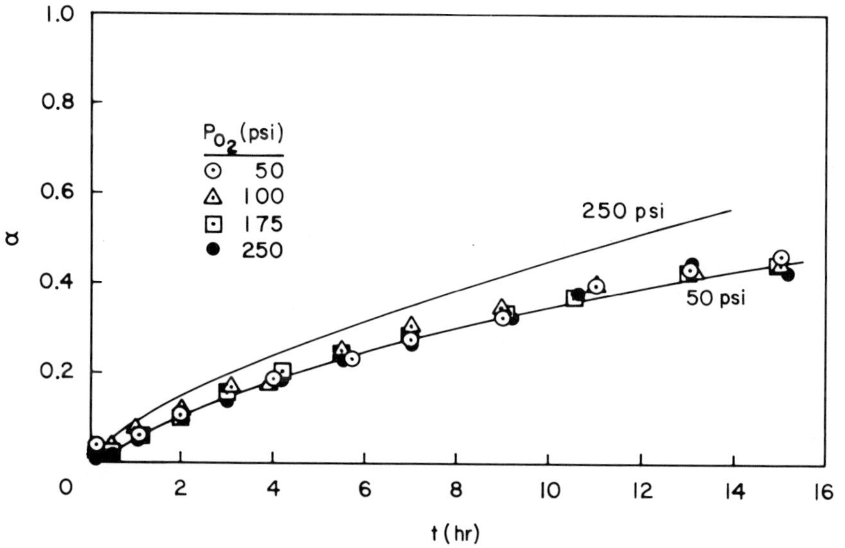

Figure 8. Effect of oxygen pressure on the leaching of uranium ore for $CO_3^=$ and HCO_3^- concentrations of 0.1 and 0.04 M, respectively. The lines show the dependence of the fraction reacted on time as predicted by the rate model.

Using the previously developed electrochemical model applied to cylindrical geometry, it may be shown that,

$$1 - (1-\alpha_r)^{1/2} = k_r \; [HCO_3^-]^{1/2}[\frac{KP_{O_2}}{1+KP_{O_2}}]^{1/2} \; t \qquad (22)$$

where α_r is the fraction of the reduced uranium (U^{+4}) leached in time t and k_r is the rate constant for the $[CO_3^=]/[HCO_3^-]$ ratio used. If k_r' is the intrinsic rate constant, $k_r = k_r' \; M/r_o \rho G$. Using the relationship,

$$\alpha = f_o \; \alpha_o + (1-f_o) \; \alpha_r \qquad (23)$$

α-times curves were calculated. A good correlation was obtained using the values K = 0.0188 psi^{-1}, k_o = 0.624 m^{-1} ℓ hr^{-1}, and k_r = 0.192 $m^{-1/2}$ $\ell^{1/2}$ hr^{-1}.

The curves shown in Figures 5 through 8 are the calculated curves at 250 psi and 50 psi for the four concentrations shown using Equations (20), (22) and (23). Figures 9 and 10 illustrate the calculated and experimental results at 250 and 50 psi oxygen pressure for four carbonate concentrations at constant pH. Using 3 psi oxygen pressure, the curves of Figure 3 are the calculated leaching results under nitrogen. The correlation, in all cases, appears satisfactory.

Using the above rate constants, the kinetics of leaching for a totally reduced ore were calculated and are presented in Figure 11. The importance of carbonate concentration is apparent. For example, the rate of leaching at 50 psi and 0.5 M HCO_3^- are comparable to results obtained for 250 spi and 0.2 M HCO_3^-.

Figure 12 illustrates calculated leaching curves for a totally oxidized ore. The curves illustrate the rapid initial dissolution of uranium followed by slow release at increasing values of α. These results may explain, in part, the attenuated release of uranium in flow systems.

SUMMARY

Leaching results obtained on a low-grade uranium ore indicate that the electrochemical mechanisms observed for pure UO_2 are applicable to such low-grade ores. Also, the results obtained were empirically correlated assuming cylindrical geometry. Oxidized fractions were observed to follow a diffusion model and cylindrical geometry. Leaching results for partially oxidized ore, in this case 40% oxidized, were well correlated by a rate expression involving the sum of rates including the diffusion mechanism for the oxidized ore

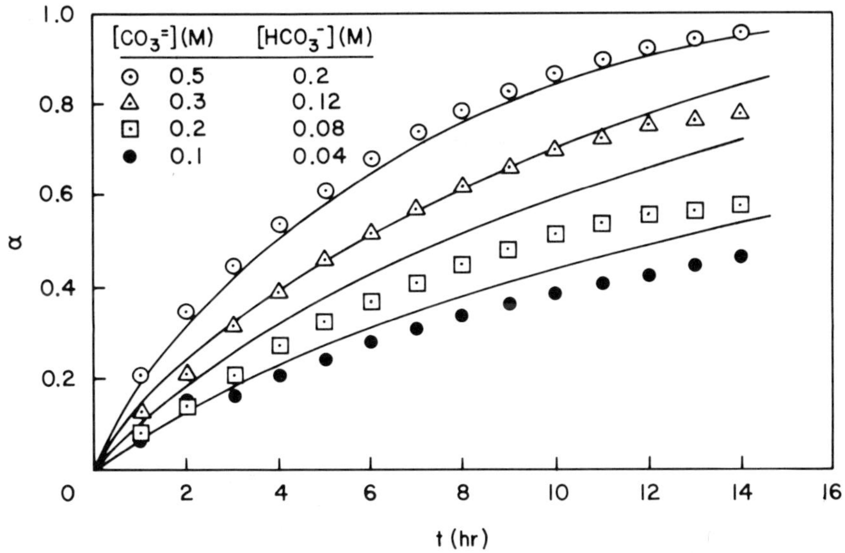

Figure 9. Effect of the concentrations of $CO_3^=$ and HCO_3^- on the leaching of uranium ore for an O_2 pressure of 250 psi. The lines show the dependence of the fraction reacted on time as predicted by the rate model.

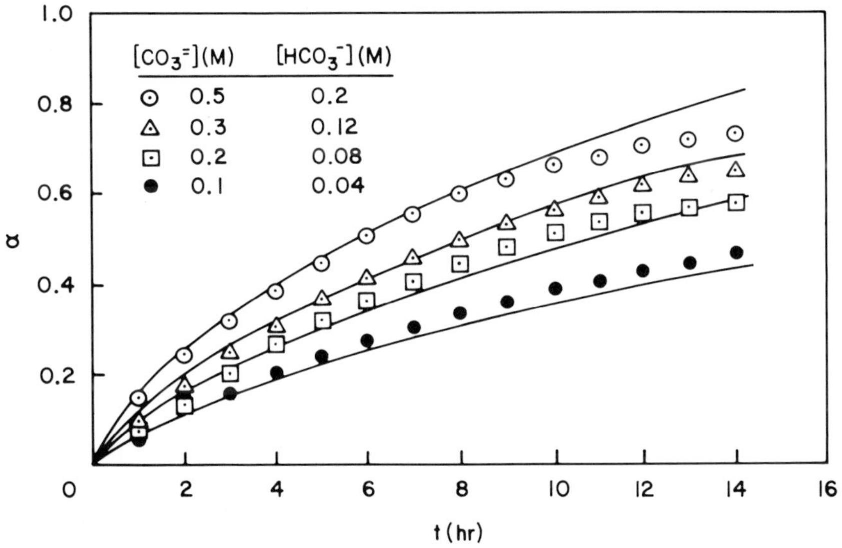

Figure 10. Effect of the concentrations of $CO_3^=$ and HCO_3^- on the leaching of uranium ore for an O_2 pressure of 50 psi. The lines show the dependence of the fraction reacted on time as predicted by the rate model.

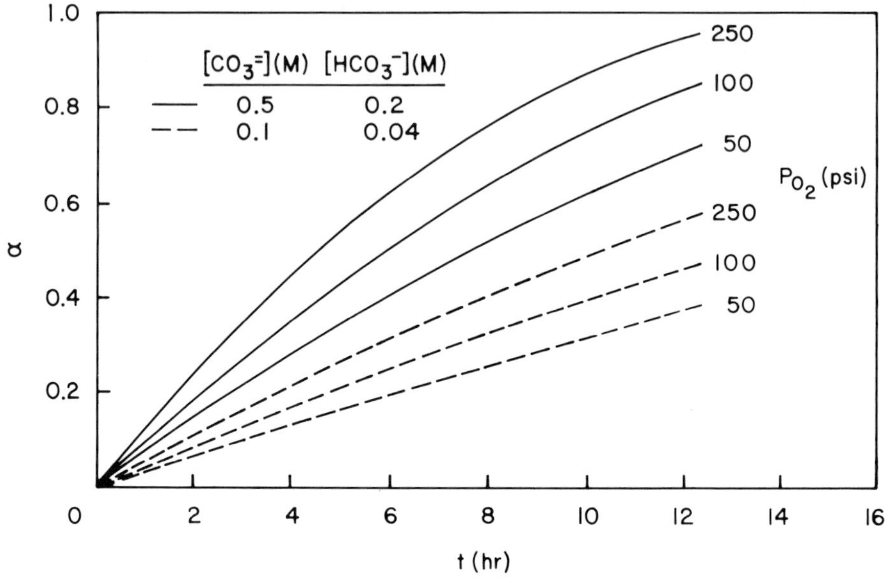

Figure 11. Dissolution of uranium as a function of time predicted by the rate model.

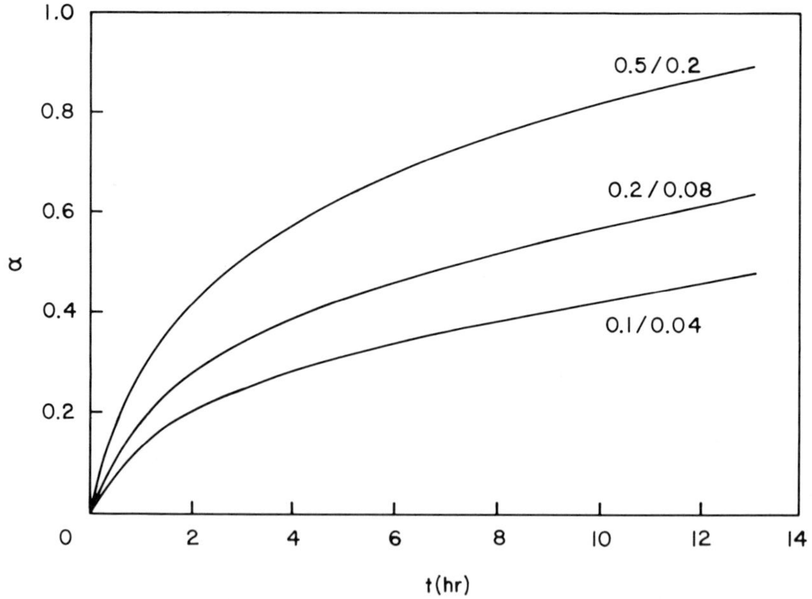

Figure 12. Effect of the concentrations of $CO_3^{=}$ and HCO_3^{-} on the rate of leaching of a completely oxidized ore.

and the surface electrochemical reaction mechanism for the reduced fraction. The overall results indicate the great importance of the carbonate concentration in the rate of uranium extraction.

ACKNOWLEDGMENTS

The authors are indebted to Gulf Research and Development who provided the funding for this research. Thanks are extended to the Reserve Oil and Gas Company for supplying the uranium ore used. Ms. C. A. Petrache is associated with the Atomic Energy Commission of the Philippines and is visiting the Department of Metallurgical Engineering at the University of Utah under sponsorship of the International Atomic Energy Commission.

BIBLIOGRAPHY

1. NURE Uranium Geology Symposium, Publication No. GJBX-1278, 1978, Department of Energy.

2. Mining Engineering, October 1978 pp. 1410, 1422. Article adapted from the paper presented by John J. Bockert and Carl E. Gerity at 1978 South Texas Uranium In Situ Seminar.

3. Wadsworth, M. E., "Physico-Chemical Aspects of Leaching," a paper given at the In-Place Leaching and Solution Mining Short Course, Mining Engineering Department, Mackay School of Mines, University of Nevada, Reno, November 1975.

4. Nicol, M. J., Needes, C., Finkelstein, N. P., "Electrochemical Model for the Leaching of Uranium Dioxide," in Leaching and Reduction in Hydrometallurgy, ed. A. R. Burkin, IMM, pp. 1-19, 1975.

5. Hiskey, J. B., "Kinetics of Uranium Dioxide Dissolution in Ammonium Carbonate," IMM Trans. C, Vol. 88, pp. C145-152, 1979.

6. Pearson, R. L. and Wadsworth, M. E., "A Kinetic Study of the Dissolution of UO_2 in Carbonate Solution," Trans. AIME, Vol. 212, pp. 294-300, 1959.

7. Vetter, K. J., Electrochemical Kinetics, Academic Press, pp. 642-644, 1967.

Chapter 14

RECOVERY OF URANIUM FROM LIGNITES

Fred J. Hurst

Oak Ridge National Laboratory
Oak Ridge, Tennessee

In 1934, Irvin Lavine (1), University of North Dakota, stated that the low-rank coals (lignite and subbituminous) of this country represent a vast tonnage of fuel of commercial value and predicted that some day they will play an important role in our economic welfare. Lavine also pointed out that one of the major problems with efficient utilization of the low-rank coals is their high moisture content. Following the discovery of uranium-bearing lignite in the Williston basin of North and South Dakota in 1948 by D. G. Wyant and E. P. Beroni, detailed investigations were conducted (primarily during 1950 to 1954) by members of the U. S. Geological Survey to evaluate the geologic factors and economic possibilities of the radioactive lignites as a source of uranium. The information gained in these studies and summarized in several papers indicated that a large potential reserve of uranium is contained in the subbituminous coal and lignite deposits of the Dakotas and Wyoming (2,3,4). Most investigators concluded that an epigenetic emplacement of uranium in the coal by means of circulating groundwater and the more extensive concentrations of uranium in coal are apparently related to the permeability of the rocks enclosing the coal beds. Breger (5) and others have shown that uranium is readily adsorbed by coal and is held irreversibly as a disseminated adsorbed constituent.

Briefly, lignite beds containing 0.1 to 10% uranium have been discovered in the Dakotas and Wyoming. The distribution of uranium was found to be erratic and was usually concentrated in thin layers, which were characteristically 10 to 14 in. thick and occasionally 30 to 36 in. thick. Lignite beds more than 3 ft. thick generally contain more uranium at the top of the bed than at the bottom. Special strip mining techniques are required to mine these thin uraniferous layers.

Paralleling these investigations by the USGS, there was a concentrated effort by Battelle Memorial Institute (6), National Lead Company's Winchester Laboratory (7), and Oak Ridge National Laboratory (ORNL) (8) to characterize selected lignite samples and test their amenability to various recovery procedures. Unfortunately, this work was halted before recovery processes were completely developed and reported because of the discovery of large amounts of relatively low-cost uranium in western U.S. sandstone deposits.

During 1963 to 1967, several companies (9,11,12), including Kermac, Union Carbide, Susquehanna Western, and Mines Development, Inc., recovered about one million pounds of U_3O_8 from uraniferous lignite in the Williston Basin. Production ceased because of limited uranium resources and technical and environmental problems.

In 1969, M. Steinberg (10), Brookhaven National Laboratory, estimated that about 240,000 tons of uranium, at an average concentration of 20 ppm, are contained in 11,970 million tons of western subbituminous coal and lignite that will be consumed by the year 2000. He further estimated that a total of 21 million tons of U_3O_8 are contained in these deposits. Steinberg suggested that earlier explorations were made with the view of obtaining rich uranium ore with the extraction of power from coal being a secondary consideration. He recommended that a thorough survey and more extensive exploration be made of the coal geology in the western areas to confirm the resource information available for uranium and coal.

In a recent report by the Electric Power Research Institute (EPRI) on nonsandstone uranium deposits, one chapter is devoted to uraniferous coal resources (11). They concluded that by-product production of uranium from the burning or gasification of coal and lignite is economically borderline; recovery is technically problematic, and the extent of the resource is relatively insignificant. Most known uranium-coal deposits are low in either uranium or heating value and are isolated from production centers. They show an updated version of major uraniferous coal deposits which includes locations, tonnage, percent of U_3O_8 in the coal, and pounds of U_3O_8 in the ash. The data indicate that about 95,000 tons of U_3O_8 are contained in 3,730 million tons of coal.

Additional updated information on uranium in coal deposits lists numerous samples from the USGS national coal resources data system and includes a large bibliography (12). They conclude that only 1 to 3% of the domestic uranium requirement can be met by processing coal ash, provided processing technology can be developed for uranium recovery at acceptable costs.

There is a renewed interest in lignites and possibly uranium as a source of fuel (11,13). This paper summarizes the experimental work carried out at ORNL during 1952 to 1954 that deals with the sulfuric acid leaching of lignite ash. A section on the possible utilization of sulfur dioxide in flue gases should be of value to those companies planning to process lignites.

LIGNITE

Description of Samples

Our investigations were done on selected samples obtained in 1951 to 1952 by USGS personnel from the Sentinel Butte region of North Dakota, the Slim Buttes region of South Dakota in Harding County, and the Red Desert area of Wyoming in Sweetwater County, North of Wamsutter. The samples were taken at least three feet from any exposed edge or outcropping that would have been subject to weathering or leaching by wind and rain. Because the highest concentration of uranium was in the top portion of the bed, only the top portion of the seams were included in the bulk samples after first removing the final overburden by hand. Thus, it was fairly certain that the samples contained no extraneous mineral matter and that the complete upper portion of the top seam was included.

The bulk samples were crushed to minus 4 mesh, quartered, riffled, and submitted for analysis. Portions were air dried and ground to minus 20 mesh for leaching and roasting studies.

Proximate and ultimate analyses were made to characterize the lignite samples as to their rank (Tables 1 and 2). Compared to a typical lignite, the mositure and ash contents were somewhat high, and the fixed carbon and volatile matter were low. Also, the samples were low in carbon and hydrogen and high in oxygen. According to Table 3, the tar yield and heating value of the samples were on the low side.

The uranium content (dry basis) varied from 0.047% for the North Dakota sample to 0.007% for the subbituminous Wyoming sample. The group of east-side South Dakota samples was generally lower in uranium content than those from the west side, that is 0.007 to 0.013% as compared with 0.010 to 0.032%.

The noncarboniferous material was mainly composed of analcite, jarosite, gypsum, clay minerals, and quartz. The major inorganic constituents in the samples are discussed later in the section pertaining to the ash.

Distribution of Uranium

Numerous tests were made to determine whether the uranium was associated with the organic matter or with any of the mineral

Table 1. Proximate Analyses of Lignite Samples[a]

Lignite No.	Percent			
	Moisture	Ash	Fixed Carbon	Volatile Matter
		Sentinel Butte		
ND-1	42.5	17.5	22.0	18.0
		Slim Buttes - East		
86	40.0	27.2	17.6	15.2
92	48.8	12.3	14.8	24.1
121	46.5	16.8	16.7	20.0
141	41.3	11.8	23.2	23.7
		Slim Buttes - West		
95	40.0	20.0	21.7	18.3
97	44.0	13.5	22.6	19.9
101	46.7	11.0	19.9	22.4
103	41.3	14.2	23.7	20.8
109	38.6	18.3	23.1	20.0
		Wyoming		
Wyo-1	28.1	12.4	24.1	35.5
Wyo-2	28.8	20.0	19.8	31.4
		Typical Lignite[b]		
	36.0	7.0	30.0	27.0

[a]As-received basis.

[b]See Ref. 1.

Table 2. Ultimate Analysis of Lignite Samples[a]

Lignite No.	Percent				
	Carbon	Hydrogen	Oxygen	Nitrogen	Sulfur
		Sentinel Butte			
ND-1	51.0	5.0	42.3	0.7	1.0
		Slim Buttes - East			
86	60.6	3.9	31.9	2.2	0.6
92	59.2	3.1	35.6	1.6	0.5
121	66.5	3.9	27.3	1.6	0.7
141	59.2	2.9	35.7	1.7	0.5
		Slim Buttes - West			
95	60.7	3.5	33.5	1.2	1.1
97	59.8	7.8	33.0	1.7	1.7
101	63.9	3.7	30.1	1.5	0.8
103	52.5	3.3	42.2	1.3	0.7
109	57.2	2.6	35.2	1.2	3.8
		Wyoming			
Wyo-1	59.7	3.3	33.5	2.0	1.5
Wyo-2	59.7	3.3	33.0	2.1	1.9
		Typical Lignite[b]			
	68.0	5.5	24.5	1.0	1.0

[a]Moisture and ash-free basis.

[b]See Ref. 1.

Table 3. Potential Fuel Values of the Lignites[a]

Lignite No.	Uranium (%)	Sulfur (%)	Heating Value (btu/lb)	Tar (gal/ton)
		Sentinel Butte		
ND-1	0.047	0.7	5400	7.0
		Slim Buttes - East		
86	0.0075	0.3	4835	2.7
92	0.012	0.4		6.3
121	0.013	0.5	7085	5.4
141	0.008	0.4	6645	8.2
		Slim Buttes - West		
95	0.032	0.7	5935	2.0
97	0.010	1.3	6640	5.9
101	0.015	0.6	6998	2.5
103	0.019	0.5		
109	0.021	2.6	5695	2.5
		Wyoming		
Wyo-1	0.0075	1.4	8750	7.6
Wyo-2	0.0079	1.5	7550	6.0

[a]Dry-basis

constituents in the lignite. The most graphic test (Table 4) shows very clearly that the uranium is not associated with any of the major mineral constituents but definitely concentrates in the organic-rich fraction. To make these tests, a petrographer* handpicked small amounts of the major minerals from a large batch of air-dried lignite and analyzed them for uranium.

Table 4. Uranium Content of Mineral Fractions

Constituent[a]	Percent
Jarosite	0.002
Coarse-grained clay	0.003
Fine-grained clay	0.003
Gypsum, "satin-spar"	0.002
Gypsum, dehydrated	0.002
Lignite, carbonaceous[b]	0.055
Head sample, approximate	0.044

[a]None of the samples were completely free from carbonaceous matter.

[b]The ash content was not measured. Similar samples have shown about 15% ash whereas the head material is about 30%.

It was later established that 93% of the uranium in a typical lignite taken from the Mendenhall strip mine, Harding County, South Dakota, is present in the form of organouranium complexes or ionic organouranium compounds that are soluble at pH <2.18 (14). Only about 1% of the uranium is held by ion exchange, and over 98% of the uranium is soluble in hot 6 N HCl.

Beneficiation

Several techniques, namely, screening, sink-float, air classification, and flotation were tested in an attempt to concentrate the uranium or remove part of the reagent consuming materials. None of these methods showed any promise.

*Dr. T. N. McVay, Bureau of Mines, Tuscaloosa, AL

Direct Treatment of Raw Lignite

About 75% of the uranium could be extracted from raw (minus 14 mesh) lignite at ambient temperatures with mineral acids. At 80°C recoveries were 10 to 15% higher. Figure 1 shows the data for sulfuric acid, which was slightly better than nitric, hydrochloric, or phosphoric acids. In these single-batch tests, large quantities of acid were required for maximum solubility (∿1000 lb H_2SO_4 per ton of lignite). This consumption could be reduced considerably by using a cyclic leach process, but acid requirements based on uranium recovered were still excessive for most of the samples tested (Table 5). For example, 2,000 to 2,700 lb H_2SO_4 were consumed per pound of uranium dissolved from the Slim Buttes-East and Wyoming samples (15). These data, plotted in Figure 2, show the effect of lignite grade on acid consumption. As the concentration of uranium in the lignite decreased below ∿0.15 lb/ton (0.0075% uranium), the acid requirement increases dramatically.

Thickening and filtration of raw-lignite leach slurries were difficult. When the acid supply was either insufficient or excessive, filtration was very slow. With intermediate quantities, filtration was quite rapid. With insufficient acid, the extracted metals reprecipitated as gelatinous hydroxides, whereas at high concentrations of acid, the organic matter was partially decomposed. The leach solutions were, in all cases, discolored by organic matter and, unless this material was removed by activated carbon, it would cause serious problems during recovery.

Treatment of Char

The dissolution of uranium from samples of char, prepared by retorting lignite under reducing conditions at 500 to 800°C, with mineral acids was only 5 to 10% (Table 6). Recoveries were not improved by the addition of oxidants during leaching. However, if the char was given an oxidizing roast at 500°C prior to leaching, a carbon-free ash was produced from which 80 to 85% of the uranium was extracted by leaching with water.

Although uranium in raw lignite is almost completely soluble in both dilute acid and carbonate solutions (data not included), almost complete removal of carbon is required for optimum leaching efficiency of burned lignite. This is also true for the extraction of uranium from carbonaceous material associated with sandstone ores (16,17).

PREPARATION AND LEACHING OF ASH

Major emphasis was placed on developing methods to recover uranium from lignite ash rather than raw lignite. Roasting concentrates the uranium significantly, thereby reducing the size of the

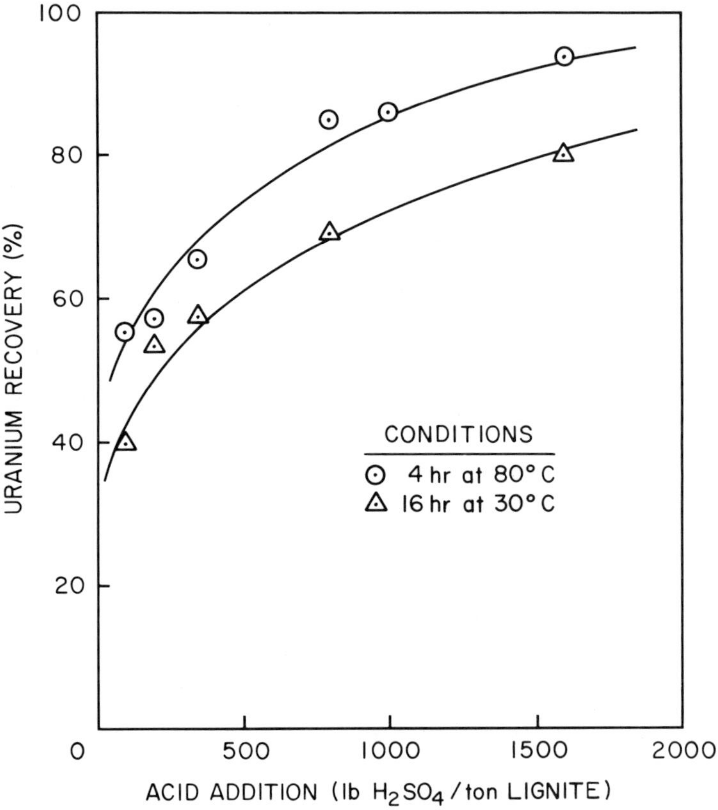

Figure 1. Direct leaching of air-dried (5% H_2O) raw lignite 109 at 20% pulp density.

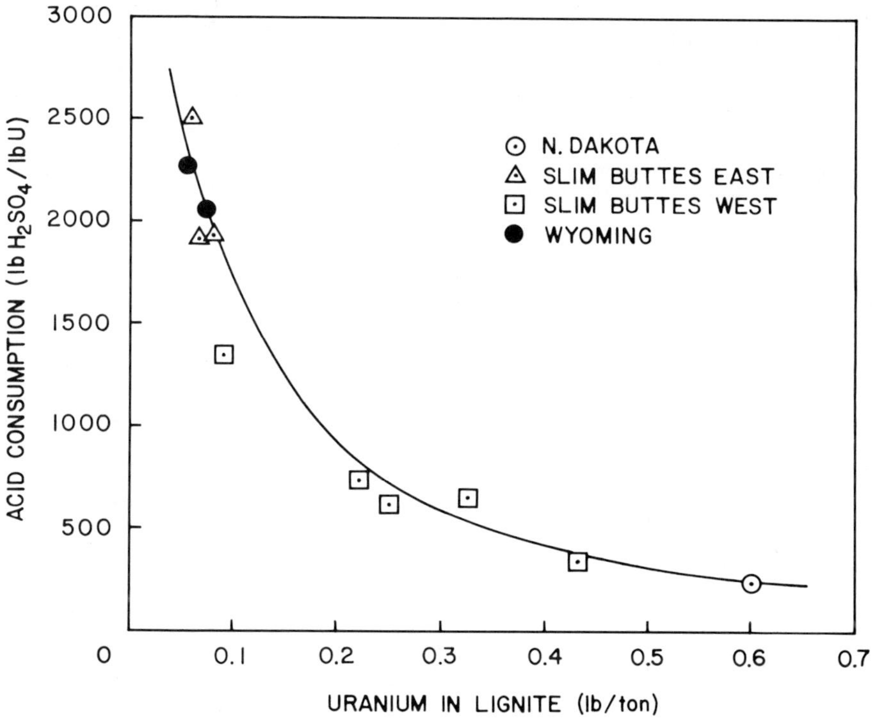

Figure 2. Effect of uranium concentration on acid consumption for raw lignite (conditions in Table 5).

Table 5. Uranium Extraction from Raw Lignite with Sulfuric Acid.

Lignite No.	Uranium (%)	lb/ton	H_2SO_4 Consumed lb/ton ash	H_2SO_4 Consumed lb/lb uranium
		Sentinel Butte		
ND-1	71	0.60	140	235
		Slim Buttes-East		
86	67	0.07	190	2715
92	69	0.11	280	2545
121	70	0.13	250	1925
141	69	0.12	230	1915
		Slim Buttes-West		
95	83	0.43	140	325
97	69	0.09	120	1335
101	84	0.25	150	600
103	87	0.22	160	725
109	86	0.33	210	635
		Wyoming		
Wyo-1	42	0.06	136	2265
Wyo-2	51	0.07	144	2055

Leach conditions: -14 mesh partly air-dried (∿20% moisture); lignite leached at 20% pulp density for 4 h at 80°C with 12.5% H_2SO_4 (1000 lb H_2SO_4 per ton of lignite).

Table 6. Uranium Extraction from Lignite Char.

Retorting Temp (°C)	Leach Agent	Uranium Extracted (%)
500	H_2SO_4	8
500	$H_2SO_4 + KMnO_4$[a]	10
Char Roasted Under Oxidizing Conditions at 500°C		
500	H_2O	86
600	H_2O	81
700	H_2O	82
800	H_2O	84

Retorting conditions: -14 mesh lignite (ND-1) heated for 2 h under reducing conditions in a 300-ml Fischer retort.

Leaching conditions: -100 mesh retorted lignite or roasted char leached at 10% pulp density for 4 h at 80°C with H_2O or 5% H_2SO_4.

[a]1 g $KMnO_4$/10 g char.

leaching facility for an equivalent uranium throughput, and reduces the acid requirement. Removal of organic matter also greatly simplifies recovery of uranium from the leach liquors.

A standard roasting or burning procedure was adopted that gave definite and reproducible material in our laboratory tests. However, a relationship to a given fire-box temperature in a large furnace was never established.

In our tests, the ash samples were usually prepared by spreading raw lignite in a thin (0.25 to 0.5 in.) bed on a tray in a muffle furnace with the controls set at the desired ashing temperature. Measurements with an imbedded thermocouple during burning indicated that combustion caused the temperature of the burning lignite to exceed that of the furnace. For example, as shown in Table 7, at a furnace setting of 400°C, measured bed temperatures approached a maximum of 565°C. Some particles may have reached higher temperatures as evidenced by a bright red glow which indicated temperatures of ∿700°C. Unless otherwise noted, the temperatures in this paper are furnace settings. A burning or roasting time of 1.5 to 2 h in air was required to burn off the carbon at 400 to 800°C. Longer times were required with deeper beds.

Table 7. Furnace vs Bed Temperatures of Burning Lignite.

Furnace (°C)	Bed (°C)
400	565
500	625
600	745
700	796
800	900

In screen analysis and air-classification studies of lignite ash produced under oxidizing conditions at temperatures of 500, 800, and 1000°C, the uranium (and molybdenum) tended to concentrate in the fine sizes of the 500°C ash. As the temperature was raised to 800°C, increasing percentages of both elements appeared in the coarser sizes. At 1000°C the ash had aggregated to a large extent. The data suggest that the uranium held originally in the lignite by organic matter, upon burning, becomes an inherent part of the rather uniform fine-ash fraction. As the burning temperature is increased, the uranium becomes refractory by physical or chemical interaction with the fine mineral constituents. Simultaneously, the fine particles begin to aggregate, distributing more and more to the coarser fractions as the temperature and heating time increase.

Most of the lignites tested formed an acidic ash when roasted at a low temperature (400 to 500°C), and ∿75% of the uranium in this ash was water soluble (Table 8). Sulfur oxides, which are formed from the oxidation of sulfides in the raw lignite, may react to form sulfate salts such as ferric sulfate. A water leach will dissolve these salts and become acidic due to hydrolysis of the metal ion. In fact, dilute solutions of ferric sulfate have been successfully used for leaching constituents from ores in several hydrometallurgical processes (18,19).

As the temperature was increased to ∿800°C, the acid-forming sulfate salts decomposed, and the ash became basic, but the uranium was still readily soluble in dilute acid (Table 8). At 1000°C, the ash became more basic, and the uranium became more refractory due to the formation of stable silicates, and recoveries dropped to ∿50% with dilute acid. These refractory silicates are not attacked by the common leaching agents. Once formed, only complete destruction (e.g., with fluoride compounds) will open up the mineral and allow recovery of the uranium.

As the lignite is burned, sulfate is lost from the material through both decomposition and reduction as shown in Table 9. The reactions that occur during roasting of lignite are similar to those which occur during the roasting of oil shale (20). A very important difference, however, in a low-temperature roast (400 to 500°C) of shales and lignites is that the length of the roast time required in the lignite treatment to form an acidic ash in which the uranium can be extracted into water is only 1.5 to 2 h as compared to 10 to 15 h for shales. This is encouraging from a process standpoint and offers an attractive processing route for recovery of uranium from uraniferous lignites that should be explored further.

When the lignite roasting temperature was increased to 800°C, the acidic constituents were decomposed, and the ash became basic. However, 70 to 80% of the uranium could still be extracted into dilute sulfuric acid from any of the lignites even though large quantities of acid were consumed in some cases. The approximate quantities of acid consumed by each sample under conditions selected for nearly maximum uranium recovery are shown in Table 10. As with the raw lignites, the greatest acid consumers were the Slim Buttes-East and the Wyoming samples. These data are plotted in Figure 3 to show the effect or ore grade on acid consumption. As the concentration of uranium in the lignite decreased below about 0.6 lb per ton of ash, the acid requirement increased dramatically.

Table 11 shows the important inorganic constituents in the 800°C ash samples in which the uranium has been concentrated by a factor of

Table 8. Effect of Roast Time and Temperature on Extraction of Uranium from Lignite.

Roast Conditions				
Time (h)	Temp (°C)	Leach Agent	Final pH	Uranium Extracted (%)
3	400	H_2O	3.2	68
2	500	H_2O	3.0	68
5	500	H_2O	2.7	68
16	500	H_2O	2.7	76
4	800	H_2O	7.8	0
4	800	H_2SO_4	0.5	82
6	800	H_2O	8.0	0
6	800	H_2SO_4	0.4	76
2	1000	H_2O	11.0	0
2	1000	H_2SO_4	1.4	45
4	1000	H_2O	12.0	0
4	1000	H_2SO_4	0.4	55

Roast conditions: Raw lignite (ND-1) placed in a muffle furnace at 400°C and heated rapidly to the desired temperature.

Leach conditions: -20 mesh ash leached at a pulp density of 10% for 1 h at 80°C with H_2O or 5% H_2SO_4.

Table 9. Loss of Sulfate During Roasting

Furnace temp (°C)	time (h)	SO_4 in ash (%)
		37[a]
500[b]	16	30
800	2	22
900	1	20
1000	0.75	18
1000	1.75	14
1100	1	11

[a]Calculated from head analysis.

[b]Placed in furnace at 500°C; others were placed in furnace at 400°C and the temperature was raised to that indicated.

about 4 from 0.06 to 0.6 lb/ton in the air-dried raw lignite to 0.24 to 2.7 lb/ton in the ash. Also, significant concentrations of molybdenum and aluminum are present that may be recovered as by-products, thereby improving the economics of uranium recovery.

The metal cations contributing to the consumption of acid were determined for most of the samples. The quantities of these elements dissolved and their equivalent quantities of sulfuric acid are shown in Table 12 for representative samples from each location. The calculated consumptions of acid are in good agreement with titration measurements considering that no allowance was made for consumption by calcium minerals which reacted but remained in the residue as calcium sulfate (15).

Aluminum compounds were by far the largest consumers of acid. In the examples shown, 57 to 86% of the consumed acid can be accounted for by the reaction with aluminum. A significant amount of the aluminum present in the ash was dissolved and could be recovered as a by-product. For example, 50 and 84% of the aluminum was dissolved from ash samples Wyo-2 and Wyo-92, respectively. This aspect is discussed in more detail in the section on by-product recovery.

If uraniferous lignite were to be used as a power plant fuel, the burning temperature would obviously be maintained at the highest level, possibly in the range of 1200 to 1400°C. Thus, a few scouting tests were made for short periods at temperatures as high as 1400°C. The results described in Table 13 show that 70 to 80% of the uranium

Table 10. Uranium Extraction from Lignite Ash Formed at 800°C.

Lignite No.	Extraction		H_2SO_4 Consumed	
	Uranium (%)	lb/ton	lb/ton ash	lb/lb uranium
		Sentinel Butte		
ND-1	80	2.7	200	74
		Slim Buttes - East		
86	71	0.24	510	2125
92	80	0.77	1000	1300
121	80	0.68	1000	1480
141	66	0.50	760	1520
		Slim Buttes - West		
95	79	1.40	210	150
97	77	0.67	400	600
101	80	1.28	410	320
103	80	0.98	290	295
109	80	1.09	300	275
		Wyoming		
Wyo-1	71	0.43	928	2180
Wyo-2	65	0.26	887	3410

Roast conditions: Ash samples were prepared by placing a thin ¼ to ½ inch bed of raw lignite on a tray in a muffle furnace for 2 h at 800°C in an air stream.

Leach conditions: -20 mesh ash leached at a pulp density of 10% for 1 h at 80°C with 5% H_2SO_4 (1000 lb H_2SO_4 per ton of ash).

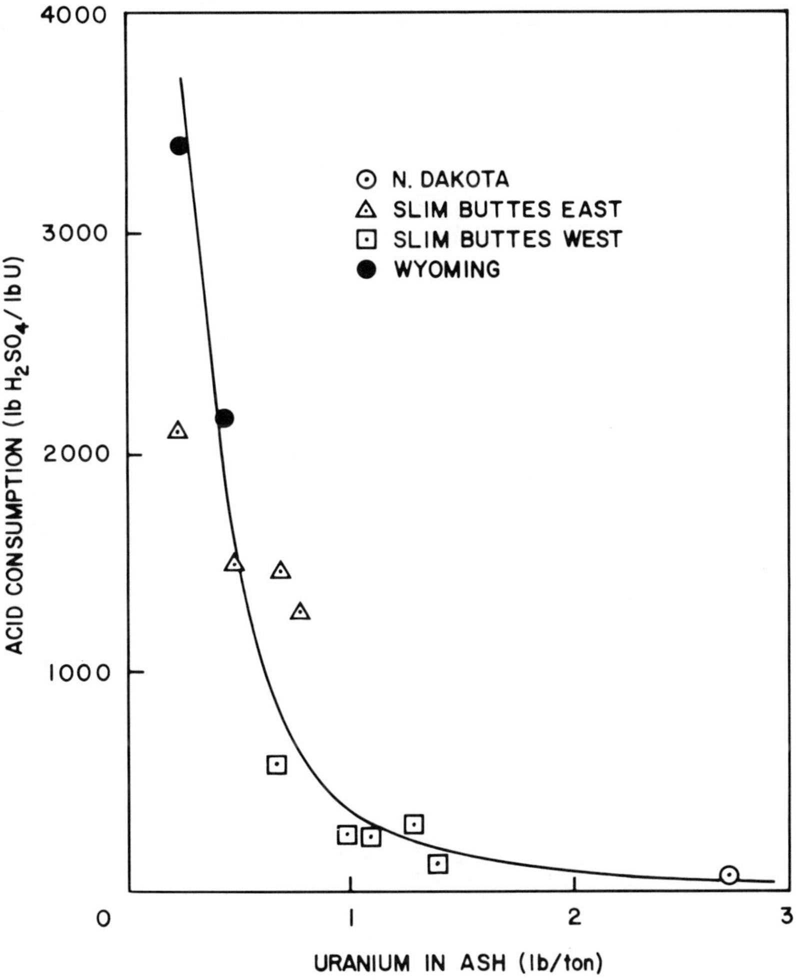

Figure 3. Effect of uranium concentration on acid consumption in the leaching of ash (conditions from Table 10).

Table 11. Chemical Analyses of Lignite Ash Samples

Lignite No.	Chemical Analysis (%) U	Mo	Fe	Al	Ca	Mg	Na	K	SO_4
				Sentinel Butte					
ND-1	0.17	0.29	17.1	4.6	7.4	0.7	1.8	0.5	25.7
				Slim Buttes - East					
86	0.017	0.12	10.5	8.6	2.8	1.0	2.4	1.1	6.5
92	0.048	0.14	11.5	7.8	7.2	1.2	1.6	0.4	4.6
121	0.042	0.14	10.5	8.4	7.4	1.4	1.8	0.9	5.8
141	0.038	0.17	10.7	4.8	8.1	0.9	4.7	0.5	18.2
				Slim Buttes - West					
95	0.090	0.17	13.9	9.3	7.6	0.8	1.8	0.9	18.0
97	0.043	0.07	7.5	6.6	6.3	1.8	3.9	1.0	18.4
101	0.080	0.27	9.7	4.6	9.3	2.4	5.0	0.7	26.8
103	0.062	0.17	12.9	4.6	10.8	1.0	1.6	0.6	25.7
109	0.069	0.31	12.3	6.0	9.8	1.2	1.8	0.6	23.4
				Wyoming					
Wyo-1	0.030		6.0	14.1	6.4	1.5	0.3	0.3	12.1
Wyo-2	0.020		6.5	14.7	4.9	1.1	0.3	0.5	8.2

Roast Conditions: 15 g of 3/8-in. air-dried lignite in a quartz tray (bed depth, ¼ in. to ½ in.; surface area, 65 $in.^2$) was inserted in a muffle furnace at 400°C. The temperature was rapidly increased to 800°C until ignition was complete (∿1.5 h).

Table 12. Impurities Dissolved and Acid Consumed by a Sulfuric Leach of Lignite Ash[a]

Impurities Dissolved	ND-1		92		109		Wyoming-2	
	lb/ton	lb H_2SO_4 (ton)	lb/ton	lb H_2SO_4 (ton)	lb/ton	lb H_2SO_4 (ton)	lb/ton	lb H_2SO_4 (ton)
Al	19	102	131	715	30	163	140	762
Fe	9	26	17	45	10	26	25	66
Ca	13	32	20	50	12	29	19	47
Mg	1	4	6	25	3	12	6	24
Na	12	26	23	50	28	60	<1	<1
K	5	6	4	5	5	6	4	5
Total		196		890		296		904
Acid Consumption by Titration		200		1000		285		887

[a]See Table 10 for roast and leach conditions.

Table 13. Uranium Extraction from Lignite Ash Formed at High Temperatures

Burning Temperature (°C)	Burning Time (s)	Uranium Extracted (%)	H_2SO_4 Consumed lb/ton ash	H_2SO_4 Consumed lb/lb uranium
		Slim Buttes - East, No. 86		
1200	15	75	280	2000
	30	82	430	1650
	60	71	330	1370
	300	58	235	1100
1300	15	81	475	1800
	30	75	420	1500
	60	69	315	1400
	300		Sample Molten	
1400	15	78	480	2000
	30	77	370	1400
	60	69	280	1275
		Slim Buttes - West, No. 109		
1200	15	84	300	320
	30	77	220	260
	60	76	175	225
	300	74	340	320
1300	15	84	320	300
	30	85	275	270
	60	78	240	245
1400	15	81	320	280
	30	82	270	220
	60	81	260	210

Leach conditions: ash leached at 10% pulp density for 3 h at 80°C with 5% H_2SO_4.

can be extracted with dilute sulfuric acid from ash subjected to temperatures as high as 1400°C if the exposure time was only 15 to 60 s. Samples heated for 5 min at 1200°C were sintered, and the uranium became more refractory (~60% recovery). When the temperature was raised to 1300°C, the sample became molten after a few minutes. These results were very encouraging because they indicated the possibility of flash burning of the lignite at a high temperature (such as that reached by a power plant fuel) without causing the uranium to become refractory. To test this theory, a small gas-fired burner was designed and built.

Raw, oven-dried (110°C) -100 and -200 mesh lignite was fed to the burner at approximately 400 g/h by a screw feeder into the top of a 5-in. porcelain tube that was 3-ft long. The tube, surrounded by fire brick, was heated with one to three gas burners (No. 4A tip for National 3A blow pipe). The resulting ash was sucked through a water-cooled, stainless stell cooling chamber into a vacuum cleaner from which it could easily be removed for chemical analysis and testing. Three thermocouples inside the tube (top, middle, and bottom) and three in the air space between the tube and the fire brick were used to measure the burning temperature and to aid in controlling the burning. Temperatures were recorded on a 12-point Brown potentiometer.

Air or air and oxygen mixtures were fed with the lignite through the screw feeder. In addition to aiding the ignition, the air stream was important in feeding or dispersing the lignite in the tube. A mirror at the bottom of the cooling chamber was used to observe the burning of the lignite. The majority of the lignite appeared to burn in a jet 4 to 8 in. long as it entered the tube, and secondary burning occurred spasmodically in a swirling motion throughout the upper half of the tube. A certain amount of the ash stuck to the tube walls, and over a period of several runs, it built up as a slag that dripped from time to time into the Buchner funnel.

The tube was usually preheated to ~200°C less than that desired because the introduction of the lignite caused an immediate increase in temperature which made up the deficit. Burning temperatures to 1000 to 1400°C were easily attained and controlled.

Representative samples of the lignites were burned in the furnace at 1000 to 1400°C in air and at 1300°C in a 50% air-50% oxygen stream. In general, 70 to 80% of the uranium from all of the ash samples was readily soluble in dilute sulfuric acid (Table 14). The acid requirements were quite high, mainly due to the loss of sulfur during burning and partly to the formation of soluble iron and aluminum compounds. As described below, this may be overcome to a large degree by recovery and recycle of the sulfur compounds from the off- or flue gases.

Combustion was essentially complete at 1000 to 1400°C. The lower limit of burning is probably 1000°C as shown by a higher percentage

Table 14. Uranium Extraction from Flash-Burned Lignite Ash

Lignite No.	Burning Temp (°C)	Ash (% SO_4)	Extraction Uranium (%)	Extraction Molybdenum (%)	H_2SO_4 Consumed lb/ton ash
		Slim Buttes - East			
86	1100	3	81	59	632
	1300	2	62	75	448
	1380	1	84	7	1000
92	(double leach)		89	78	1500
		Slim Buttes - West			
97	1100	12	74	39	756
	1300	8	76	39	846
	1300[a]	12	72	59	686
101	1100	10	77	11	1022
	1300	8	74	1	922
	1300[a]	19	84	81	790
109	1000	17	80	51	614
	1100	17	80	56	572
	1200	14	81	59	600
	1300	11	82	66	615
	1300[a]	17	84	69	400
	1400	5	76	65	830
		Wyoming			
Wyo-1	1150	4	68		690
Wyo-2	1150	3	56		640

Leach conditions: -100 mesh ash leached at 10% pulp density for 4 h at 80°C with 5% H_2SO_4 (~1000 lb H_2SO_4 per ton of ash).

[a]50% air--50% O_2 during burning; others burned in 20% excess air.

of carbon (∿1 to 3%) in the ash and as evidenced by visual observation of the burning. At 1400°C, ∿15% of the recovered ash was slag. This slagging was thought to be caused by an accumulation of dust on the burner walls, which became molten and dripped from the burner, rather than by actual fusing of the free falling particles. The upper limit for burning, 1300°C, is apparently optimum.

The amount of oxygen present during burning is very important, not only to the rate and degree of combustion but also to the formation of acid consumers in the ash. For example, as the temperature was raised from 1000 to 1400°C (in air), the percentage of sulfate in sample-109 ash dropped from 17 to 5%, and the acid consumed during the subsequent leach increased from 614 to 830 lb H_2SO_4/ton. When the sample was burned under more oxidizing conditions (50% O_2-50% air), the sulfate was more stable, and most of it remained in the ash (17% compared to 11% in air at 1300°C). In this case acid consumption was only 400 lb/ton.

Molybdenum extraction values are also included in Table 14. Higher recoveries were obtained at the higher burning temperatures and from the most completely oxidized ash. The data also suggest that molybdenum requires more free acid for dissolution than does the uranium.

Under the same conditions, some of the lignite samples did not burn as completely as others. The temperature and degree of oxidation need to be adjusted for the type of lignite being burned to optimize recoveries of uranium and potential by-products (molybdenum, aluminum). A major problem for a large commercial roaster or burner is control of the combustion temperature and rapid dissipation of the heat to minimize sintering and slagging.

UTILIZATION OF SULFUR DIOXIDE FROM FLUE GAS

Flue gases from the burning lignite normally contain significant quantities of sulfur dioxide. In the samples tested, the amount of SO_2 released during burning was, on the average, equivalent to ∿300 lb H_2SO_4 per ton of ash. It is known that SO_2 in aqueous solution may be oxidized to sulfate by air in the presence of metal ions (19,20). Thus, we made a few scouting tests in which synthetic flue gas, prepared by mixing air-N_2-SO_2-CO_2 in the proportions usually found in typical flue gases, was bubbled through a slurry of ash and water for various times and temperatures. These preliminary tests showed that SO_2 was readily oxidized and converted to sulfuric acid on contact with the slurry (Table 15).

A 1 h addition of synthetic flue gas containing SO_2 equivalent to 1000 lb H_2SO_4 per ton of ash to a leach slurry consisting of 1150°C Wyo-1 ash and water at room temperature produced a filtrate pH of 2.5 and a solubilization of 25% of the uranium. The sulfate content of the

Table 15. Extraction of Uranium from Flash-Burned Lignite Ash Using Synthetic Flue Gas

Gas (%)				Leach		Filtrate			Uranium Extracted (%)
Air	N_2	CO_2	SO_2	Time (h)	Temp (°C)	pH	SO_4 (g/L)	lb H_2SO_4 / ton of ash	
83	0	17	0.25	1	30	2.5	7.5	300	25
				5	30	2.2			35
				1	95	1.3	10.5	420	54
				2	95	0.73	22.0	880	54
				3	95	0.85	22.0	880	65
				5	95	0.60	44.5	1780	63
				6	95	0.50	38.0	1520	60
100	0	0	0.25	6	95	0.46			56
83	0	17	0	6	95	7.5			7
0	83	17	0.25	6	95	4.0			10
10	73	17	0.25	6	95	1.15			52

Ash: -100 mesh Wyo-1 ash burned at 1150°C in flash burner.

Leach Conditions: Synthetic flue gas (as shown) was bubbled at a total flow rate of ~4.9 liters per min. through a slurry consisting of 200 ml of H_2O and 10 g of ash.

filtrate was 7.5 g/l or ~300 lb H_2SO_4 per ton of ash. Prolonging the leaching time from 1 to 5 h resulted in only a slight increase in uranium solubilization to 35%, whereas raising the temperature to 95°C produced a sizeable increase in sulfuric acid consumption (~400 lb/ton) and uranium dissolution (54%). A 3 h leach at 95°C resulted in 65% dissolution of the uranium which is equivalent to that obtained by leaching the sample with sulfuric acid (Table 10). Additional 6 h, 85°C tests confirmed that air (or oxygen) is necessary to convert the SO_2 to acid. In these tests only 30 to 40% of the SO_2 was converted to sulfuric acid, and the control and measurement of the SO_2 flow was not as good as desired.

A few additional tests were also made using very accurate methods for measuring flow, particularly that of the SO_2 which was measured with a calibrated, pressure-differential device. Sodium hydroxide scrubbers containing hydrogen peroxide were used to measure any SO_2 that escaped the leach vessels so that a material balance over the system could be obtained.

These data (Table 16) show that about 40% of the SO_2 was absorbed in a single leach when the quantity of SO_2 (7 cc/min x 60 min) evolved during the burning of Wyo-1 at 1150°C was passed through the leach slurry. In this test, ~14% of the uranium was solubilized, and an equivalent of 135 lb H_2SO_4 per ton of ash was analyzed in the filtrate. An additional leach of the ash with 300 lb H_2SO_4/ton increased the recovery to 53%.

Increasing the SO_2 flow by 50% while maintaining a constant air flow rate at 450 cc/min did not result in an improvement in SO_2 conversion or in uranium extraction. However, when the amount of air was increased, and the leaching time was extended to allow more time for oxidation, 58% of the SO_2 was absorbed and over 40% of the uranium was solubilized. It is rather apparent that these tests were limited by inadequate oxidizing conditions. Also, SO_2 absorption tends to decrease as the pH decreases.

So that process efficiency could be improved, a series of tests was made in which the flue gas was cycled through five consecutive leach slurries. In these tests, 96% of the SO_2 was absorbed, and ~50% of the uranium was solubilized without additional acid.

These tests, although limited in scope, were very encouraging and indicate that a large part of the acid requirement for uranium recovery may be obtained from the lignite itself. To determine whether these reactions can be utilized on a practical operating basis with actual flue gases will require much additional work. Such a process would also help relieve the environmental disposal of sulfur dioxide in the plant.

Table 16. Absorption of Sulfur Dioxide in a Leach Slurry

Gas Flow (cc/min) Air	Gas Flow (cc/min) SO_2	Filtrate pH	Filtrate SO_4 (g/L)	Filtrate lb H_2SO_4/ton ash	SO_2 Absorbed (%)	Uranium Extracted (%) 1	Uranium Extracted (%) 2[a]
450	5.0	3.3	2.9	118	48	12	49
450	7.0[b]	3.0	3.3	135	39	14	53
450	10.5	3.1	3.3	135	27	11	52
4050	16.5	1.9	9.3	380	48	42	56
4050	16.5[c]	2.2	11.4	466	58	44	

Ash: -100 mesh Wyo-1 burned at 1150°C in flash burner.

Leach Conditions: Synthetic flue gas containing air-N_2-CO_2-SO_2 was bubbled through a slurry consisting of 200 ml H_2O and 10 g of ash for 1 h at 85°C. Total gas flow was ∿4.9 L/min. CO_2 flow was 850 cc/min. N_2 flow may be obtained by difference.

[a]The tailing from the flue gas treatment was given an additional 4 h leach at 80°C with 5% H_2SO_4 (300 lb H_2SO_4/ton ash).

[b]Represents the quantity of SO_2 evolved during burning.

[c]Ash leached 2 h after gas flow was stopped.

PRODUCT RECOVERY AND BY-PRODUCT POSSIBILITIES

Recovery is not complete until the uranium is recovered from the leach solutions and converted to a relatively pure U_3O_8 product that is acceptable to uranium refineries that produce UF_6. Our work on lignites was halted before we had an opportunity to study this part of the problem. However, it was studied at several other laboratories (6,7) and during the intervening years, a very large amount of information has been accumulated on the recovery of uranium and other metal values from similar solutions using solvent extraction and ion exchange techniques. The front end, which includes mining, ore preparation, burning, leaching, thickening and/or filtration of the leach slurry is considered to be the most cost-intensive part of a process scheme. Recovery of uranium from clean liquors should be relatively simple and inexpensive.

The quantities and composition of coal and lignite residues or ash make this material a potentially valuable resource for numerous important metals in addition to uranium. These include molybdenum, aluminum, and iron plus a number of other metals that are present in trace quantities. Several investigators (21-25) are currently evaluating various methods for recovering these products from commercial coal wastes, and a similar program has begun on Chattanooga oil shale leach solutions (26). The major goal of these studies is to determine the economics of by-product metal recovery. Based on the cost of energy generation from lignite at $0.30/10^6 Btu and current spot prices for uranium, molybdenum, aluminum, and iron, the value of 1 ton of ash is $119 (Table 17) or ~$12 per ton of raw lignite "as-received." This does not include any value for trace elements or sulfur that may be recovered during burning and used for leaching the metal values. Although total recovery, as shown in Table 17, cannot be achieved, such an analysis indicates some of the advantages of co-recovery of thermal energy and metal values.

SUMMARY AND CONCLUSIONS

Our studies showed that uranium in raw lignite is associated with the organic matter and is readily soluble in acid (and carbonate) solutions. However, treatment of raw lignite did not, and does not now, appear to be attractive because reagent consumptions are high, liquid-solid separations are difficult, and the liquor is fouled by organic matter. Beneficiation techniques were not successful for concentrating the uranium or removing part of the reagent-consuming materials.

Once the lignite was heated by roasting, retorting, or charring, the uranium became much less soluble in both acid (and carbonate) solutions, and essentially complete removal of carbon was required to convert it back to a soluble form. Proper burning (with respect to time, temperature, and oxidizing conditions) tends to improve

Table 17. Products in Lignite

Product	Content[a]	Ash ($/ton)
Heating value, Btu/lb	6550[b]	24.00[c]
Uranium, lb/ton	1.4	45.00
Molybdenum, lb/ton	3.5	32.00
Aluminum, lb/ton	150	8.00
Iron	215	10.00
Total		119.00

[a]As oxide except for heating value; based on average values from Tables 1, 3, and 11.

[b]Dry basis.

[c]Based on an average ash content of 16.3%.

acid-leaching efficiency, that is, it reduces the reagent consumption and concentrates the uranium, thereby reducing plant size for comparable uranium throughput, and it eliminates organic fouling of leach liquors. Restrictions are necessary during burning to prevent the uranium from becoming refractory. In general, although a prolonged roasting time has a beneficial effect on the uranium extractability of a low-temperature (500°C) ash and no appreciable effect on a medium-temperature (800°C)* ash, it has a definite detrimental effect on a high-temperature (>1000°C) ash. When the lignite was roasted or burned at 1000 to 1100°C for 0.5 to 1.5 h, the uranium extractions were erratic. The general effect of increased temperatures (for rather long periods of time) was to decrease the availability of the uranium.

The most encouraging results were obtained by flash-burning lignite at 1200 to 1300°C and utilizing sulfur dioxide released during burning to supplement the acid requirement.

*These results do not agree with current information, which states that uranium recoveries are impaired at temperatures of 600°C, Ref. 11.

The major acid consumers were aluminum and iron. At the time this study was made, these constituents were considered waste products; today, because of changing economics, such material is a potential resource, and co-recovery could result in significant cost savings.

It should be emphasized that the material tested in this study may not be completely representative of the average mineable material desired for power production. Ash can vary within rather wide limits, not only in its composition but also in its chemical behavior, depending on its history, the type of lignite, roasting time and temperature, previous dressing operations, etc. Before successful production of power can be coupled with the economical recovery of uranium and other potential by-products, additional information is needed concerning the properties of ash produced in commercial lignite-fueled power, gasification, and liquefaction plants. Ideally, a balance in operating conditions is needed to obtain maximum heating values or gaseous or liquid products without causing excessive uranium refractory.

ACKNOWLEDGMENTS

This work was done during 1952-1954 under the direction of K. B. Brown and includes contributions by J. E. Beard, C. F. Coleman, F. J. Hurst, R. S. Lowrie, R. C. Nelson, F. A. Schimmel, and F. G. Seeley.

The research was sponsored by the Division of Chemical Sciences, Office of Basic Energy Sciences and the Supply Analysis Division, U.S. Department of Energy under Contract W-7405-Eng-26 with the Union Carbide Corporation.

The author thanks Vivian Jacobs for editorial assistance and Nancy Pope for help in manuscript preparation.

BIBLIOGRAPHY

1. Lavine, I., "Progress in Low-Rank Coals," Ind. Eng. Chem. 26 (2), 154-164, 1934.

2. Geological Survey Professional Paper 300, Geology of Uranium and Thorium, International Conference, 1955.

   - Vine, J. D., "Uranium-Bearing Coal in the United States," pp. 405-411.

   - Denson, M. N. and Gill, J. R., "Uranium-Bearing Lignite and Its Relation to Volcanic Tuffs in Eastern Montana and North and South Dakota," pp. 413-418.

- King, J. W. and Young, H. B., "High-Grade Uraniferous Lignites in Harding County, South Dakota," pp. 419-431.

- Pipiringos, G. N., "Uranium-Bearing Coal in the Central Part of the Great Divide Basin Sweetwater County, Wyoming," pp. 433-438.

- Masursky, H., "Trace Elements in Coal in the Red Desert, Wyoming," pp. 439-444.

3. "Uranium in Coal in the Western United States," Geol. Surv. Bull. 1055, 1954.

4. "Geology of Uranium and Thorium, Vol 6," Proceedings of the International Conference on the Peaceful Uses of Atomic Energy, Papers 49, 55, 56, 57, 286, and 287, 1956.

5. Breger, I. A., U. S. Geological Survey, Personal Communication, July 21, 1981.

6. Ewing, R. A. et al., "Recovery of Uranium from North Dakota Lignites - Final Report," BMI-237, 1950.

7. Beverly, R. G. (ed.), "Summary Report," Raw Materials Development Laboratory, Winchester, Mass., and Grand Junction, Colo., WIN-115, 1959.

8. Brown, K. B. et al., "Recovery of Uranium from Lignites, Progress Reports," ORNL-1308, -1384, -1446, and -1569, 1952-1953.

9. "Calcining Helps to Unlock Uranium in Lignite," Chem. Eng. 33-34, Feb. 1, 1965.

10. Steinberg, M., "Value and Prospects of Recovery of Uranium from U.S. Coal Deposits," 65th National Meeting of the AICHE, Cleveland, Ohio, May 4-7, 1969.

11. "Uraniferous Coal Resources," Sect. 6 in U_3O_8 Production Analysis for Nonsandstone Uranium Deposits, EPRI EA-733, March 1980.

12. Facer, F. J., Jr., "Uranium in Coal," GJBX-56(79), May 1979.

13. "Lignite Revival Takes Texas by Storm," Chem. Eng. 75-77, March 27, 1978.

14. Breger, I. A., Deul, M. and Rubinstein, S., "Geochemistry and Minerology of a Uraniferous Lignite," Econ. Geol. 50 (2), 1955.

15. Ingles, J. C., "Determination of Free Acid in Leach Liquors," Dept. of Mines and Technical Surveys (Canada), NP-1598, March 18, 1950.

16. Nicols, I. L., Lawrence, A. G. and Seidel, D. C., "Extraction of Uranium from Carbonaceous Sandstone Materials," Bureau of Mines, RI 8393, 1979.

17. Milligan, D. A., "Oxidizing Roasting Enhances Extraction of Uranium from some Refractory Ores," Eng. Min. J. 178 (9), 114-118 1977.

18. Dutrizac, J. E., and MacDonald, R. J. C., "Ferric Ion As a Leaching Medium," Miner. Sci. Eng. 6, (2), 59-100, 1974.

19. Neytzell-de Wilde, F. G. and Taverner, L., "Experiments Relating to the Possible Production of an Oxidizing Acid Leach Liquor by Auto-oxidation for the Extraction of Uranium," Proceedings of the Second United Nations International Conference on the Peaceful Uses of Atomic Energy, Vol. 3, P/1116, pp. 303-317, 1958.

20. Brown, K. B., Schmitt, J. M., Hurst, F. J. and Crouse, D. J., "Recovery of Uranium from Oil Shales," Part I, Extraction of Uranium from the Shale Gangue, ORNL Report Y-564, Feb. 10, 1950.

21. Egan, B. Z., Oak Ridge National Laboratory, Personal Communication, June 18, 1980.

22. Canon, R. M., Oak Ridge National Laboratory, Personal Communication, Aug. 11, 1980.

23. Egan, B. Z., Seeley, F. G. and Kelmers, A. D., "Chemistry of the Calsinter Process for Recovery of Aluminum from Fly Ash," Preprint No. 80-47, AIME Annual Meeting, Las Vegas, Nev., Feb. 24-28, 1980.

24. Gilliam, T. M., and Canon, R. M., "Removal of Metals from Coal Ash," Proceedings of the 15th Intersociety Energy Conversion Engineering Conference, Seattle, Wash., Aug. 17-22, 1980.

25. Scott, T. R., "Alumina by Acid Extraction," J. Met. 14 (2), 121-125, 1962.

26. Ryon, A. D., Oak Ridge National Laboratory, Personal Communication, Aug. 5, 1980.

Uranium II

Session Chairman

Charles J. Gessner
South Dakota School of Mines
and Technology
Rapid City, South Dakota

Chapter 15

URANIUM EXPLORATION IN THE CHORD PROJECT

Richard N. Grigsby

Union Carbide Corporation
Casper, Wyoming

INTRODUCTION

The Chord property was acquired in 1975 from Roy Chord, an early prospector and miner of the area. He had staked numerous claims on the southern flank of the Black Hills in the early 1950's, and by the late 50's, many small open pit and underground mines had been or were in operation. By the mid 1960's small tonnages of high-grade oxidized uranium ore had been mined from the property.

Starting in 1975, Union Carbide conducted a series of exploration programs to determine the remaining economic potential. The previously mined areas were re-examined in order to define additional oxidized mineralization. Meanwhile, a wide-spaced drilling program was initiated to determine the potential for deeper unoxidized deposits. Through the exploration program of 1979, measured minable reserves in excess of two million pounds of contained uranium had been discovered.

GEOLOGY

The Edgemont District is located on the southern end of the Black Hills Uplift. A succession of pre-Cambrian to Cretaceous Formations is exposed within the core and across the flanks. The uranium host is the lower Cretaceous Inyan Kara Group which is divided into two formations, the Lakota and Fall River Formations (Figure 1).

The basal Lakota Formation rests unconformably on the Jurassic Morrison Formation and consists of fluvial channel sandstones separated both horizontally and vertically by overbank sediments of thin interbedded, fine-grained sandstones and siltstones. Individual sandstone channels vary from less than fifty to several hundred feet in width and from less than five to more than fifty feet in thickness. The channels can be traced along their axis for several miles.

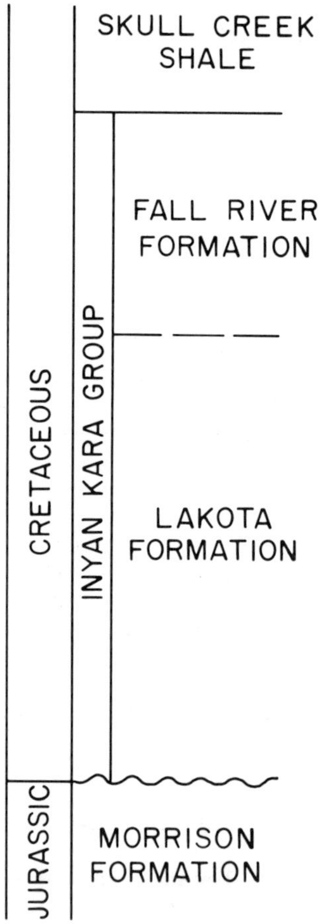

Figure 1. Stratigraphic section of the Inyan Kara Group.

The upper part of the Lakota Formation is comprised of lacustrine shales and mudstones with localized limestone and sandstone.

A transgressive disconformity separates the Lakota and the Fall River Formations, indicating a hiatus between the deposition of Lakota and Fall River sediments. The Fall River Formation is typical of uniform marginal marine deposition. Fluvial channel sandstones can be easily mapped over large areas. They appear as more of a blanket deposit and incise less into underlying sediments. Thin interbedded sandstones and siltstones are common and more widespread than in the Lakota Formation. The upper part of the Fall River Formation is a transition zone into the upper Cretaceous black shales.

The Inyan Kara Group was deposited in a structural and depositional basin which was periodically accentuated, particularly during early Lakota time. Changes in the base level of the drainage system resulted in a crudely cyclic sequence of stream, swamp and lake deposits. Both formations are comprised of reworked sediments. Material for the Lakota Formation originated from highlands in present day New Mexico, Arizona and Oklahoma (Figure 2), while the Fall River Formation was partially derived from highlands in present day eastern South Dakota and partially reworked Lakota (1).

Local structures consist of several south-plunging anticlines and monoclines and two northeast-southwest trending structural zones (Figure 3). The Chord property lies within the southern-most Long Mountain structural zone, which is made up of several small magnitude faults with displacements rarely exceeding twenty feet.

URANIUM GEOLOGY

Uranium deposits are restricted to channel sands of the Lakota and Fall River Formations and the lower interbedded sandstones and siltstones of the Fall River Formation. The deposits are proportional to the size of the sandstone channel; thus, larger channels generally contain more economic deposits.

The uranium mineralization occurs as typical "roll front" deposits. A "roll front" or geochemical cell is an irregular "C" shaped interface between altered and unaltered portions of a sandstone along which uranium has been deposited. The interface is produced by the migration of uranium-bearing oxidizing fluids down a hydrostatically controlled gradient and through a carbonaceous and pyritic, water-saturated, reduced sandstone. The oxidizing solutions alter the host rock and the uranium is deposited on the interface ahead of the altered portion.

Alteration products are usually recognizable in drill samples. Kaolinization of feldspar, chloratization of biotite, and the oxidation or absence of primary iron minerals such as pyrite and

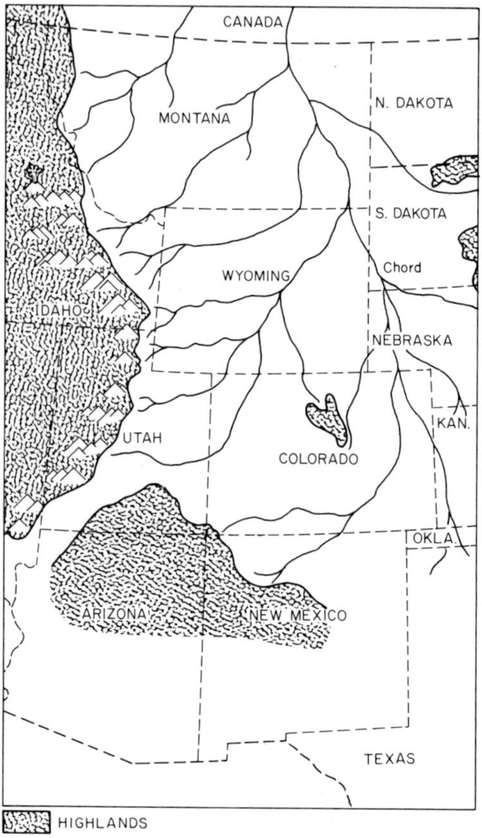

Figure 2. Early Cretaceous drainage, Ref. 2.

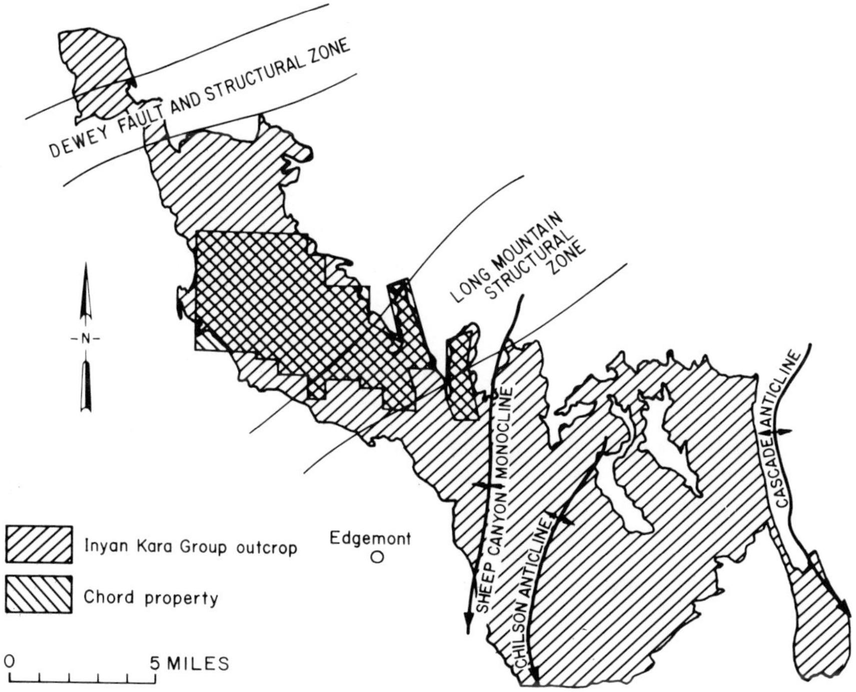

Figure 3. Structural map of the southern Black Hills, Ref. 1.

magnetite are usually the most useful alteration products (Figure 4). Iron oxide staining is an important product when working in an area where the geochemistry of the system permits the formation of limonite and/or hematite following the destruction of the primary iron minerals.

In the southern Black Hills, alteration products are less abundant and, consequently, of less value as exploration guides because the host formations are genetically orthoquartzites with less than five percent feldspar and one percent heavy minerals. Mica is generally not present except for minor amounts of muscovite in the Fall River Formation. Fortunately, iron oxide staining does occur and aids in identifying the location of geochemical cells or "roll fronts."

The source of uranium is not clearly established. A theory put forth by geologists from the U. S. Geological Survey (3) contends that the uranium traveled up through a network of breccia pipes originating in the Minnelusa Formation.

This writer believes that the host formations provide the best readily available source with a possible secondary source being the tuffaceous White River Formation. Renfro (4) stated that the unaltered Inyan Kara sandstones contains an average of 14 ppm U_3O_8, while equivalent altered sandstone averages 5 ppm U_3O_8. The 9 ppm loss could easily account for the uranium concentrated in the deposits in the Black Hills.

Tuffaceous sedimentary rocks have long been considered as a source of uranium. The White River Formation at one time unconformably covered the Inyan Kara Group. Uranium could have been leached from the tuffaceous material and transmitted into the Inyan Kara sandstones by ground water.

Both oxidized and unoxidized deposits are found on the property. Oxidized deposits are found within the zone of surface oxidation and contain uranium-vanadium mineralization (e.g. carnotite). The sandstone hosts are stained by limonite derived from surface oxidation obscurring redox boundaries. In some cases a change from limonite to hematite staining will denote the redox boundary.

Disequilibrium is an important factor in the oxidized deposits. It occurs when the amount of uranium is not in proportion to the amount of radiation produced by its daughter products. In the case of the oxidized deposit, acidic fluids tend to remove the uranium and carry it to a chemically more favorable location for precipitation. Vanadium reduces the effect of acidic solutions by creating insoluble uranium-vanadium compounds. Disequilibrium within an ore deposit occurs when vanadium concentrations are not sufficient to hold all of the uranium within insoluble compounds during the leaching by the acidic fluids. Oxidized deposits on the Chord property have chemical to radiometric ratios ranging between 0.81 to 0.41.

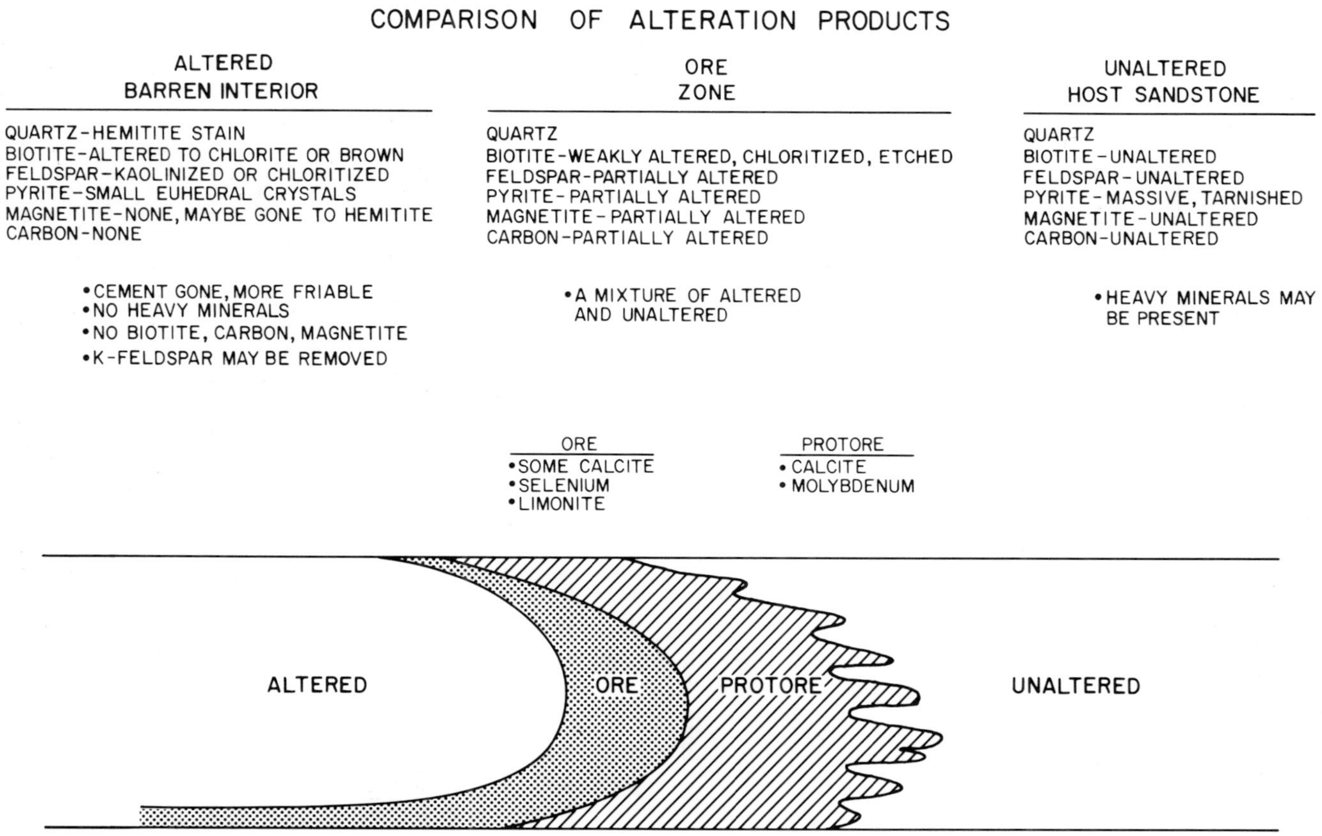

Figure 4. Idealized roll front.

The unoxidized deposits are usually deeper and are found below the zone of surface oxidation. Uraninite and coffenite are the primary uranium minerals with only minor concentrations of vanadium being found. Disequilibrium is not a factor in these deposits. Mineralization is usually in equilibrium or slightly in favor of the chemical assay (Table 1).

Table 1. Characteristics of Oxidized and Unoxidized Zones

	Oxidized	Unoxidized
Location	Above Zone of Surface Oxidation	Below Zone of Surface Oxidation
Mineralization	Secondary Uranium-Vanadium Minerals, eg. Carnotite	Primary Uranium Minerals eg. Urananite, Coffenite
Staining	Iron Staining in Altered and Unaltered Sandstone	No Staining in Unaltered Sandstone
Disequilibrium	Mineralization out of Equilibrium in Favor of Radiometric Assay	Mineralization in Equilibrium or out of Equilibrium in favor of the Chemical Assay

Some deposits display characteristics of both oxidized and unoxidized deposits. They usually occur near the lower limits of surface oxidation or where perched ground water or large amounts of carbon trash have preserved a strongly reducing environment within the oxidized zone. These deposits tend to be relatively small due to the limited extent of the reducing control.

The location of the "roll fronts" is controlled by structural confinements and sedimentary changes within the host formation. Localized folding that creates a flattening of dip or a small syncline tends to reduce or pond the fluid movement. In Figure 5, notice that the mineralization is found at or near the dip change.

Sedimentary controls, such as thinning and silting of the sands at the channel margin, affect the lateral migration of the mineralizing fluids. Also, higher concentrations of potential reductants (carbon) occur along the channel margin creating a more favorable environment for the deposition of uranium. In Figure 6 the localization of the mineralization is controlled by the thinning of the upper units. The trend from which this section was taken parallels very closely the mapped location of the channel margin.

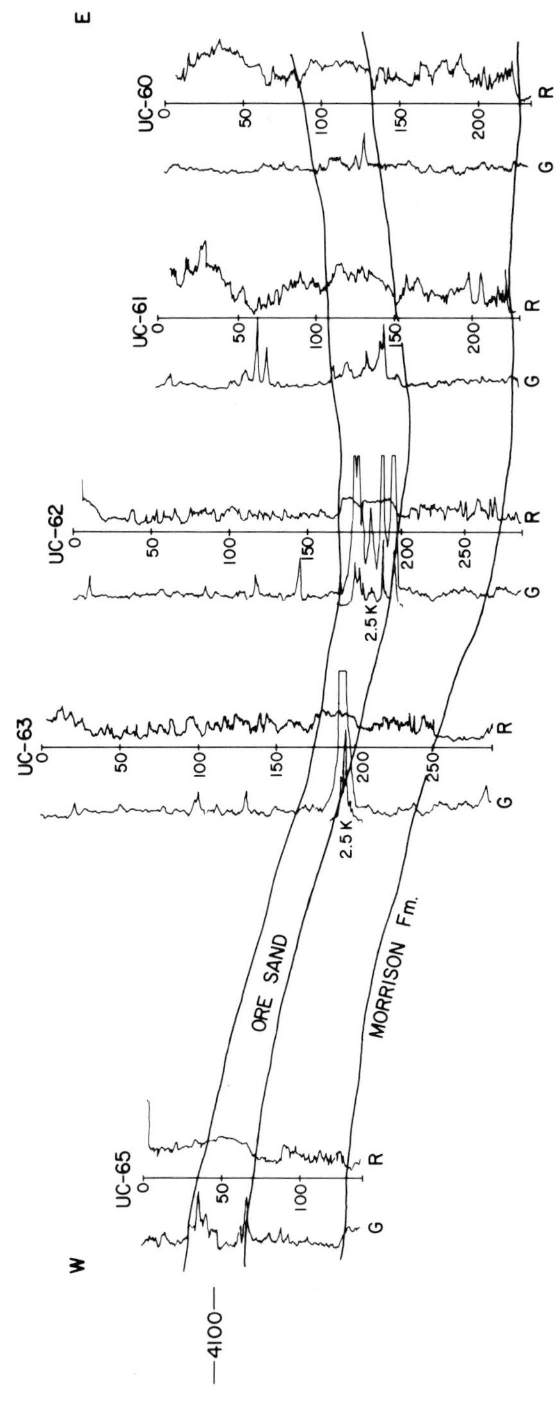

Figure 5. Cross-section showing structural control of ore deposition. (G-gamma and R-resistivity)

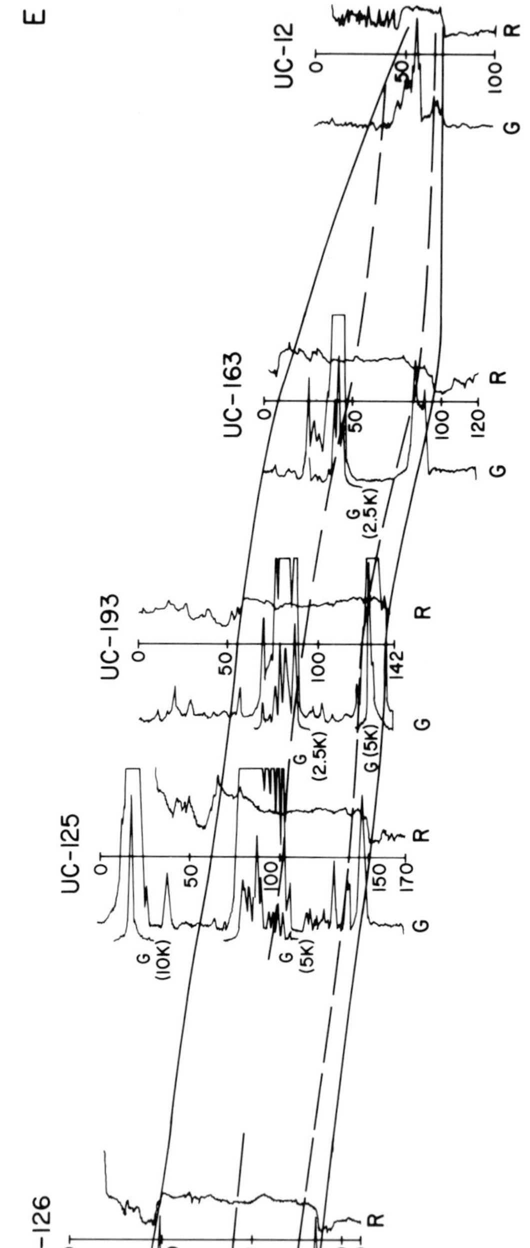

Figure 6. Cross-section showing structural control of ore deposition. (G-gamma and R-resistivity)

Faulting does not appear to be a major factor in controlling deposition of the uranium except possibly on a local level. The major effect faulting may have had was to deflect and disrupt the stream channels during the deposition of the host sediments creating impediments to ground water flow for the later deposition of the uranium.

EXPLORATION PRACTICES

During the exploration of the property, the "roll front" model was used as an exploration guide. Alteration products that may be seen and gamma log characteristics, along with the geology of the host formation, outlined the most favorable areas.

Gamma log characteristics used in uranium exploration were developed on uranium deposits found within the Wyoming Basins and are based on the shape of the "roll front" and position where the drill hole intersects the front. An idealized "roll front" is cresent shaped. The thickest, strongest mineralization (nose) occurs in reduced ground with generally weaker, thinner mineralization (tails) in oxidized ground.

The intersection of drill holes with different parts of a front produces distinctive signatures on the electric log and creates a picture of the mineralized zone. This can indicate whether or not the system contains significant uranium mineralization and where the next exploratory hole should be located.

Figures 7 and 8 are examples of this type of interpretation. In Figure 7, an ideal situation has been diagrammed showing the expected configuration of the gamma curve for different locations within the front. Figure 8 shows this same interpretation using actual drill hole data from the Chord property. Note the zonation of the uranium mineralization from weak tails to stronger tails, to thick strong mineralization at the nose, to thick weak mineralization in the protore zone. Also, note that the mineralization is often stronger in the tails immediately behind the nose than within the nose itself. Hematite staining and feldspar alteration are seen in the samples in the altered zone and disappear at the nose. At the same time grain size decreases towards the unaltered zone indicating that permeability affected deposition.

Figure 9 shows this same interpretation but with the complexity of multiple fronts within a single geochemical cell. This is more the rule than the exception when dealing with the larger channels found within the Lakota Formation.

The use of this technique, along with lithologic information and an understanding of the geology of the host, can and does expedite exploration for deposits in the area. It is more difficult when

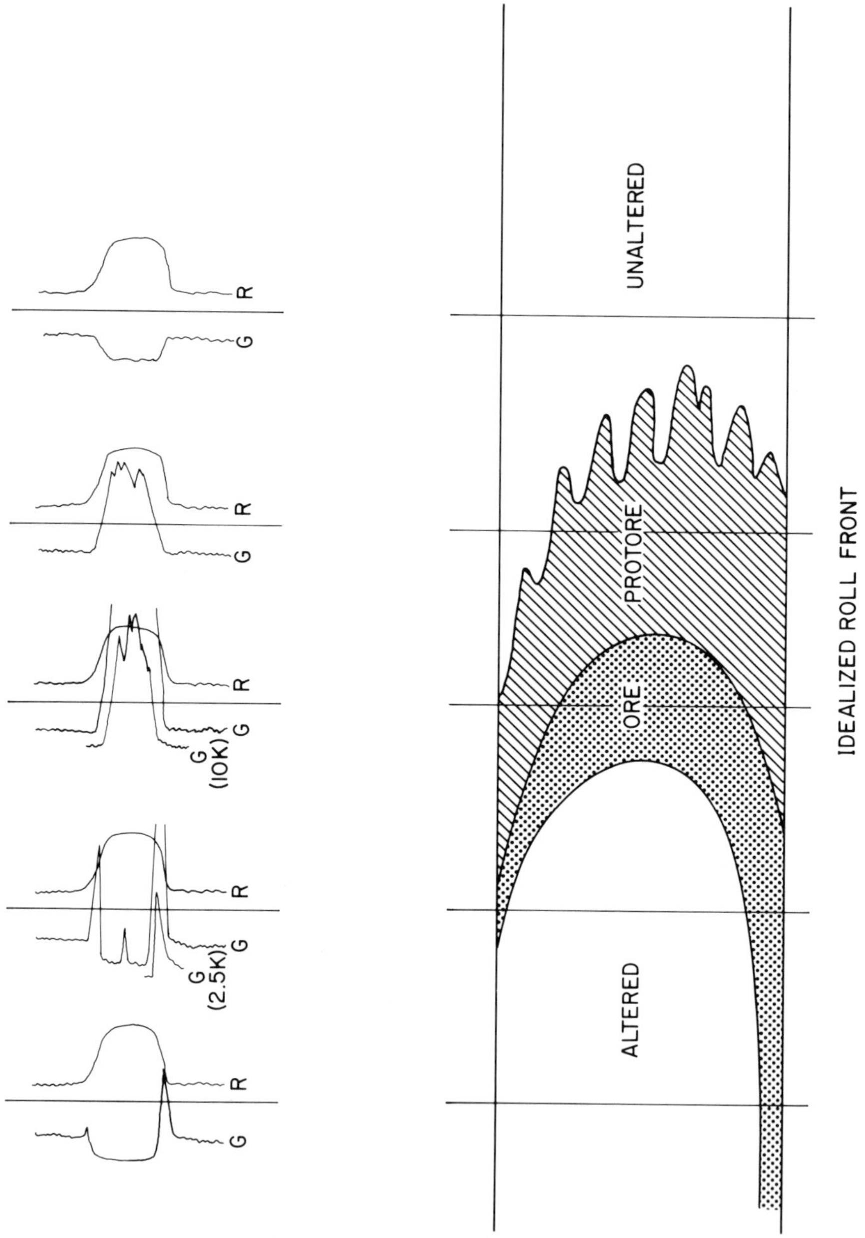

Figure 7. Interpretation of gamma log characteristics for an idealized roll front deposit.

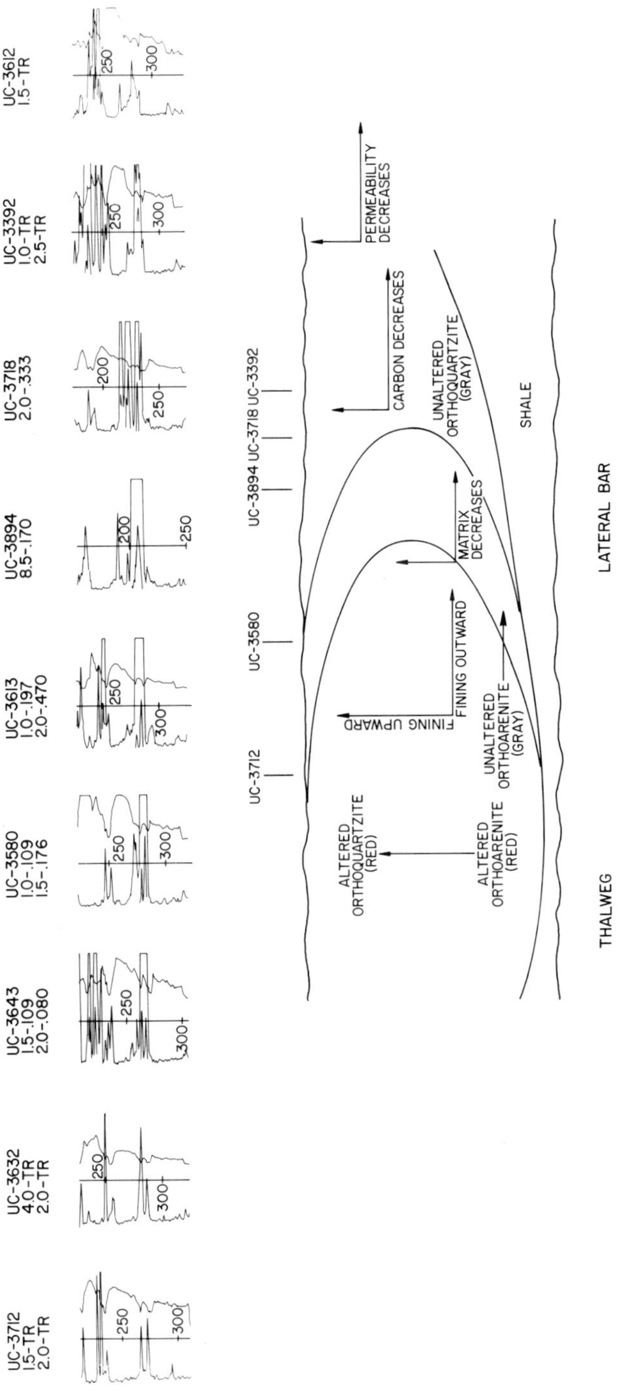

Figure 8. Diagrammatic lithographic cross-section of Number 4 trend, Viking area.

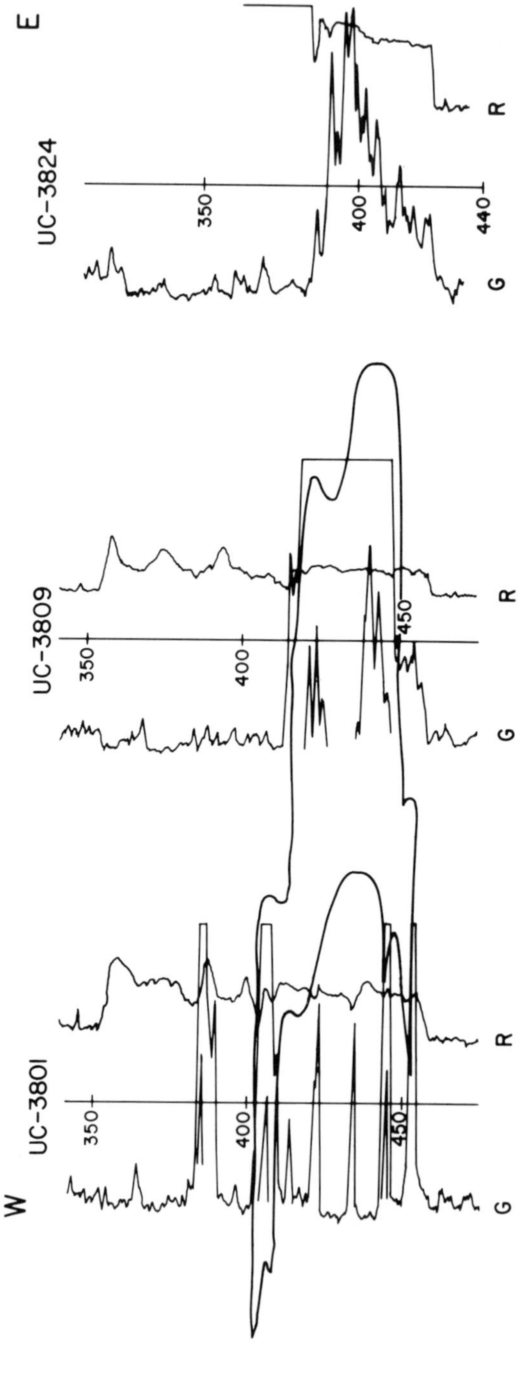

Figure 9. Cross-section showing the roll front interpretation of the unoxidized front of the October-Jinx ore body.

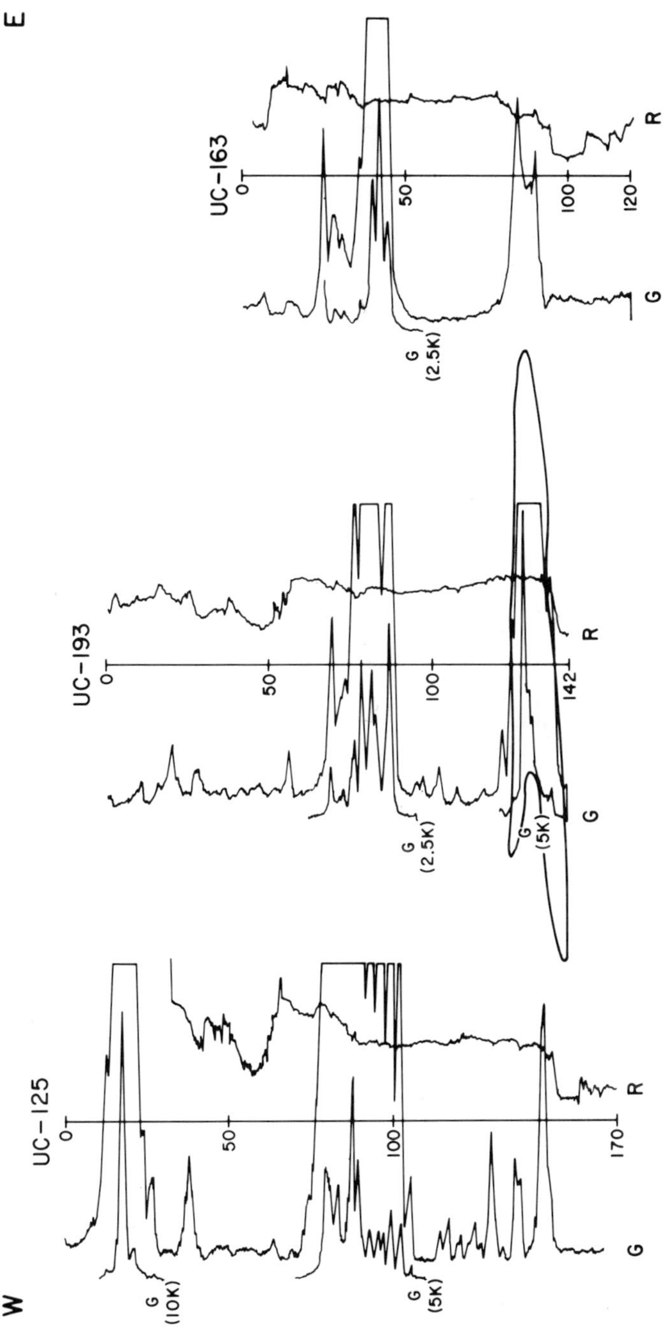

Figure 10. Cross-section showing roll front interpretation of the Long Mountain oxidized front.

looking for deposits in the zone of surface oxidation. Iron oxide staining is obscurred by the surface oxidation to the point that minor alteration features and electric log characteristics are the only remaining guides. Figure 10 shows that gamma log interpretation can still work on oxidized deposits.

BIBLIOGRAPHY

1. Gott, G. B., Walcott, D. E. and Bowles, C. G., "Stratigraphy of the Inyan Kara Group and Localization of Uranium Deposits, Southern Black Hills, South Dakota and Wyoming," U. S. Geological Survey, Prof. Paper 763, p. 24-27, 1974.

2. McGookey, D. P., Haun, J. D., Hale, L. A., Goodell, H. G., McCubbin, D. G., Weimer, R. J. and Wulf, G. R., "Geologic Atlas of the Rocky Mountain Region," Rocky Mountain Association of Geologists, p. 196, 1972.

3. Bowles, C. G., "Theory of Uranium Deposition From Artesian Water in the Edgemont District, Southern Black Hills," Wyoming Geologic Association Guidebook, p. 125-130, 1968.

4. Renfro, A. R., "Uranium Deposits in the Lower Cretaceous of the Black Hills," Wyoming Uranium, Wyoming University Contr. Geology Vol. 8, No. 2, Pt. 1, p. 87-92, 1969.

5. Von Holdt, R. H., "Annual Chord Project Report," Union Carbide Corporation Project Report, 1975.

6. Mark, D. L., Grigsby, R. N. and Pate, S. M., "Annual Chord Project Report," Union Carbide Corporation Project Report, 1977.

7. Grigsby, R. N., Mills, M. M. and Pate, S. M., "Annual Chord Project," Union Carbide Corporation Project Report, 1978.

8. Grigsby, R. N., "Annual Chord Project Report," Union Carbide Corporation Project Report, 1979.

Chapter 16

URANIUM ORE CONTROL USING COMPUTER MAPPING

J. F. Crouch* and R. L. Sandefur**

*Pathfinder Mines Corporation
Riverton, Wyoming

**Utah International, Inc.
San Francisco, California

INTRODUCTION

Roll front uranium has unique geologic characteristics (1,2) which require specialized mining procedures. Roll front uranium occurs in arkosic sandstones with the ore zone typically less than 100 feet in thickness with up to 400 feet of overburden. The areal extent of ore is often a few thousand acres. Within the ore zone, ore grade material occurs in C and S shaped rolls, which are a few feet in width and height but may run laterally for several hundred feet. Since the total mine plan may include mining all ore, a roll front uranium mine may cover a few thousand acres. For both economic and environmental reasons, a roll front deposit is mined as a series of phases or sub-pits (3,4) as illustrated in Figure 1. A typical sub-pit is 400 to 500 feet across. Sub-pit 1 is stripped initially and the overburden is spoiled on the surface. Depending on haul distance and overburden characteristics, stripping is done with trucks and shovels (or front-end loaders) or scrapers.

When ore is encountered in sub-pit 1, mining begins, and the stripping equipment is moved to sub-pit 2, with sub-pit 2 overburden also being spoiled on the surface. When ore is encountered in sub-pit 2, sub-pit 1 mining should ideally be complete so the mining equipment can move to sub-pit 2 and sub-pit 3 overburden can be spoiled in the mined out sub-pit 1 area. Ore is mined selectively in a series of 12 to 24 inch cuts by a 2-1/2 yard backhoe on the basis of radiometric probing. For effective drainage, ditches must be maintained on the boundary of sub-pits, and mining lifts are kept as level as possible, with the orebody slightly elevated as part of general good mining practice. Waste removal in the ore zone is done by self-loading scrapers near the orebodies and by push-pull or push loaded scrapers away from the orebodies. Each self-loading scraper cut is

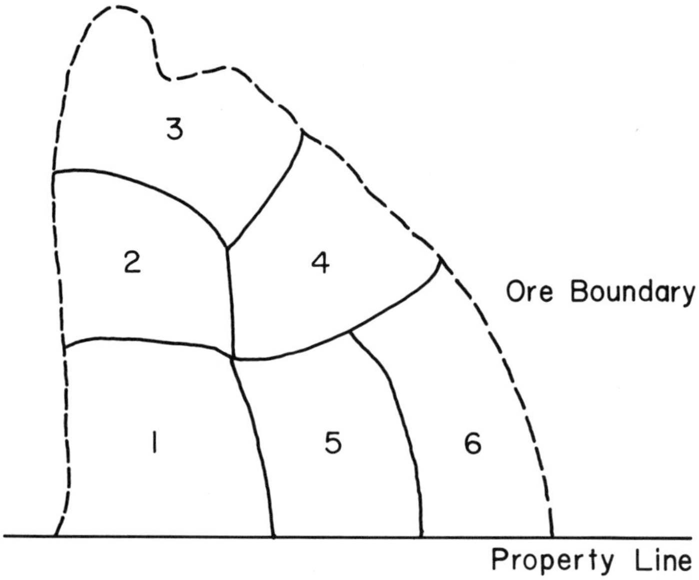

Figure 1. Typical plan sub-pit layout.

only a few inches deep and, in stripping around ore, a radiometric check for ore is made after each scraper pass.

Although grade and thickness are very erratic within a lift, there is a definite structure to the ore rolls. On the basis of data available from radiometric probing in the drill holes (Figures 2 and 3), this structure is not obvious except to those with considerable roll front experience. Even to those with considerable experience, there is often a plethora of possibilities for the structure of the rolls and interpretations often show individual biases.

HISTORICAL DEVELOPMENT

During 1973, problems with the availability of technically-oriented people for the ore control positions in Pathfinder's Wyoming uranium operations caused concern with the possible difficulty of maintaining the quality of mining operations. The basic problem was the fact that the people available for the ore control type of job were inexperienced and generally had a weak educational background. These people simply could not grasp the complexity of uranium geology or the roll front type of orebody. Consequently, their efforts in the area of ore control were not on the level that Pathfinder's uranium operations desired.

In examining the overall problem, it was felt that a pictorial product would bring better end results than intensive geologic training. The old saying of "a picture being worth a thousand words" was what appeared to be a good possible solution for the problem at hand. This general idea was manipulated into a project to create some sort of pit mapping that was extremely simple to understand, covered various pit levels, and was not overly difficult to produce. Thus, in 1974, an effort was made by Utah International (later Pathfinder) to develop a computerized procedure for mapping roll fronts.

The first step in the project was a very systematic and careful effort to map the orebodies in pits that were currently producing. Mapping was carried out using an alidade and plane table. All of the mining cuts were surveyed, and items such as orebody grade, fresh leach contacts, and ore thickness in the cut were mapped. This mapping continued through each mining cut through entire pits. The end result of the orebody mapping was a series of maps in horizontal section with no more than two-foot vertical separation between maps through more than 60 vertical feet of solution front orebodies.

With the completion of major portions of the plane table mapping, the problem then became that of developing a type of computer mapping or modeling that would create a pictorial of the orebody that was a true and realistic representation.

All drill hole information at Lucky Mc was available on magnetic

+ 6729.8

6377.0	2.1	0.04
6368.1	1.9	0.10
6335.1	1.6	0.06
6290.5	1.4	0.08
6270.0	2.0	0.04

+ 6738.0

6375.3	7.5	0.05
6347.9	3.5	1.30
6339.5	1.5	0.60
6328.8	1.5	0.05
6288.4	1.3	0.15
6314.5	2.4	0.04

+ 6737.7

6277.0	1.3	0.04
6380.2	1.3	0.06
6368.2	2.7	0.07
6365.5	1.2	0.18
6333.3	1.1	0.18
6325.0	2.2	0.06
6285.7	1.2	0.06

+ 6737.8

6544.9	1.1	0.04
6541.1	1.9	0.18
6539.2	0.8	0.21
6440.8	0.9	0.28
6439.9	1.3	0.32
6432.9	1.0	0.05
6429.2	1.3	0.13
6396.3	2.2	0.08
6361.8	1.8	0.06
6331.6	1.3	0.07
6309.2	1.6	0.07

+ 6736.8

6369.4	2.0	0.04
6367.4	2.0	0.06
6347.0	2.4	0.06
6342.2	3.8	0.04
6338.4	2.0	0.42
6316.1	2.3	0.05
6296.1	1.5	0.21
6291.4	1.4	0.08
6279.5	1.5	0.05

Figure 2. Plan view showing the elevation of the top of each drill hole. Data in the three columns indicate elevation of the top of each ore body (left), thickness (center), and grade thickness (right).

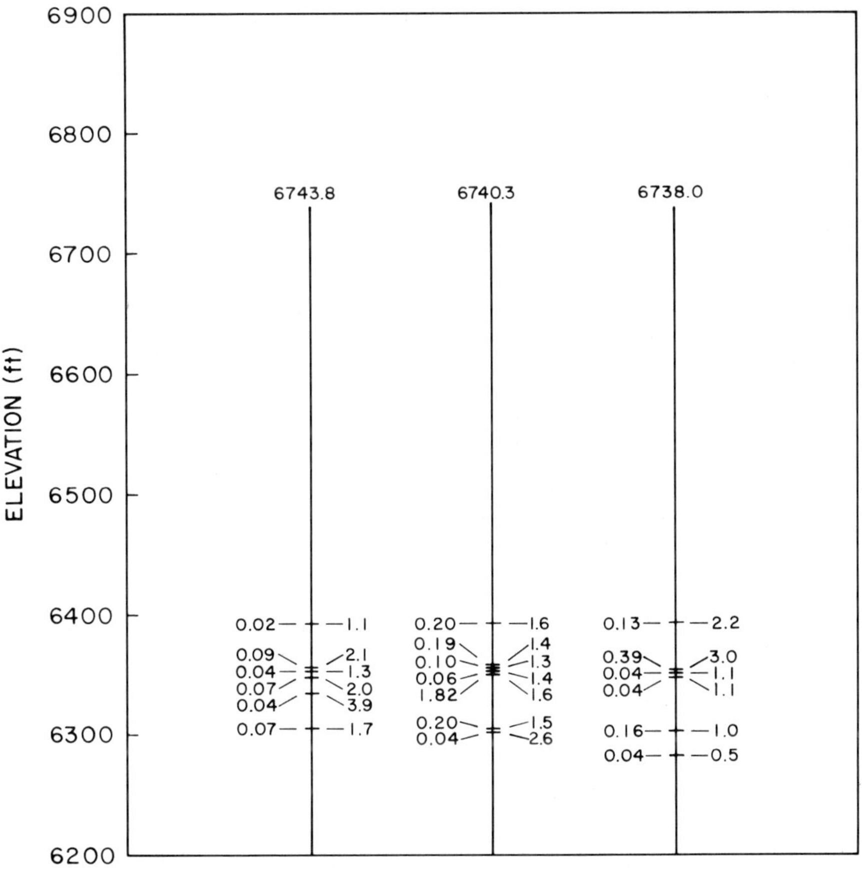

Figure 3. Analytical data for three drill holes indicating grade thickness (left) and slice thickness (right) for each intersection with an ore body.

computer tape. This information is tied strictly to elevation control and contains information on hole collar elevation, top elevation of each ore intercept, ore intercept thickness, ore intercept grade, bottom hole elevation, and drill hole coordinates.

Only two dimensional computer mapping procedures are generally available. Based on past very successful top-of-coal seam structure mapping and successful coal thickness mapping, the NUMAPROX function of IBM's STAMPEDE (5) package was selected for initial attempt. NUMAPROX assigns values to points of a square grid from irregularly spaced data using weighted least surfaces for grid squares containing data points and first order finite differences for other grid squares. Thus, NUMAPROX produces a smooth minimum curvature contour map which honors the data. STAMPEDE provides several other useful computer mapping functions. The STAMPEDE package is distributed for the cost of the source tape ($25). STAMPEDE is coded in FORTRAN II but contains a few IBM assembler bit manipulation routines which would have to be recoded to run on machines which cannot decode IBM assembler instructions.

There were a number of parameters to be chosen for the mapping procedure: variable to be mapped, slice thickness, and mapping grid size. A number of variables were possible candidates for mapping: ore grade, ore thickness and ore grade thickness. Grade thickness was selected for a number of reasons. Within the range of slice thickness which seemed reasonable for mapping (1 to 5 feet), ripping and mining by backhoe mixes ore and waste. Grade thickness within a slice is proportional to area under the gamma log track within the slice increment. Grade thickness multiplied by ore density gives pounds U_3O_8 per square foot. Even though grade thickness seemed to be the proper variable for mapping, there was still a question of cutoff grade to be used in calculating grade thickness. Grid size in the STAMPEDE manual was suggested as about 1/2 of the average minimum interhole separation. It is Lucky Mc practice to drill on a 70.7 foot square grid at 45° to the north, so a range of from 5 to 50 feet was considered.

A lengthy series of error plots were produced using variations in slice thickness, cutoff grade, mapping grid size, and contour interval. Each set of maps was compared to the actual orebody mapping done in the pit (Figures 4 and 5). This method of map comparison and subsequent adjustment of mapping led to the first regular auto plot mapping which was done on two-foot slices, a 0.05% grade thickness, a 12.5 mapping grid and with a 0.05% grade thickness contour interval. The good agreement between the plane table maps and computer maps gave even more useful results when several slices were taken together; see Figure 6.

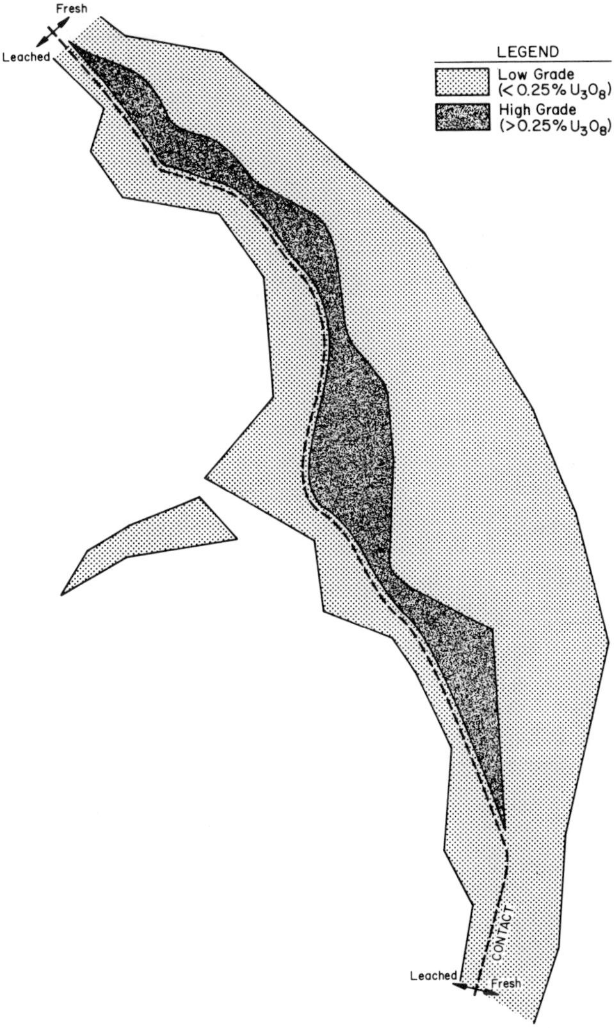

Figure 4. Lucky Mc Pit slice map showing areas of high grade and low grade ore as determined by analysis of the mined ore.

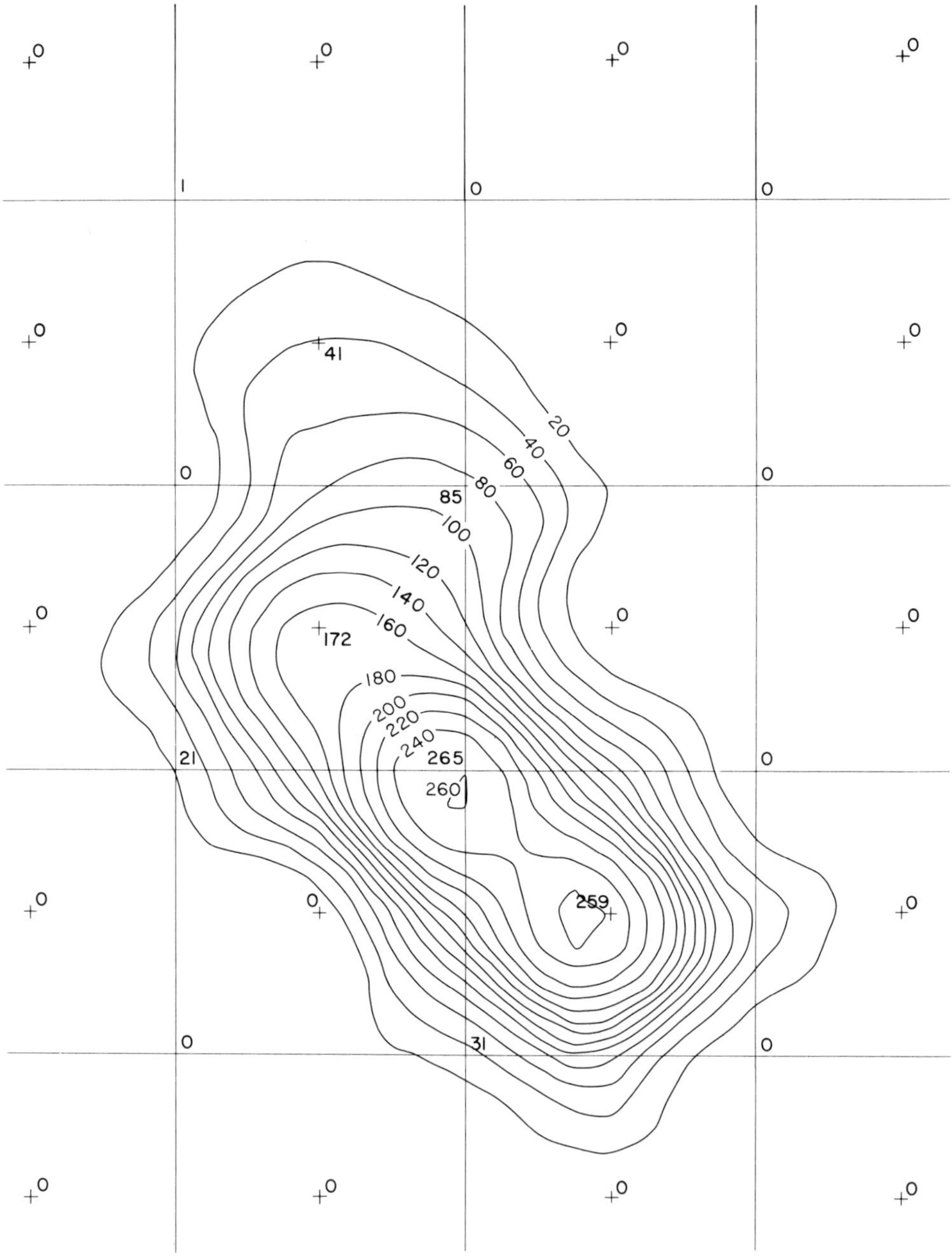

Figure 5. Predicted grade thickness for the orebody shown in Fig. 4 obtained from drill hole data.

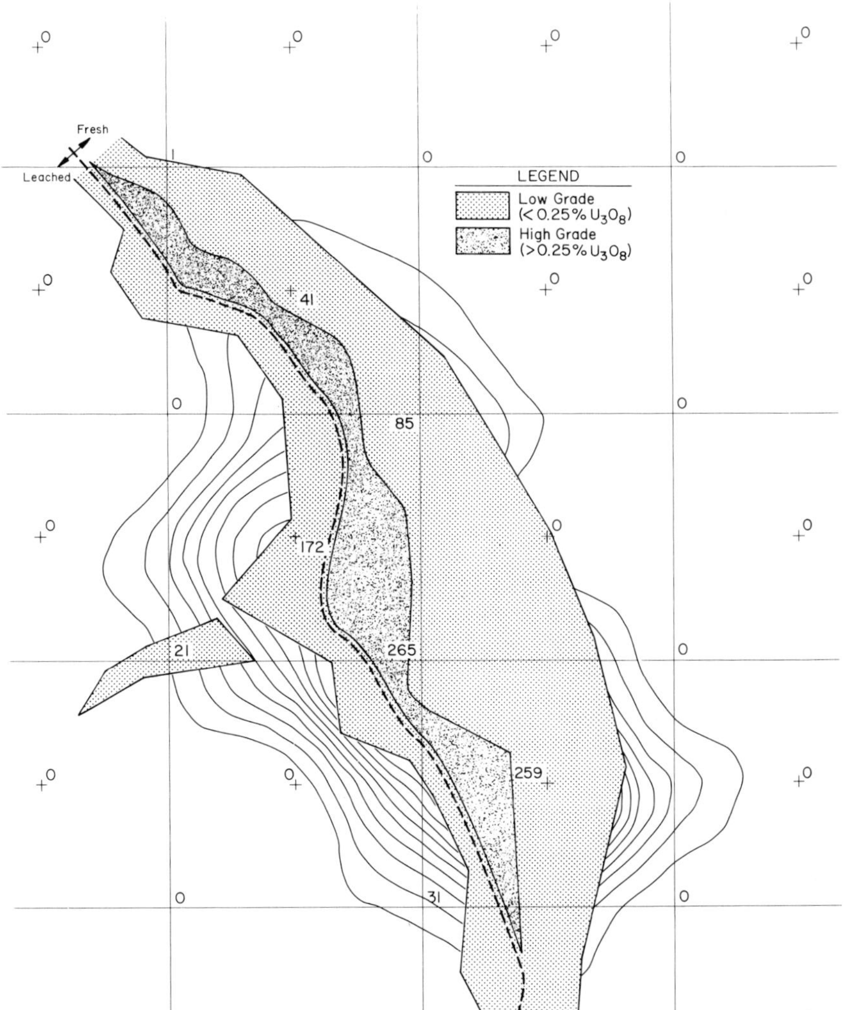

Figure 6. Comparison of the predicted and measured grade thickness data for the information presented in Figures 4 and 5.

CURRENT PRACTICE

The auto plot slice mapping is currently done using a 40 scale map. These individual maps are reduced photographically to a size that can be made into a booklet form, generally the longest dimension being a maximum 16 inches, but many of the booklets are in the 8½x11-inch size. The booklet format uses semi-transparent overlays so the ore controlman can see the orebody and how it changes in 6 to 10 feet of depth.

In the Pathfinder operations, ore controlmen as well as stripping and mining foremen are provided with the auto plot booklets. In the case of stripping foremen, this mapping shows them where they can safely strip and where they must exercise extreme caution. Of course, the mining foremen and ore controlmen use the slice maps as an aid in the actual mining operation.

Knowing the structure of the rolls before mining provides information on locating ramps and ditches on the boundary of sub-pits. Additionally, information of preferred stripping directions can be obtained.

It is also possible to produce vertical sections (Figure 7) on the basis of slice maps. These sections are useful for designing pit toes and walls and for planning the type of underground mining done at Lucky Mc where rat hole drifts are driven from the pit wall along high grade rolls. Because most of Pathfinder's ore is mined by open pit methods, vertical sections are seldom produced.

Besides being a very useful mining tool, other uses have been developed from the mapping system. Exploration and development work have benefited from the program by using the system to project ore trends from sparse and side spaced drilling.

Since grade thickness within a slice is proportional to pounds U_3O_8 per square foot, it is possible to calculate ore reserves within a slice by integrating (computer planimetering) the slice maps. Pounds of U_3O_8 obtained by this method agree within 5 percent, and are well within the geostatistical estimation variance for a slice. Whether or not integration of slice maps provides a better ore reserve estimate is a moot point. If the mapping procedure is for ore thickness, tons of ore may also be calculated by planimetering the thickness slice map.

INCONSISTENCIES IN SLICE MAPS

Even though the slice maps provide a good picture of the ore stringers, maps sometimes show unreasonable features. The first of these is edge effect which sometimes occurs at the boundary of the NUMAPROX grid. Edge effect is due to instability in the finite

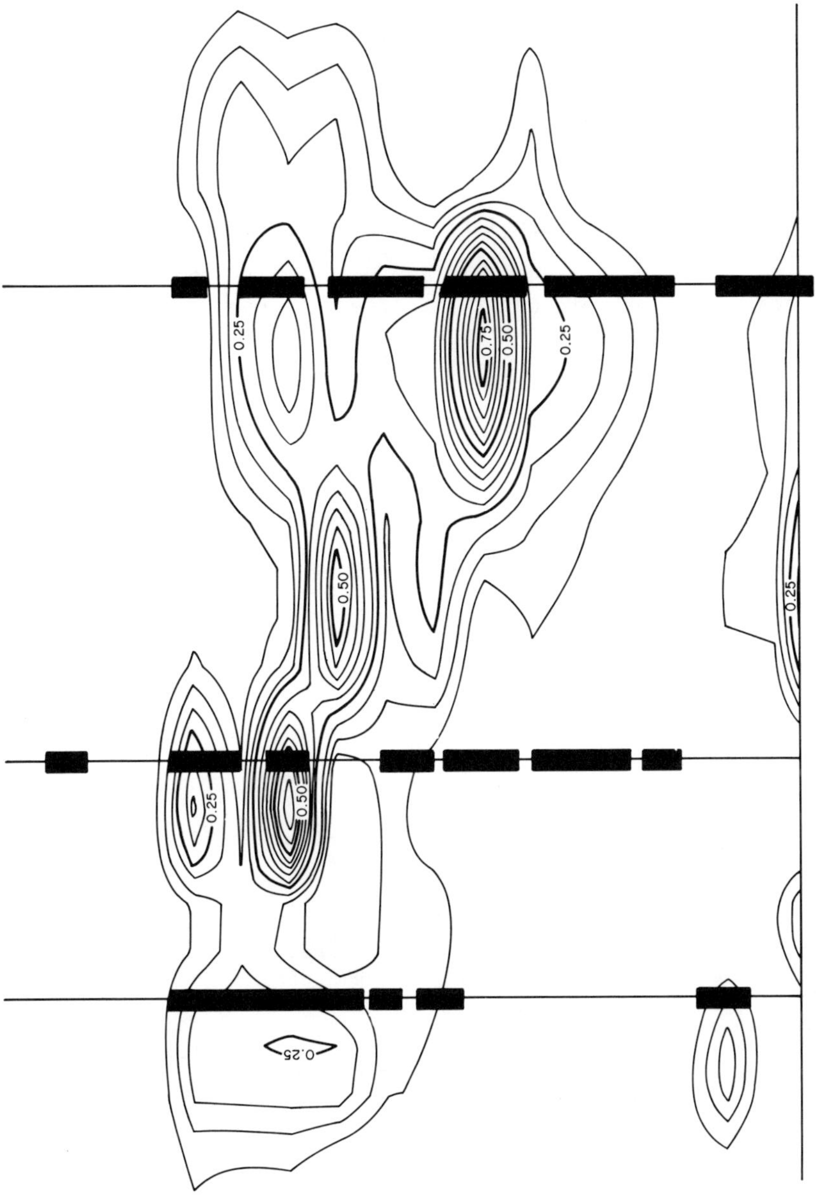

Figure 7. Predicted grade thickness variation for a vertical section of the Lucky Mc Pitt.

difference phase of NUMAPROX. Edge effect can be eliminated by additional holes (either dummy or real) on the boundary.

The second and more serious inconsistency is that maps sometimes show negative grade thickness values. When NUMAPROX fits the minimum curvature surface to the data, there is no guarantee that negative values will not occur. Two techniques were tried to overcome this. The first method was to use the SMOOTHER function of STAMPEDE, which averages various numbers of grid squares. This procedure eliminated most negative values but spread the ore zones too much.

The second technique was to take the logs of values before NUMAPROXING and then exponentiate before contouring. This was successful, but there was always a question as to what value to assign as the log of zero and with a minimum contour of 0.05% ft., the logged exponentiated values were not that different from the untransformed maps, so the problem of negative values has little or no effect on the projected geometric structure of the rolls.

Later geostatical (6,7) work has indicated that attempting to assign thickness or grade thickness values to points away from the drill hole has estimation error of at least ± a factor of 4 at 95% confidence. Thus, the slice map is not reliable for indicating grade thickness values at a point. However, the slice maps do show the geometry of the rolls to sufficient accuracy that the slice mapping system is an integral and important part of a successful open pit uranium mining operation.

BIBLIOGRAPHY

1. Harshman, E. N., "Uranium Ore Rolls in the United States," Uranium Exploration Geology (Proc. IEAR Vienna, 1970), IAEA, Vienna, 219, 1970.

2. Anderson, D. C., "Uranium Deposits of the Gas Hills," Contrib. to Geol. Wyo. Uranium Issue, 8 2 101, 1969.

3. Crouch, J. F., "Mining Technique at Lucky Mc Uranium Corporation's Lucky Mc Mine," Proceedings of First Conference on Uranium Mining Technology, Kim, Y. S., Editor, Mackay School of Mines, Reno, Nevada, 1977.

4. Anderson, D. C., "The Lucky Mc Mine," Case Studies of Surface Mining, Hartman, H. L., Editor, AIME, pp. 275-286, 1969.

5. IBM Corporation, STAMPEDE Program Documentation, Program Number 360D-17.4.001, IBM Corporation.

6. Sandefur, R. L. and Grant, D. C., "Preliminary Evaluation of Uranium Deposit. A Geostatistical Study of Drilling in Wyoming Solution Fronts," Exploration for Uranium Ore Deposits, IAEA, 695-714, 1976.

7. Sandefur, R. L. and Grant, D. C., "Applying Geostatistics to Roll Front Uranium in Wyoming," Engineering and Mining Journal, 181, No. 9, February 1980.

Chapter 17

EXXON MINERALS COMPANY, U.S.A.
HIGHLAND URANIUM OPERATIONS
GEOLOGY AND MINING METHODS

Ronald A. Murdock

Exxon Minerals Company, U.S.A.
Casper, Wyoming

INTRODUCTION

Exxon's Highland Uranium Operations in eastern Wyoming consists of an open pit, an underground mine and a mill complex. The geology of the deposit is reviewed in this paper, and an overview of the mining operations at both the surface and underground mines is presented. Topics discussed include geology, stripping of overburden and mining of ore in the open pit, and underground development and stoping methods.

GEOLOGY

The Highland uranium deposit was discovered in 1968 by Exxon Minerals Company, U.S.A., in Converse County, Wyoming, 35 miles north of Douglas. The mine is located at an elevation of about 5,200 ft in a high-plains, sagebrush-covered prairie typical of eastern Wyoming. The deposit is near the southern edge of the Powder River Basin in the Paleocene Fort Union Formation.

The Fort Union Formation consists of 1,900 to 3,200 ft of sandstone, siltstone, mudstone, claystone and coal beds. Siltstone, mudstone and claystone are collectively referred to as "shale" at Highland. There are many alternating layers of shale and sandstone in the Fort Union Formation; but in the Highland area, there are only three principal sandstones, known as the Highland Sand Group, which host the uranium. The combined sand thickness is about 150 ft.

Roll-Front Deposits

The uranium at Highland is in roll-front deposits. In this case ground water carries oxidized uranium in solution, and the uranium precipitates upon encountering a reducing environment. The resulting ore mineralization at Highland is in the form of coffinite $(USiO_4)(OH)$ and uraninite (UO_2) which coats sand grains in the sandstones.

A typical roll-front deposit is crescent shaped in cross section. In the Highland deposit, the roll-fronts in the three sands may be found stacked together, or they may be apart. The center of the roll-front is oxidized and generally barren of ore. The area ahead of the roll-front is also lacking in ore.

Drilling

Delineation of a uranium ore body is done by extensive drilling. The holes are logged using a gamma ray detecting instrument to locate the ore zones. There are over 16,000 holes at Highland with a total drilled depth of about 8.7 million feet or 1,650 miles.

SURFACE MINING

Surface mining consists of two separate operations: (1) stripping of overburden and (2) mining of ore. Each operation uses its own unique equipment fleet and requires a different operational plan.

Stripping

Contract stripping of overburden began at Highland in 1970 and has evolved into the present Exxon-owned truck/shovel operation for removing primary waste. To date, over 75 million bank cubic yards (bcy) of primary waste have been moved. Currently, the stripping fleet moves about 1.5 million bcy of overburden per month.

The overburden consists of weakly cemented shales and sandstones overlying the ore sands. The total thickness of the overburden ranges from 250 to 375 ft. This material has an average compressive strength of 500 to 1,000 psi and can be dug by a shovel. Calcareous boulders are occasionally encountered. Regular blasting is performed with ANFO to ensure higher productivity and less wear on the equipment.

The stripping fleet consists of three electric shovels equipped with 15-cu-yd buckets and thirteen 100-ton diesel-electric, end-dump trucks. A trailing electric cable carries 4,160 volts to power each shovel. Rubber-tired dozers clean up the immediate areas where the shovels are working.

Slope Stability and Floor Conditions

Water causes operational problems for the stripping fleet and enters the pit through the permeable sandstones and as precipitation. Slope stability and poor floor conditions are two major concerns resulting from excess water within the pit.

Wet shale causes slope-stability problems as the shovel digs along the face. The wet, water-softened shale squeezes out from between sandstone layers towards the free face and leaves the sandstone unsupported. The sandstone then fails in tension and causes localized wall failures.

Excess water also creates poor floor conditions. A wet pit floor is a difficult working surface because the sand is very soft and slippery. Equipment traction can be a problem under these conditions. To minimize the impact of water-related slope stability and floor problems, several water-control techniques are used.

Water Control

Water control is managed by: (1) reducing water flow into the pit, and (2) diversion and pumping water from the pit. To reduce water flow into the pit, dewatering wells are drilled around the perimeter of the pit prior to beginning excavation. The dewatering wells pump water to the process-water storage pond for use in the mill. This reduces the water flow through the sandstone into the pit with a significant reduction of pore pressures within the pit walls. Reduction of total water flow decreases the amount of water that must be handled inside the pit, and reduction of pore pressure reduces slope failure.

To control the water that does enter the pit, backhoes excavate trenches along the toe of the pit wall. As a shovel advances along the face, backhoes extend the trench through the newly excavated area. Water collects in these ditches and is routed to a low elevation central sump within the pit. This water is also pumped to the mill pond to be used as process water.

Secondary Waste Removal

Due to the irregular nature of the uranium mineralization within the roll front, there is a considerable amount of internal waste that must be removed after the top of ore has been reached. This material, identified as secondary waste, may lie above, below, or around a mineable ore pod. The proximity of the ore requires that equipment be capable of selectively removing the secondary waste without removing ore at the same time. At Highland a contractor-owned fleet of scrapers removes the secondary waste. These scrapers include push-pull and self-loading types with capacities from 17 to 28 bcy.

Grade control personnel, using probes, outline the ore pods around which secondary-waste is removed by scrapers. A bulldozer usually rips the areas to be worked by the scrapers to assist scraper loading. Scrapers will often load themselves but may be assisted by another scraper or pushed by one or two bulldozers. The scrapers travel a short distance to dumps in previously-mined areas and then return to complete the cycle.

OPEN PIT MINING

Grade-control technicians locate and delineate ore pods by using hand-held Geiger counters and T-probes. The grade-control technicians must also use geological maps and visual inspection of the sands to help define ore. Small colored flags indicate the boundary between waste and ore and differentiate between low-grade material and high-grade ore. The ore pods are outlined carefully as the scrapers remove the secondary waste around the pod. Upon completion of this waste removal, the pod stands about three feet above the level of the surrounding ground prior to mining.

The ore is mined by backhoes with 2-3/4-cu-yd buckets loading into Exxon-owned 35- and 50-ton diesel trucks. The backhoe climbs on top of the raised ore pod and will dig the ore pod down to the level of the adjacent pit floor.

After the backhoe completes loading the truck, the truck travels to an assay-probe tower to determine the grade of the mined ore. The tower uses an auger to sample the ore in the truck in three different locations. After being sampled, the truck continues on to the ore pod.

The sample taken from the truck is pulverized, dried, and analyzed by beta and gamma counts to compute a grade in % U_3O_8 in less than a minute. The tower operator is then able to radio the truck driver and tells him on which stockpile to dump the load.

UNDERGROUND MINING

In August 1973, Exxon contracted the construction of the Buffalo vertical mine shaft and erection of a suitable steel head frame. The shaft is 14 ft in diameter and lined with 1 ft of reinforced concrete. The head frame is 75 ft high. The shaft contains 3 compartments for a 12-ton service cage and two counterbalanced 80-cu-ft ore skips. The ore skips are hoisted by a 400-hp double-drum hoist. A separate 200-hp hydraulic hoist powers a single deck service cage.

Underground Development

The underground mine ore is accessed on two levels. A lower track-haulage level is at a depth of 600 ft and is equipped with rail haulage and drainage for water control. The ore-mining level is at 550 ft and is laid out for trackless operations. A 10% incline from the haulage drift accesses the ore level. Broken ore moves from the ore-mining level to the haulage level through cribbed raises developed on approximately 250-ft centers. Slushers move the ore from scram drifts above the track level to load rail cars. The rail haulage transports the ore to the station for hoisting to the surface.

Track Level Development

Exxon took over development of the underground mine in October 1975. A total of 2.6 miles of drift for the track level was driven during the following three years. Two basic mining methods were used: (1) conventional drill and blast, and (2) a continuous shield mining system.

The conventional drill-and-blast method accounted for about half of the haulage level. Blastholes were drilled with jacklegs and shot with stick powder. Mucking machines loaded the broken rock into 110-cu-ft rail cars pulled by 8-ton diesel locomotives. The inside drift dimensions are 8-1/2 x 9 ft where timber sets are used and 9 x 9 ft where bolts and mesh are used. Progress was slow and hampered by loose, wet running shales.

An electric/hydraulic shield miner is capable of driving a 10-ft circular opening within the protection of a steel shield. Teeth on an excavator bucket rip the face, and the bucket moves the broken muck to a conveyor in the center of the machine. The conveyor then discharges the muck into rail cars positioned at the end of the conveyor. Circular ring steel sets on 4 ft centers with 4-in. oak lagging provide a finished 9-ft diameter, completely enclosed drift. The shield is advanced by six hydraulic jacks that push against the previous steel set. The rate of advance using the continuous shield miner proved to be considerably greater than the conventional drill-and-blast cycle. Advance rates as high as 48 ft in two shifts were achieved with two 5-man mining crews. The maximum completed shield drift footage in one month was 762 ft.

Trackless Level Development

The trackless level, or the mining level, was planned to be developed using a continuous drum-head miner. Rubber-tired two-cu-yd load-haul-dump units (LHDs) removed the muck from behind the machine. Shortly after starting in 1975, several drawbacks quickly became apparent. The soft, wet floor caused the miner to bog down. Timber flooring was required to keep the miner out of the mud. It was difficult to keep ground support near the front of the miner. Caving problems required forepoling with rebar and rail. Because of these disadvantages, use of the continuous miner was discontinued in late 1976 after a total advance of 870 feet.

Development of the trackless level continued using conventional drilling and blasting with timber and steel sets for support. In early 1977 "friction rock stabilizers" ("Split Sets") and wire mesh were introduced underground and now account for the majority of ground support. Rubber-tired one- and two-cu-yd LHDs are used for muck removal for both mining and development.

Ground, Radon and Water Control

Early project planning and development identified three major problem areas to be controlled for a successful underground project: very weak ground, radon concentrations, and water.

Ground Control

Ground conditions at Highland led to experimentation with several ground-control techniques. The rock at the underground mine consists of wet, weakly-consolidated sandstones and shales similar to that found in the surface mine. Floor heave of three feet in three months is not uncommon in the underground mine. Steel floor beams imbedded in one foot of concrete were tried in an attempt to reduce floor heave. Other techniques tried for ground control included standard rock bolts with wire mesh, steel sets, timber sets, epoxy resin bolts, and friction rock-stabilizer bolts. Many of these attempts gave unsatisfactory results and only three of the techniques currently in use at Highland will be described. These methods are: (1) circular steel sets, (2) timber and steel sets, and (3) friction rock-stabilizers bolts.

The three-piece, circular-steel sets with a 9-ft diameter proved to be the most stable system for the track level. These sets were installed on 4-ft centers and used 4-in. oak lagging to provide a completely enclosed drift. These solved the problem of floor heave on the track level. Timber and steel sets are used throughout the ore mining level. Timber sets are installed on 5-ft centers. Timber dimensions are 10 x 10 in. Two- and three-in. fir is used as lagging. Most drift intersections in shale require 10WF49 steel sets. Similar steel sets are used wherever severe ground conditions are encountered.

After unsatisfactory experiments with rock bolts, Highland adopted the friction rock stabilizer bolts. Friction bolts are used with 10-gauge wire mesh to provide protection on all the ribs and back to prevent small-scale slabbing. These bolts show a 3- to 5-ton pull test compared to less than 2 tons for conventional expansion-shell rock bolts.

Radon Control

A positive-pressure ventilation system provides the primary control of radon buildup at Highland. Highland is believed to be the only uranium mine in the United States completely on positive pressure ventilation. The pressure is 7- to 9-in. water gauge.

A ventilation shaft, a production shaft, and four exhaust bore holes provide the primary openings for the ventilation system. The ventilation shaft is 9 ft in diameter, concrete lined, and 650 ft deep and is located 1000 ft from the production shaft. Four other 5-ft diameter exhaust holes lined with full-hydrostatic steel casing are located around the perimeter of the mining zones.

Air-flow direction and rates are regulated to control radon buildup in active mining areas. Two 350-hp fans on the surface provide 350,000 cfm of fresh air to the mine. Air is circulated from the ventilation shaft to the track level and then up to the stopes and out the exhaust holes and the production shaft. Air doors and pneumatically-controlled covers over ore passes prevent recirculation of fresh air flows. Two natural-gas heaters can increase the intake air temperature by 60°F during the winter.

Inactive areas are sealed with bulkheads. Bleeder ventilation tubing extends from the bulkheaded areas to exhaust drifts. Positive suction is used to reduce the radon concentration and pressure in bulkheaded areas. This standard practice at Highland is a result of a joint study between Exxon and the Spokane Research Center of the U.S. Bureau of Mines. Test results indicate that the bleeder lines can reduce radon buildup in the bulkheaded areas by approximately 80%.

Water Control

Preliminary drilling indicated that the ore sands were water saturated and that dewatering measures would be necessary. A ring of 16 dewatering wells, ranging from 490 to 745 ft deep, were drilled around the outline of the underground ore body. These 20-in. diameter wells, with 10-in. steel casings, formed a 6,000-ft diameter ring around the mine. A 30- to 50-hp pump powered each well. In 1973, flow rates from individual wells ranged from 45 to 125 gpm and have decreased to the current flow rates of zero to 30 gpm.

Additional drain holes are drilled from development drifts to dewater the ore sands in the mine. Longhole fan drilling from the track level provides local drainage from the ore horizons to the lower haulage drift. The drilling was done on 150 to 200 ft centers. Each fan consisted of four or more holes with hole lengths up to 300 ft. The holes making up the fan were drilled on an incline from 20 to 50° from the horizontal and are cased with 1-1/4-in. PVC perforated pipe to keep the holes open and to allow radiometric probing of the ore body. The cased longholes are the source of most of the water now being pumped from underground with an average discharge of 150 to 200 gpm.

Stoping Methods

Two basic ore-mining methods are used at Highland: a modified room-and-pillar retreat method and a sandfill stoping method. The choice of stoping methods depends upon ore thickness and ground conditions. Ore thicknesses range from several inches to over 30 ft, and the minimum mining height is 8 to 9 ft.

Room and Pillar

The modified room-and-pillar retreat method is used for ore zones less than 15 ft high. Development is normally driven to the boundaries of the deposit. Pillar extraction is done on the retreat in a shortwall method with rubber-tired LHDs loading out and transporting ore to ore passes. Split sets and 10-gauge wire mesh provide primary ground support. Timber stulls on 5-ft centers provide secondary support as needed.

Several different pillar dimensions and mining strategies have been used. Initially, pillars as large as 250 ft long and 120 ft wide were developed. Headings were generally 9 x 9 ft. Drift intersections became unstable at the edge of the cuts and required support.

Larger pillars were then quartered into 50 x 50-ft pillars. Each of these pillars were further quartered into 20-ft square stumps. Two drill-blast passes were used to extract the stumps. Support problems caused delays and lowered productivity with some modification of pillar dimensions.

The current pillar dimensions are 150 ft long and 100 ft wide. These pillars are mined using a short-wall retreat method which is the best mining system used.

Sandfill

Originally, stopes over 15 ft high used square-set timbering for support. However, horizontal movement of the timber sets due to the crumbling of the sand from under the side blocking even under low vertical loading on the timber became a problem. To halt this movement, many of the square sets were cribbed tight with timbers, and mining was stopped in the immediate area.

Since the square-set timbering method proved unsatisfactory in controlling the ground, Exxon modified the mining method by combining the successful friction rock bolts and wire mesh with sand backfill support. The sandfill stopes are opened by driving a heading at the top of the ore. Split sets and wire mesh are used for support. A series of underhand cuts then excavate the floor to form a slot 10 ft wide, 25 ft high and 150 ft long. Horizontal timber stulls on 10 ft centers provide support for the ribs. After the last underhand cut, the miners build a wooden retaining wall on each end of the slot to contain the sandfill.

The sandfill slurry is prepared on the surface and is gravity fed through a 3-in. plastic line to the underground stopes. The sand is screened to 100% passing 1/4 in., and the sand-to-cement ratio is maintained between 10:1 and 15:1 at 60 to 75% solids. Over 50,000 tons of sandfill have been used since March 1979. Due to the excellent support of the sandfilled areas, extraction of the ore in adjacent stopes approaches 100%.

SUMMARY

Exxon's Highland Uranium Operations consist of a series of large open pits, an underground mine, and a mill. The open pit uses Exxon-owned electric shovels and 100-ton trucks to remove sandstone and shale overburden to a maximum depth of about 375 ft. A contractor-owned scraper fleet removes secondary waste after the top of ore is reached. Exxon backhoes mine the ore and 35- and 50-ton trucks transport the ore to stockpiles at the mill after sampling and grade determination. Dewatering wells reduce water flow into the pit and improve slope stability and floor conditions.

New techniques have been tried and some introduced in the underground mine to handle difficult underground mining conditions. The use of the continuous shield mining system controlled the problem of floor heave on the track level, and friction bolts eliminated much of the need for timber sets on the ore level. Other significant developments include a surface well dewatering program, a positive pressure ventilation system, and a sandfill stoping method.

Exxon has been a pioneer in both surface and underground uranium mining in the Powder River Basin. In the last eight years, the work force has grown from 90 to over 600 people as production increased to its present levels. At Highland the development of new technology and improved mining methods has created better working conditions, increased safety, and reduced costs.

Coal I

Session Chairman
G.E. Calahan
AMAX Coal Company
Gillette, Wyoming

Chapter 18

PERMITTING PROCESS FOR COAL MINING - FEDERAL AND STATE

Lyle D. Randen

Cities Service Company
Gillette, Wyoming

With the environmental movement beginning in the late 1960's came a deluge of permit requirements for the mining industry. Coal mining has been one of, if not the most, severely impacted of the mineral industries as a result of the passage of the Federal Surface Mining Control and Reclamation Act of 1977. Numerous provisions of this Act and numerous regulations developed under authority of the Act are being challenged in differing levels of the judicial system across the United States. Such court cases have had varied success. Numerous other regulatory programs that are not quite as dramatic as the SMCRA have presented an unending regulatory problem for mining operators. My comments in this presentation relate primarily to the major Federal and State agencies and their legal responsibilities which we in the coal mining industry must address in order to receive authority to construct and operate surface coal mines in the State of Wyoming. A list of the Federal and State agencies one could deal with in an application is as follows:

1. Bureau of Land Management (BLM). BLM is responsible for deciding where and when to lease federally owned coal and for establishing terms and conditions for leases. BLM also has some control over surface reclamation on federal land, i.e. regrading to eliminate highwalls, covering of spoils, maximum extraction of coal to limit future disturbance, contemporaneous reclamation, etc. BLM is also responsible for designation of federal lands unsuitable for all or certain types of surface coal mining operations.

2. Geological Survey (USGS). USGS coal mining responsibilities apply only to Federal and Indian lands and concern compliance with production requirements including diligent development, minimum production levels, royalty payments and resource protection, and

determinations of logical mining units. Part 211 regulations have been completely rewritten to reflect USGS's responsibilities in the coal management program.

3. Fish and Wildlife Service (USFWS). The Department of the Interior has the responsibility for implementing the Endangered Species Act. The Secretary of the Interior must use his programs to further the purposes of the Act. The USFWS plays a consultation role under the Fish and Wildlife Coordination Act, i.e. all permitting processes where any threatened or endangered species are encountered must be coordinated through the USFWS.

4. Heritage Conservation and Recreation Service (HCRS). HCRS, in conjunction with an Advisory Council on Historic Preservation, administers historic preservation statutes, regulations and orders. Federal agencies must consult with the Council on permitting actions that could affect any property that is, or is eligible to be, listed on the National Register of Historic Places maintained by HCRS. Local consultation extends protection to cultural resources on non-federal lands.

5. Environmental Protection Agency (EPA)

A. The Clean Water Act

Under Sections 301, 304, and 401 of the Clean Water Act, coal mining operations must obtain National Pollution Discharge Elimination System (NPDES) permits and comply with the effluent limitations for point source discharges of pollutants to the surface waters of the United States. Section 208 of the Act provided a scheme for local control of nonpoint sources of water pollution from surface mines.

B. The Safe Drinking Water Act

The Safe Drinking Water Act authorizes EPA and approved state programs to issue permits to control underground injections (subsurface emplacement of fluids by well injections). Drinking water on mine sites may or may not be regulated depending upon its use at the site, i.e. number of service connections.

C. The Resource Conservation and Recovery Act

Depending upon the outcome of numerous studies being funded and conducted by EPA, regulations will be developed for such things as controls over environmental contamination from coal waste disposal including protection against both surface and groundwater pollution. The "Hazardous Waste Management System Regulation" requires present operators to inventory "hazardous wastes." Legislation is currently in Congress to exempt coal mines subject to OSM jurisdiction from RCRA requirements.

D. The Clean Air Act

The Clean Air Act contains three areas of concern to the coal mining industry as related to ambient air quality. They are: the primary air quality standard which provides for air quality levels that are not harmful to public health, the secondary ambient air quality standard which provides for air quality levels that are not harmful to the public welfare, and the prevention of significant deterioration levels which provides a method for insuring no significant deterioration of air quality levels and visibility depending upon area classification.

6. Department of Agriculture (USDS). The USDA Soil Conservation Service (SCS) has authority for the protection of prime farmlands and the effects of surface coal mining. SCS has the authority to establish standards for soil surveys, to review and comment on permit applications, to establish specifications for soil removal, storage, replacement and reconstruction, and to establish criteria for designation of prime farmlands. USDA (SCS) also plays a role in determining whether or not surface mining will be allowed on National Forests west of the 100th Meridian.

7. Department of Energy (DOE). The Department of Energy regulatory responsibilities do not relate directly to environmental concerns. They do play an active role in coal leasing on federal lands and issue regulations relating to competitive bidding, rates of production, diligence requirements, and establishing energy production goals.

8. Department of Labor (DOL). The Secretary of Labor (MSHA) must concur in writing with those parts of any permanent regulatory program which could affect the health and safety of miners. Also, MSHA must review and approve all water impoundment structures that impound in excess of 20 acre-feet.

9. Corps of Engineers. The Chief of Engineers must concur in writing with regulations concerning the design, modification, removal or abandonment of new and existing coal mine waste piles or impoundments. The Corps must also approve any dam or dike constructed across navigable waters and discharge of dredge and fill materials (permit). The Corps is also in charge of certain dam inspections.

10. Office of Surface Mining (OSM). In March, 1979, OSM promulgated its permanent program regulations provided for by the Surface Mining Control and Reclamation Act of 1977. Since that time, approximately 60 of these regulations have been invalidated. The regulations are some 150 pages in length and provide for a permit to be issued by OSM before surface mining can occur; establish extensive environmental protection and reclamation requirements;

provide for the identification and protection by the Department of Interior of areas of federal lands unsuitable for surface coal mining; provide that interested parties may petition to have areas declared unsuitable for surface coal mining; and absolutely prohibit surface coal mining in certain areas. The law and regulations encompass nearly all aspects of coal mining from exploration to post mining land use.

11. National Environmental Policy Act (NEPA). The National Environmental Policy Act requires that any major federal action where the environment may be affected be the subject of an "Environmental Impact Statement." This "Environmental Impact Statement" is usually prepared by the federal agency proposing the action and must be reviewed by other agencies including EPA.

As can be seen with the preceding list of agencies, the coal mining industry is probably one of these most economically and operationally-impacted by federal regulations. Following is a list of sixteen major federal laws affecting coal mining that were passed in the last ten years. They are partially the basis for the preceding regulatory agencies.

The Federal Coal Mine Health and Safety Act of 1969
The National Environmental Policy Act
The Clean Air Act of 1970
The Clean Water Act
The Clean Water Act Amendments
The Critical and Endangered Species Act
The Historic Preservation Act
The Solid Waste Disposal Act of 1971
The Federal Surface Coal Mine Health and Safety Act of 1971
The Federal Water Pollution Control Act Amendments of 1972
The Safe Water Drinking Act
The Coal Leasing Act Amendment of 1975
The Resource Conservation and Recovery Act of 1976
The Clean Air Act Amendments of 1977
The Mine Safety and Health Act Amendments of 1977
The Surface Mining Control and Reclamation Act of 1977

Of course, it was Congressional intent for the States to assume "primacy" for all of the programs and develop laws and regulations to implement each and every one of them. This has resulted in dual regulatory agencies resulting in redundancy of administration and confusion in the permitting process. Often times a State review of an application is duplicated by a Federal review of the same application, and such reviews may provide conflicting conclusions or conditions for approval.

The State of Wyoming has been very aggressive in trying to retain state control for all regulatory programs. The Wyoming Legislature in 1973 passed the Wyoming Environmental Quality Act which in turn established the Wyoming Department of Environmental

Quality. The Department of Environmental Quality is comprised of the following divisions: (1) Air Quality Division, (2) Land Quality Division, (3) Solid Waste Division and (4) Water Quality Division.

1. The Air Quality Division (AQD) is the counterpart of the Air Quality Control Section within EPA. The AQD is responsible for implementation of the primary and secondary ambient air quality standards and the PSD provisions of the Clean Air Act Amendments. The AQD must grant construction and operation permits for surface coal mines which necessitates modeling emissions through the use of a computerized air quality dispersion model.

2. Surface coal mines were regulated by the Open Cut Land Reclamation Act of 1969 prior to the passage of the Wyoming Environmental Quality Act. The Land Quality Division (LQD) was given a major role in the control of surface mining activities through the 1978 Act. The LQD could in 1975 control water flow and quality; require conservation and replacement of topsoil; require backfill plans, slope stabilization terracing and approve appropriate land use; require toxic wastes to be buried and revegetated for a beneficial use; delay mining for archeological or paleontological surveys; require baseline studies on geology, land use, soils, topography, vegetation, water resources and wildlife; and require a reclamation bond. Prior to enactment of the Federal regulations, Wyoming had one of the most comprehensive and effective mined-land reclamation laws in the country. Now, the LQD is in the difficult process of trying to revise its laws and regulations to comply with the Federal requirements of OSM. Wyoming has submitted a state program and expects approval very shortly. This also presents the coal mining industry with problems as many states (such as Wyoming) have developed regulations that "clone" the Federal provisions. The Federal Regulations suspended or invalidated through litigation need not be included in the state programs as a condition of approval. The difficulty lies in getting the regulations that have been invalidated out of the State programs once they have been approved.

3. The Solid Waste Division is charged with implementing a state level solid waste disposal program. Surface coal mines must receive a solid waste permit to dump their domestic wastes or other wastes into the pit. The Federal Government has also developed "Hazardous and Toxic Waste" Regulations that will impact the coal industry through a potential state program in the future.

4. The Water Quality Division (WQD) issues NPDES permits for surface coal mines. These permits require strict control of effluents, requiring suspended solids to be 30 mg/l. The receiving water suspended-solids levels usually range from 150-200 mg/l and sometimes reach 30,000 mg/l. This usually brings about larger land disturbances by requiring additional control ponds or larger ponds. The WQD is also responsible for reviewing plans and specifications for sewage treatment lagoons, sediment control ponds, etc.

5. In addition to DEQ, the State of Wyoming also requires a potential operator to assess the socio-economic and environmental impacts of their proposed facility on the area of influence through a process known as the "Siting Council Permit Application." This is only done for those projects estimated to cost more than $79 million. It is the charge of the "Siting Council" to insure that all front end problems resulting from such development are mitigated by the prospective operator.

Other State agencies that a surface coal mine will have to deal with include the State Engineers Office relative to water rights and design specifications for drainage control structures; the Department of Fish and Game, for threatened or endangered species and wildlife, with mitigation actions to be taken before, during, and after surface mining; the State Archeologist to insure historic and archeologic preservation; the State Mine Inspector for health and safety; and the Health Department to insure a safe drinking water supply.

Once a property for coal development has been obtained (usually a coal lease or fee coal with surface ownership or lease), the real permitting process begins. Simply put, a permit is an effort to develop a strategy for development and operation that the coal operator and regulatory agency can agree upon prior to mining. The process, however, never is simple. Basically, there are three major parts in the OSM or State Programs in the permitting process.

1. The gathering of an environmental data base of conditions at the proposed mine site prior to any site disturbances. This is done for surface and groundwater quality and quantity, air quality and meteorology, topography and geology (drilling permit required by USGS), soils and vegetation, wildlife, aquatic life, cultural resources, land use, esthetics, prime farmlands and alluvial valley floors.

2. The effect on the base data of mining operations is established through an applicant-developed detailed mine progression plan showing all facilities including roads, structures, diversions, etc.

3. Post mining restoration activities with respect to the surface and subsurface estate is the final part of the application. This section includes descriptive activities of an applicant to achieve such items as approximate original contour, hydrologic function restoration, erosion control, reclamation species diversity with antive seed mixtures, fugitive dust control programs, subsidence control monitoring programs, etc. The "reclamation bond" requirements play a large role in deliberations on what the final reclaimed surface will look like.

Realistically, an applicant for a proposed surface coal mining permit could expect the permitting process to take at least three years from the point of initiating baseline studies to the point of having all of the necessary permits in hand to begin on-site construction. A basic flow chart for developing a proposed coal mine and associated time frames are outlined in Figure 1.

As can be seen, the permitting process is becoming very time consuming, costly, and difficult. I cannot overemphasize commencing work as early as possible on permit applications. Permitting a new surface coal mine will cost over $1,000,000.

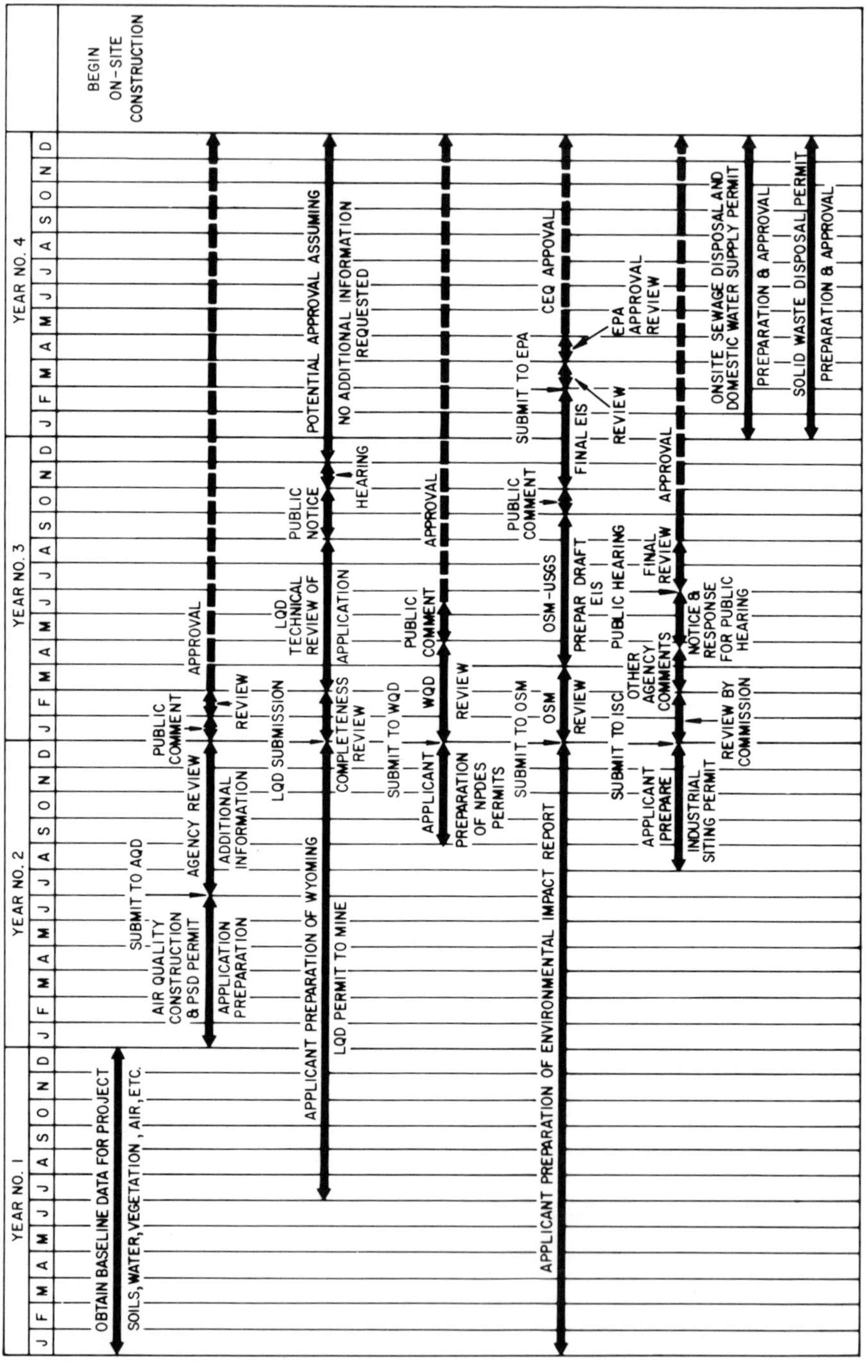

Figure 1. Length of time required to obtain state and federal permits.

Chapter 19

COAL SEAM MAPPING FOR QUALITY CONTROL

F. W. Metzger

The Carter Mining Company
Gillette, Wyoming

INTRODUCTION

Coal quality control in mining and mine planning has always been important and has become even more essential in recent years due to environmental considerations. The primary consumers of the sub-bituminous coals found in the Powder River Basin are electric utilities that require low sulfur coals to meet government air quality regulations. Electric generating plants are designed to burn coal of a specific quality, and the supplier of that coal must be able to guarantee shipments within specified quality tolerances.

Over the past several years, Exxon Coal USA and, more recently, The Carter Mining Company, a division of Exxon Coal USA, have established an evaluation and development program to accurately define the geology and quality of their coal properties. This requires an adequate drill hole density so that detailed correlations of the coal seam and the lithologic units, or benches, within the coal seam can be made. These units are recognizable in the core and also on geophysical logs.

Like any systematic approach, geologic modeling has been refined and improved through new technologies. Today, coal seam properties are mapped using computer applications. Rapid access and manipulation of these data allow for optimization of mining efforts. This presentation will demonstrate how a detailed mapping program can be important in both pre-mine development and in-pit quality control.

This procedure will be explained using data obtained from the Wyodak coal in the Powder River Basin. Figure 1 shows the location of the active and proposed mines along the Wyodak trend. The coal outcrop (or subcrop) and the 200-foot overburden line correspond to the approximate location of existing and proposed mines in this area. Strata in the Eastern Powder River Basin dip one degree to the west. Generally, the mines have opened in the areas of lowest stripping ratio, and mining has been toward the west. Current economics dictate about a 2:1 stripping ratio cutoff for Powder River Basin coals, and the western portion of most of these leases correspond roughly to this cutoff. As the demand for western coal increases and the economics of mining deeper coal improves, the mining companies will probably extend their leases westward when federal coal leasing is resumed.

In the mid to late 1960's, when the initial evaluation of the Wyodak trend began, some confusion existed in the correlation of Wyodak coal throughout the basin. Most geologists believed that when the Wyodak coal split, the two resulting coal seams were the Roland and Smith. Detailed drilling by industry and government since 1970 have confirmed that when the Wyodak seam splits, it becomes the Anderson and Canyon seams. This is the presently accepted nomenclature; however, many coal companies continue to use the old Roland-Smith terminology. In this presentation the upper and lower Wyodak will be referred to as the Roland and Smith coals. It should be emphasized, however, that all of the mining companies along the Wyodak trend mine the same coal, regardless of terminology.

RAWHIDE AREA GEOLOGY

In the area of The Carter Mining Company's Rawhide Mine, the Wyodak is split into two coals. The upper Roland seam averages about 25 feet thick, while the Smith averages about 70 feet (Figure 2). These coals are separated by a 3-10 foot parting which occasionally reaches thicknesses of greater than 100 feet. The characteristics of the coal allow the Roland to be broken into five units, or benches, and the Smith into three. These units can be clearly seen on geophysical logs, in the core, and also in the pit.

Figure 3 shows a generalized north-south cross-section through the Rawhide area. The five benches of the Roland are present in the northern portion of the area of investigation. The two lower benches gradually become more clayey in the southern portion of the area and, when a thickening or thinning of the coal is a depositional feature, a relatively uniform thickening or thinning of the benches follows. However, if the thinning is related to post-depositional erosion or weathering, the lower units maintain their thickness and the upper units are truncated.

The division into lithologic units not only aids in studying the

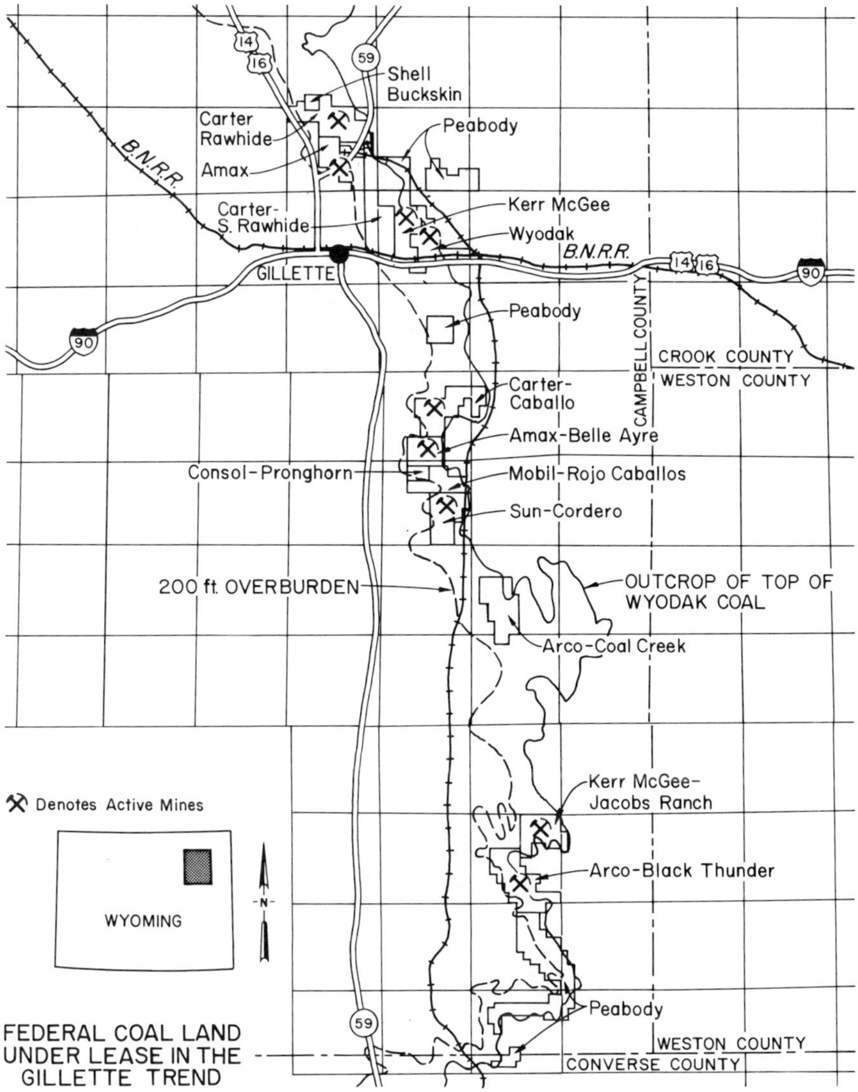

Figure 1. Federal coal land under lease in the Gillette Trend.

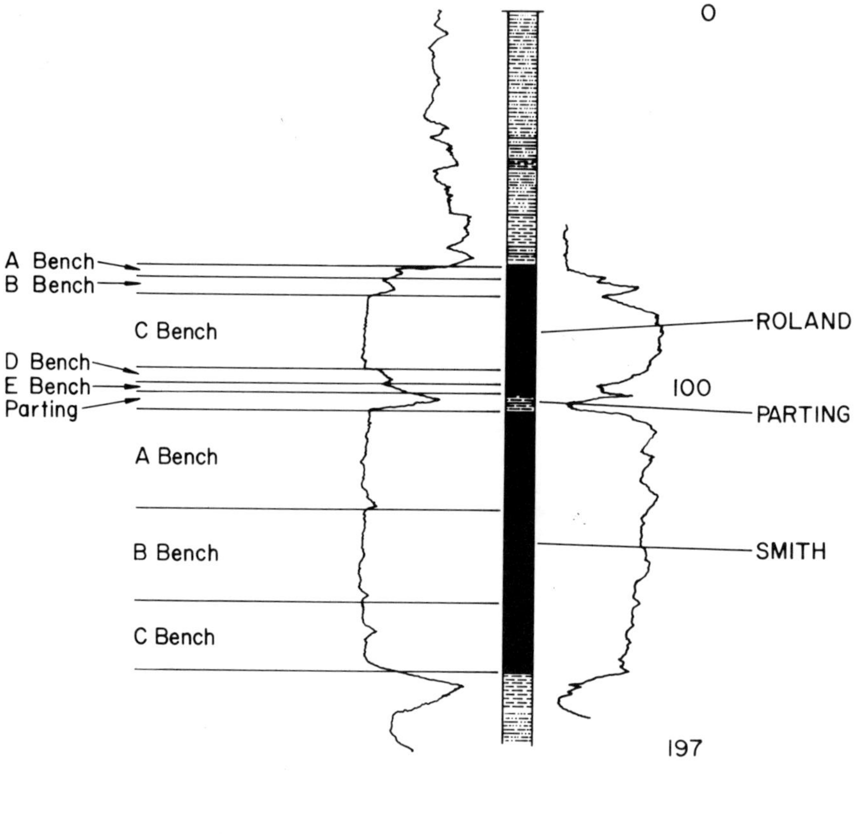

Figure 2. Geological cross-section for the Wyodak seam.

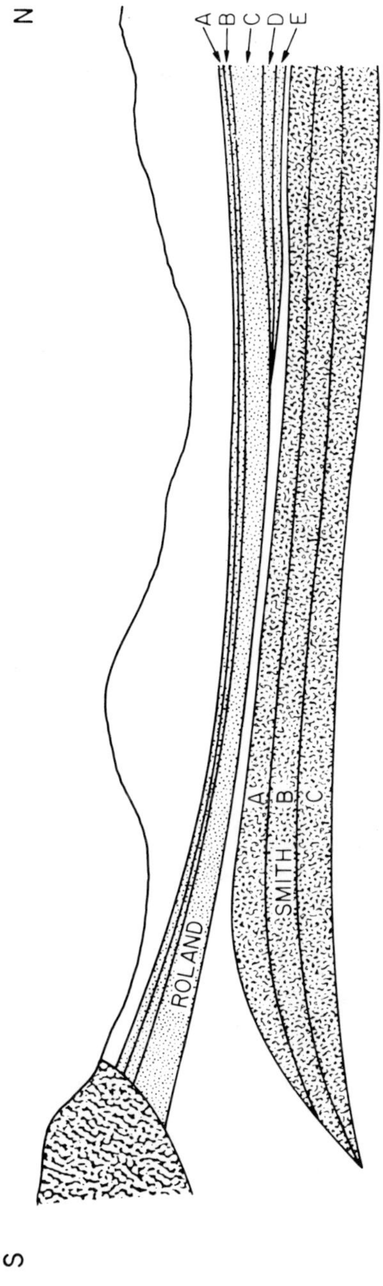

Figure 3. Generalized cross-section of the Rawhide area.

depositional and post-depositional characteristics of the coal, it also aids in building computer models of these characteristics.

DATA STORAGE AND MANIPULATION

The geologic data base contains all drill hole information. Location data, collar elevation, and depth to top and base of each lithologic bench are stored in the data base. Coal quality parameters are analyzed and stored for each lithologic bench. From this basic raw data, all other parameters such as sub-sea elevations and coal quality statistics can be calculated. This data is then accessed for mapping. The Carter Mining Company uses a complex series of computer programs to reduce the data into contour maps. These programs take randomly spaced data and generate a uniform grid spacing (Figure 4).

Coal companies usually drill two types of holes: (1) rotary drill holes which are used to define coal seam correlations and continuity, and (2) core holes from which an actual sample of the coal is retrieved and sent to a lab for coal quality determinations. Typically, a company will drill many more rotary drill holes than cores. This results in a variation in drill hole data distribution. Figure 4 illustrates how random data distribution are fitted to a uniform grid pattern. At every grid intersection, a data value is generated based on surrounding drill information. The grid points are weighed by a mathematical algorithm to account for spacing and distance from the grid intersection. With a uniform grid system and complete set of data at each point, contouring and grid arithmetic or manipulation can begin.

Gridding and mapping of each lithologic bench are the first steps in geologic model building. Geologic interpretation is required at each step to incorporate supplementary data not supplied by the drill holes. Information such as outcrops, geophysical data to define burn edges or subcrops, and geological bias to account for regional trends must be used. Without this additional information, computer models are merely mathematical representations of the data and, unfortunately, geologic bodies cannot be totally defined mathematically. The grids are mathematical models of the geology biased by a geologist.

Once the thickness and quality of each bench have been gridded and mapped, the grids are composited as shown in Figure 5. The composite coal seam thickness is simply an arithmetic addition of the bench grids. The quality grids cannot be simply added because the benches vary in thickness. If they were added, a thin poor quality bench would be overly represented in the composite. For this reason, each bench quality is multiplied by the bench thickness. The resultant grid is not mapped. It is an intermediate grid referred to as the quality factor grid. The bench quality factor grids are then added, and this sum is divided by the composite thickness. This re-

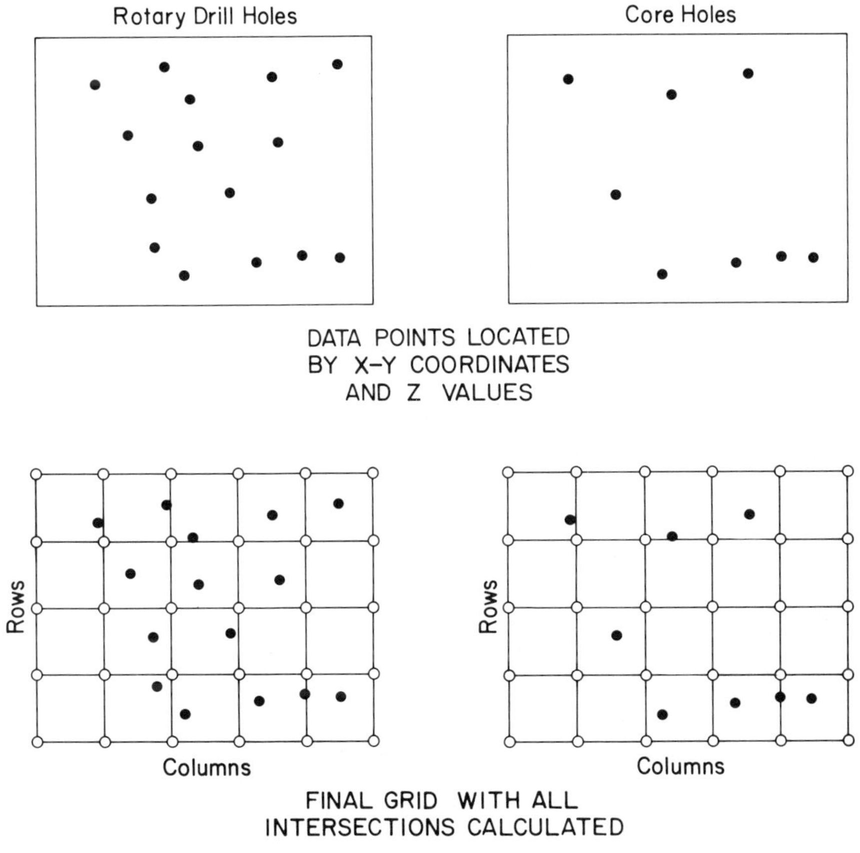

Figure 4. Grid determination from drill hole data.

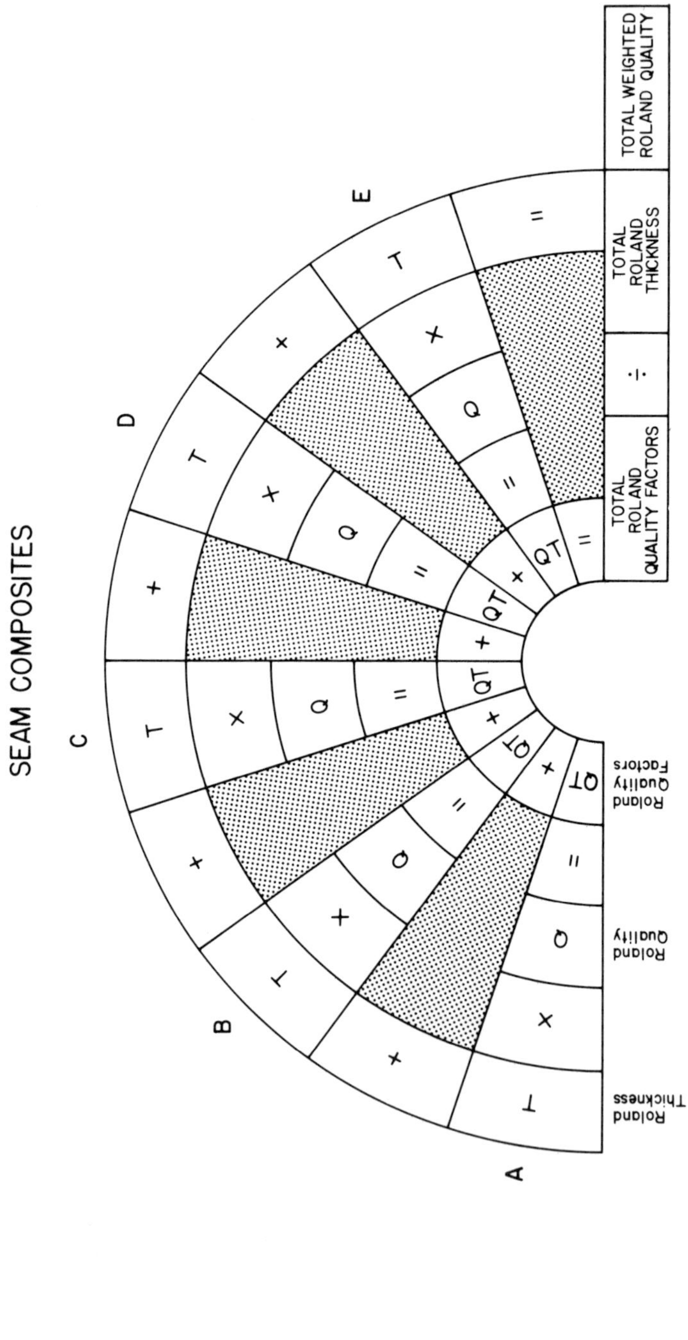

Figure 5. Calulation of seam composites.

sults in a correctly weighted composite seam quality. If other variables such as seam density are known, the quality grids should also be weighted by these factors. This is especially important in lignites where ash percentages vary greatly. The Wyodak seam is relatively uniform and, hence, thickness is the only weighting factor used.

GEOLOGIC MODEL

Figure 6 shows the Roland seam thickness in the general vicinity of the Carter Mining Company's Rawhide Mine. It is a composite of the five benches of the Roland. As previously mentioned, the D and E benches of the Roland develop in the northern portion of the Rawhide area. The thickness isopach indicates that the Roland is thicker than the average 25 feet where these two benches have developed. The depositional or subcropped edge of the Roland seam is defined by both drill hole data, surface work and geophysical interpretations. In the northwestern portion of the map, the effects of a deeply incised drainage on the coal are apparent.

Figure 7 shows the Smith coal thickness. Again, the edges are defined by both drill holes and supplemental data. The thickest Smith coal is in the southern portion of the area, while the thickest Roland occurs in the northern portion of the area. Compositing these two seams into the Wyodak seam results in a fairly uniform thickness of coal at just over 100 feet (Figure 8).

By simply subtracting the grid of the structural top of coal from the grid of topography, an overburden isopach can be generated (Figure 9). Since the dip of this coal is very slightly to the west, the overburden generally reflects topography. The hills show overburden cover greater than 200 feet and the drainages an overburden cover of less than 50 feet. In the western portion of the area, the deeply incised drainage referred to earlier is obvious.

Figure 10 shows the as-shipped BTU's for the composite Wyodak coal seam. It assumes 100 percent recovery of the coal with no rock dilution. There is a distinct gradation from low BTU to high BTU from east to west. In general, the net heat content increases down dip.

Figure 11 shows the weight percentage of ash on an as-shipped basis. As expected, low ash corresponds with high BTU.

Figure 12 shows the as-shipped weight percentage sulfur. Note that high ash coal generally coincides with high sulfur coal. Also, the total sulfur content decreases down dip.

Building a geologic model by mapping lithologic units and compositing to arrive at the full seam allows the mine engineer and quality control engineer a great deal of flexibility. A variety of

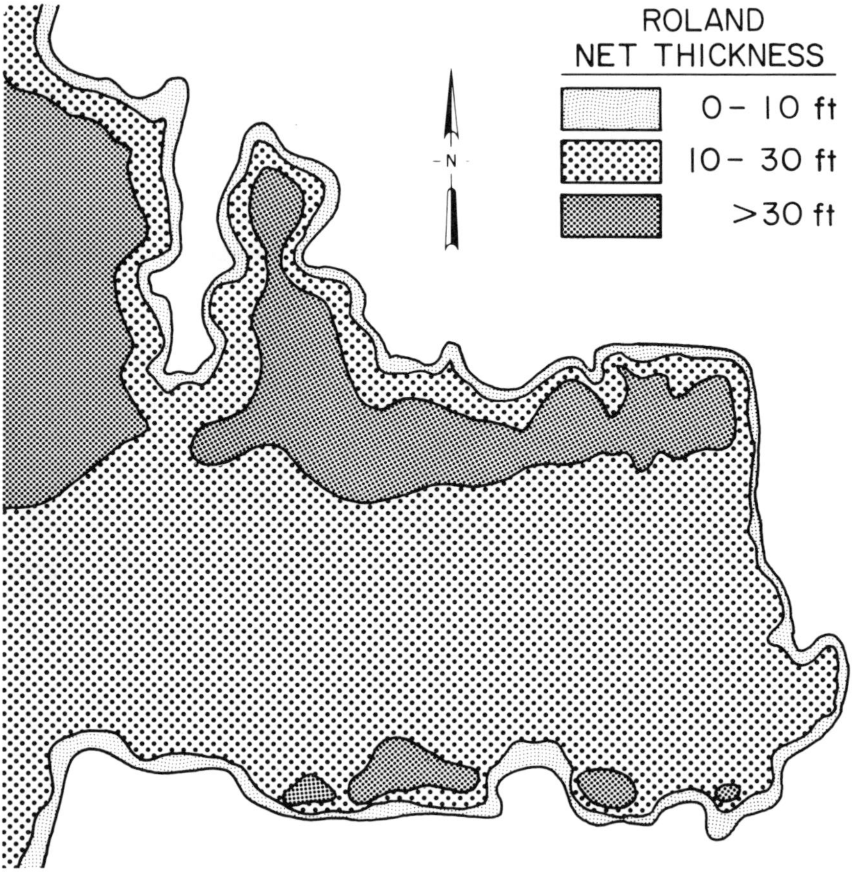

Figure 6. Thickness of the Roland seam.

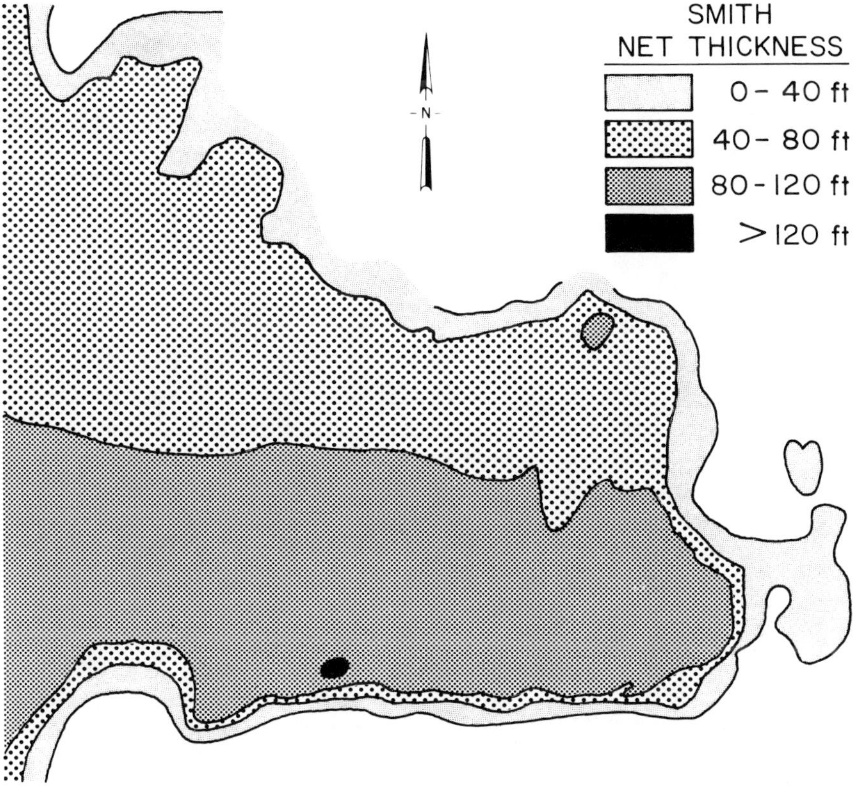

Figure 7. Net thickness of the Smith seam.

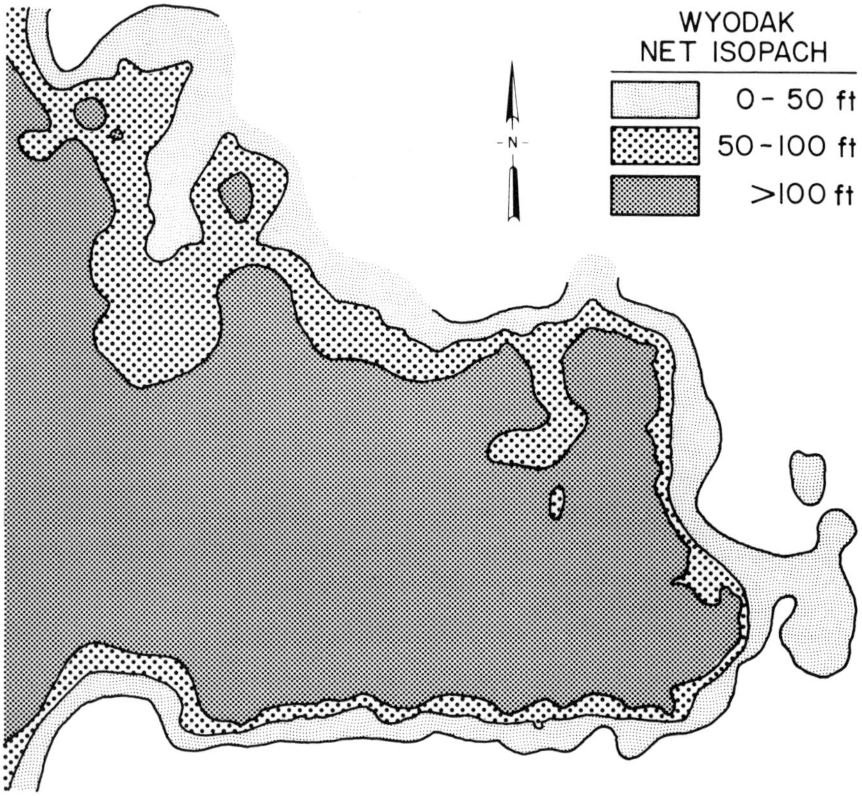

Figure 8. Thickness isopach for the Wyodak seam.

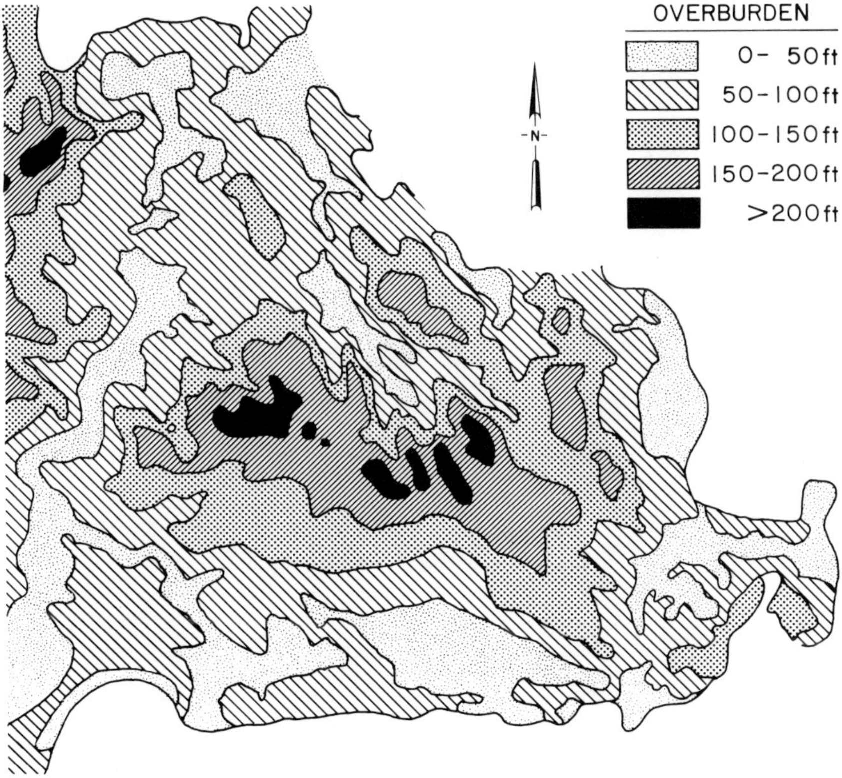

Figure 9. Overburden isopach for the Wyodak seam.

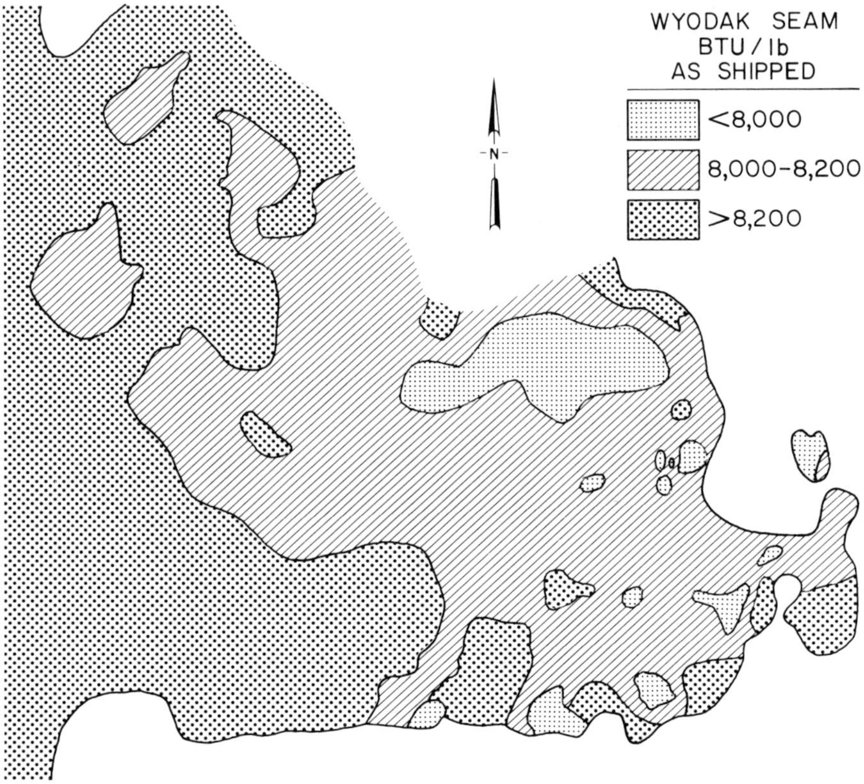

Figure 10. Fuel value of the Wyodak seam.

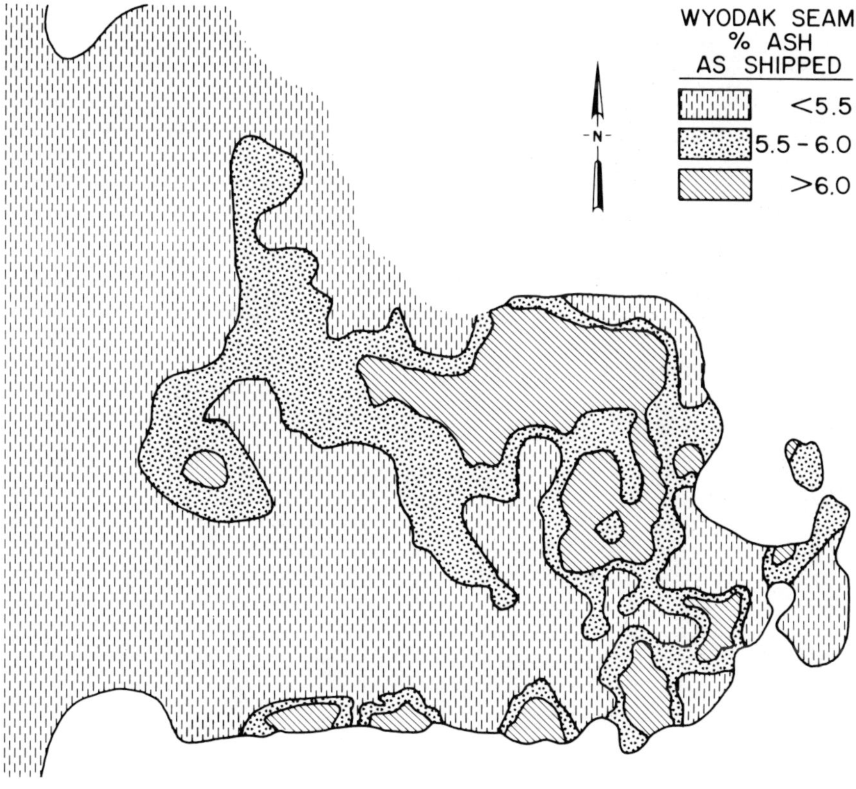

Figure 11. Ash content for the composite Wyodak seam.

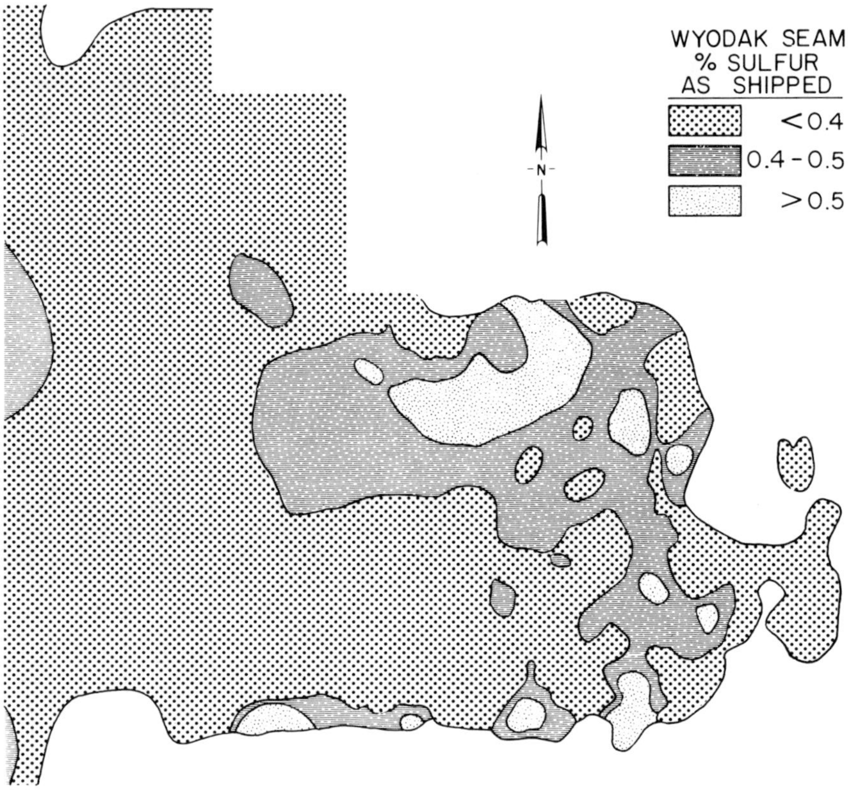

Figure 12. Sulfur content of the Wyodak seam.

different mining schemes can be simulated by adding or subtracting benches in the model. As an example, assume that the top and bottom benches of the Roland coal are not mineable (Figure 2). Calculations of reserves for such a mine plan results in a 93.5 percent recovery rate. Comparing Figure 10 to Figure 13, a significant improvement in the overall BTU content is obvious. Also comparing Figure 12 to Figure 14, a marked overall reduction in the percent of sulfur is apparent. Therefore, a slight decrease in coal recovery improves the overall block average quality significantly. With detailed mine planning, only select areas will require some coal loss, and a balance can be obtained between high quality and high recovery percentage. Incorporating blast hole sample data into the grids, the pit quality control engineer gains additional data for estimation of daily coal quality variations.

CONCLUSIONS

Once the most important coal quality parameters are mapped, it is a logical extension to map other parameters such as ash element percentages. The ash elements are calculated as a percent of the total ash and must be weighted by both seam thickness and ash percent. This weighting procedure would be very tedious and time consuming if calculated and mapped by hand. However, with the computer and accurate data files, the process is relatively easy.

Detailed mapping of coal seam lithologies with computer applications have proven very beneficial to The Carter Mining Company. It has enabled providing customers with better estimates of anticipated coal quality over the life of the mine as well as aided the mining engineers in achieving maximum recovery coupled with production of the highest quality coal possible.

ACKNOWLEDGEMENTS

Several employees of The Carter Mining Company were helpful in the preparation of this paper. Dan Duller designed and drafted the figures. T. D. Smothermon, M. R. Rayphole, J. M. Farmen, and J. D. Brown reviewed the manuscript and offered several suggestions. Mrs. B. J. Wyninger typed the manuscript.

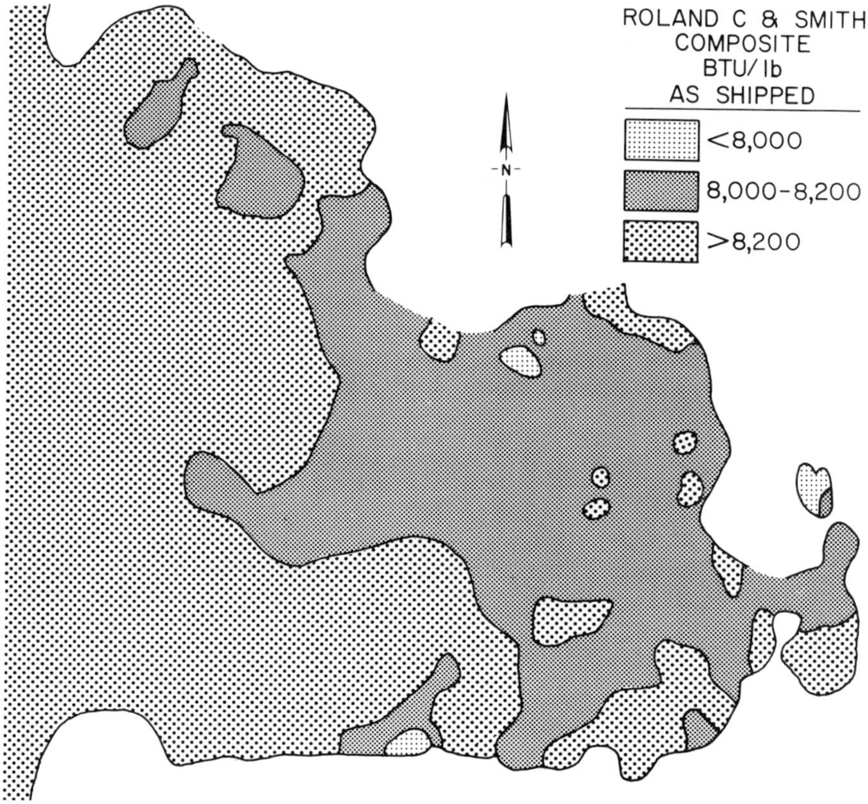

Figure 13. Fuel value of a composite consisting of the Roland C bench and the Smith seam.

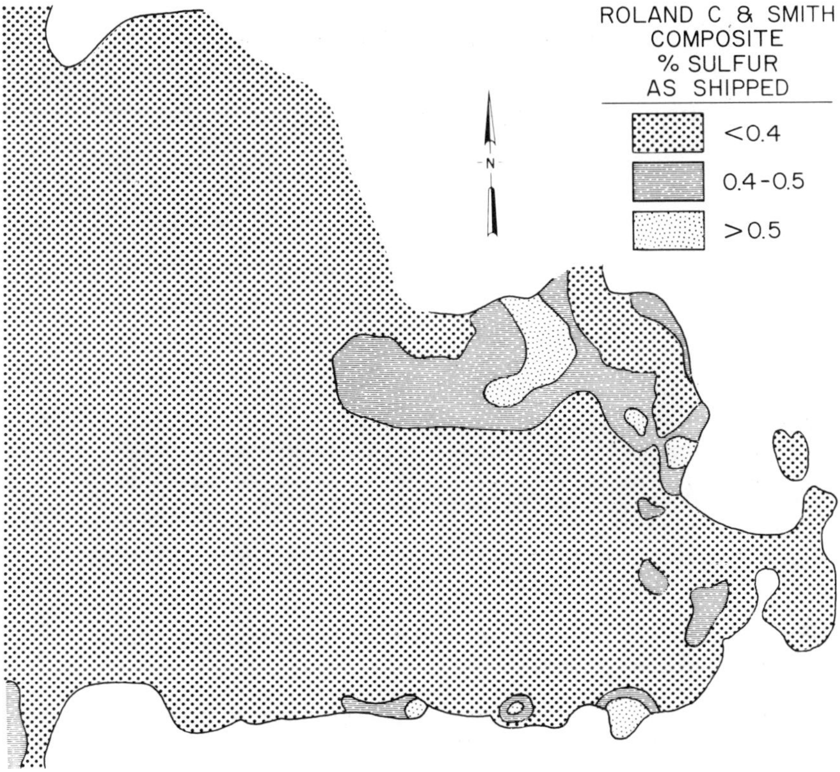

Figure 14. Sulfur content of a composite consisting of the Roland C bench and the Smith seam.

Chapter 20

THE USE OF PHOTOGRAMMETRY COUPLED WITH COMPUTER MODELING TECHNIQUES AS AN AID IN SURFACE MINE PLANNING

D. A. Farmen and P. Tilston

The Carter Mining Company
Gillette, Wyoming

INTRODUCTION

In early 1978, engineering personnel at The Carter Mining Company began using photogrammetry coupled with a computer modeling technique to aid in mine planning efforts. The decision to develop this approach was based on the following needs:

1. A method to accurately capture excavated quantities for reporting purposes,

2. A method to produce accurate up-to-date maps for planning purposes, and

3. A computer-based mine-planning technique that would be compatible with techniques used in the past.

The system developed utilizes a photogrammetric contractor to provide thousands of aerially-surveyed data points for input into the computer system. Each month the Rawhide and Caballo mines are flown by the photogrammetric contractor. The terrain of the active pit area is digitized and punched out on computer cards. This information is read into The Carter Mining Company's computer facilities, and an updated two-dimensional topographic grid model of the pit is generated. This surface is used to generate quantities moved during the month, to update isopachs, and to update quantities contained in fixed block progression polygons used for forecasting purposes.

The system has greatly reduced the amount of manhours spent in field surveying and provides a useful tool for expeditiously evaluating many different forecast cases. It also provides the engineering personnel with professional looking computer drafted maps.

This paper presents an outline of the system developed by The Carter Mining Company's engineering personnel to aid them in mine planning efforts. As such, the discussion following will cover:

1. A description of the computer modeling package,

2. A description of how the computer system is used to develop the geologic grid models,

3. A description of how photogrammetry is used to generate monthly maps and material quantities, and

4. A description of how this system can be used to evaluate many different forecast cases.

THE COMPUTER MODELING SYSTEM

The computer modeling system uses a two-dimensional gridding technique to approximate a surface from various types of mappable input data.

The system actually consists of approximately fifty basic computer programs which are interfaced and monitored by a master program. This program allows various programs and program options to be executed easily during a single computer run. Functionally, the programs can be classified into five groups as shown in Table 1. These groups are: data processing programs, data display programs, surface representation programs, surface processing and modification programs, and surface display programs.

The first group of programs enables input data such as core holes or digitized contour lines to be manipulated, edited and corrected in preparation for future transformation into grid form. Each data point record consists of X and Y coordinates and up to 97 additional words of information such as alphameric identification or numerical values. The data can be displayed for verification purposes in either printed or plotted output form by invoking the second group of programs, data display.

Following verification that the input data is satisfactory, the third group of programs can be used to generate two-dimensional approximations of the surface represented by the data. Each surface or grid is generated by a complex algorithm which assigns values to each evenly spaced grid intersection based on the values of the surrounding data points. Figure 1 outlines the gridding procedure. The grids generated by the surface representation programs can be modified further and processed by using the programs contained in the fourth group.

Table 1. COMPUTER PROGRAMS BY GROUP

I. DATA PROCESSING

Data Edit
Data Preparation
Data Correction
List Data File

II. DATA DISPLAY

Map Annotations
Print Base Map

III. SURFACE REPRESENTATION

Grid File Initiation
Numerical Approximation
Grid Partitioning
Gridding

IV. SURFACE PROCESSING AND MODIFICATION

Grid Reduction
Surface Operations
Grid Correction
Smoothing
Integration
List Grid Names
Grid To Data

V. SURFACE DISPLAY

Plotter Contouring
Cross Section
Perspective View
Plot Grid Values

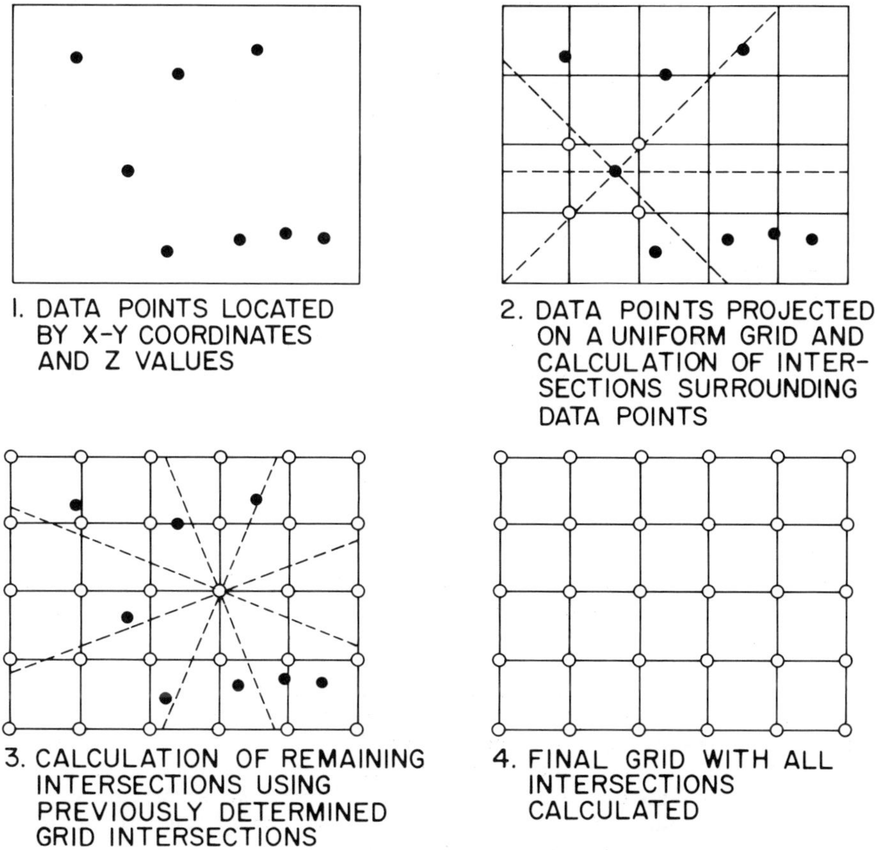

Figure 1. Generalized gridding procedure.

One of the important functions of these programs is to allow mathematical operations to be performed on one or two grids. For example, a base of coal grid can be subtracted from a top of coal grid to yield a coal isopach. Mining losses can then be subtracted from the isopach to generate a mineable coal thickness grid. Another important function of these programs is the integration routine. This program calculates volumes and areas within one or more polygons given a thickness grid as input.

The ability to display surfaces is the function of the last group of programs. These programs will generate computer maps in the form of plan views, cross-sections, or three-dimensional perspective views. Many options are available so that high quality, professional looking maps can be plotted.

Generally, the system being used can be classified into two distinct subsystems. Figure 2 shows the relationship between the data subsystem and the grid subsystem.

The remainder of this presentation will focus on how The Carter Mining Company engineering group has applied this system to its particular needs.

BASELINE GRID MODELS

The first step in establishing a computerized planning base is the establishment of grids from base line data. Drill hole data from exploration drilling programs such as drill hole location, collar elevation, depth to the different structural horizons, and hole depth are entered into a data base stored on the computer.

Grids for the top and bottom of the coal seam are established next. Drill hole locations, collar elevations, and depths to the top of coal and base of coal are retrieved from the data base. Depths to the top of the coal and bottom of the coal are subtracted from the collar elevation to establish elevations for those structures. This information is then gridded to form a top of coal structure grid and a base of coal structure grid. All grids are generated on 20-foot centers.

A topography grid can be established by either digitizing contour lines from existing topographic maps or by obtaining data generated directly from stereo models.

A topsoil thickness grid is established by digitizing soil thicknesses from soil thickness maps generated from previous soil surveys. These data are then gridded out to form a topsoil thickness grid.

From these grids a series of other grids are built. The topsoil

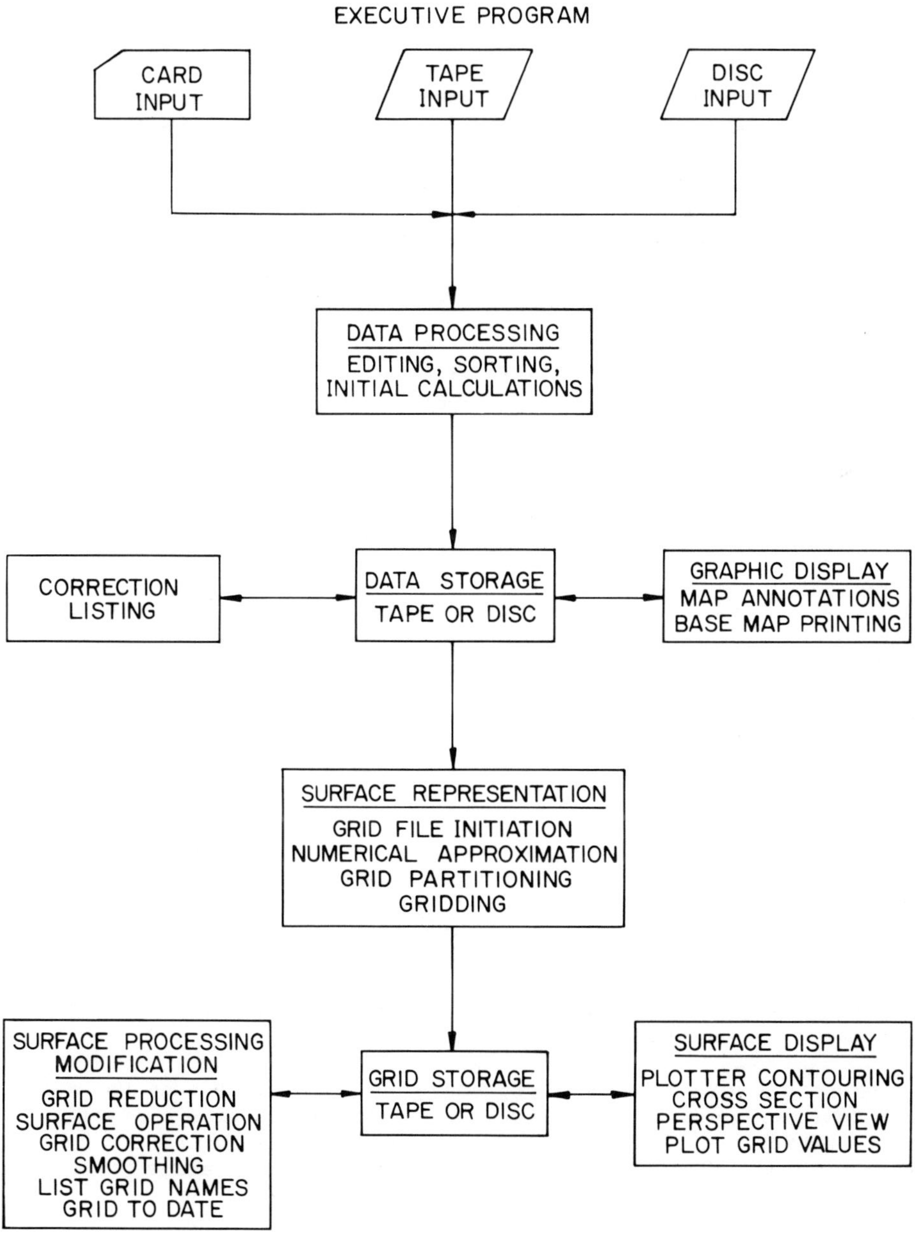

Figure 2. Flow of the computer system.

thickness grid is subtracted from the topography grid to yield a top of overburden grid; the top of coal grid is subtracted from the top of overburden to yield an overburden thickness grid, and the base of coal grid is subtracted from the top of coal grid to yield a coal thickness grid.

A backfill thickness grid is also constructed to complete the planning base. To do this it is first necessary to construct a reclaimed topography grid. This procedure is essentially a trial-and-error method. A post-mining topographic map is drawn incorporating slopes at less than 20 percent and re-establishing existing drainages. The contours from this map are digitized. These data are then gridded out to form a preliminary post-mining topography grid.

The base of coal grid is then subtracted from the post-mining topography grid to yield a thickness grid representing backfilling requirements. This backfill thickness grid is integrated within a polygon formed by the mining and disturbance limits at the edges of the pit. This yields the backfill volume (defined at the bottom by the base of coal, at the sides by the mining and disturbance limits, and at the top by the post-mining topography) required to achieve this particular post-mining topography. This volume is then reduced by the anticipated swell factor. The overburden thickness grid is integrated within a polygon formed by the mining and disturbance limits at the edges of the pit to yield an overburden removal volume. This volume is compared to the required backfill volume. The post-mining topography is then raised or lowered, and the process is repeated until the backfill volume matches the overburden removal volume.

PHOTOGRAMMETRY AND MONTHLY MAPPING PROCEDURES

Each month the Rawhide and Caballo Mines are flown by a photogrammetric contractor. The active pit area is digitized, and the location and elevation of each data point are punched onto computer cards. The engineering groups at each mine site use these data, together with supplemental survey data from the field to generate monthly quantities and an updated pit map with a minimal amount of effort. Figure 3 outlines the sequencing of the major components comprising this procedure.

The initial step in the procedure involves converting the card data which is in the form of elevations at offsets from a baseline into northing, easting, and elevation format for input to the computer program. Concurrently, a polygon is generated around the outer edge of the digitized area.

The second step involves inputting the polygon and the previous month's data file to the computer. All data points from the previous month which fall within the polygon boundary are deleted.

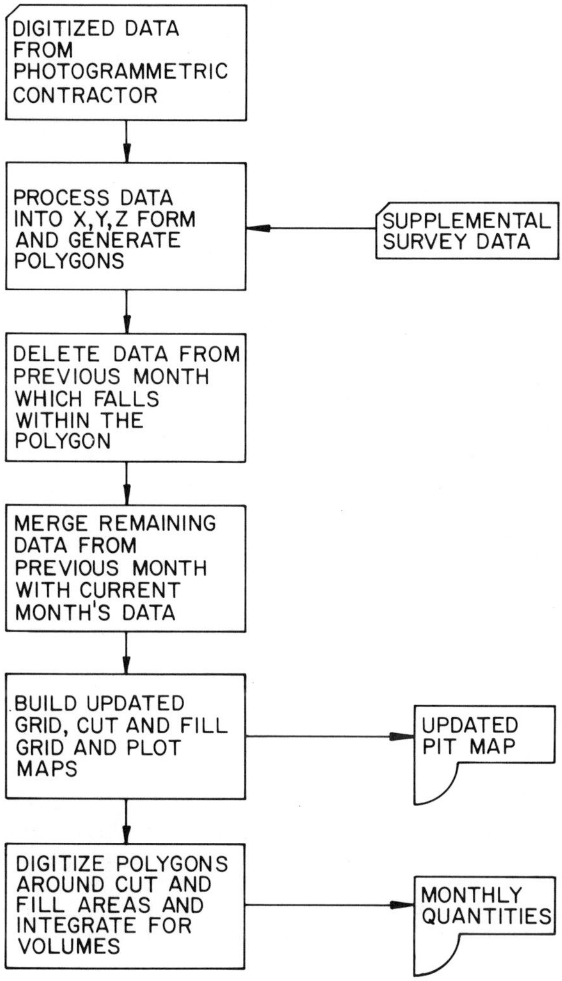

Figure 3. Monthly mapping procedure.

The next step is to insert the current month's data into the remaining data from the previous month to create a current month's base data file.

The surface representation programs are then used to build a grid of the pit surface from the base data file. This grid is subtracted from the previous month's grid to generate a cut and fill grid representing the material movement during the month. The updated pit grid is then plotted for verification purposes. Following verification, the cut and fill grid is used to calculate the quantities of coal and overburden removed and the backfill volume into which the overburden was placed.

This is done by first plotting the cut and fill grid and then digitizing polygons around the areas in coal, overburden, and backfill. These polygons are integrated over this grid to yield the cut and fill volumes.

The monthly updating procedure also gives The Carter Mining Company's mine engineering groups two additional sources of information. First, by comparing the overburden volume removed each month to the backfill volume available, it is possible to collect useful data on overburden swell factors. Secondly, surfaces representing as-mined structures at the top and base of coal can be generated from these data together with supplemental field survey data. By subtracting these structures, an as-mined coal thickness grid is obtained. This grid can be integrated and the results compared to those obtained from the geologic coal thickness grid so that recovery factors can be calculated.

Three dimensional perspective views of the mining operation can also be generated for use as visual aids. Figure 4 is an example of a three dimensional plot of an active pit.

FORECASTING

Earlier reference was made with regards to how this system aids in forecasting through the updating of quantities contained in fixed-block progression polygons. This can be extremely useful in evaluating several different production rates at any given point in time. These different cases can arise from many different reasons such as:

1. Radical variations in shipments due to transportation problems,

2. Customer lifting more (or less) coal than currently contracted, and

3. Short duration sales (spot).

Any of the above can result in a large deviation from previous fore-

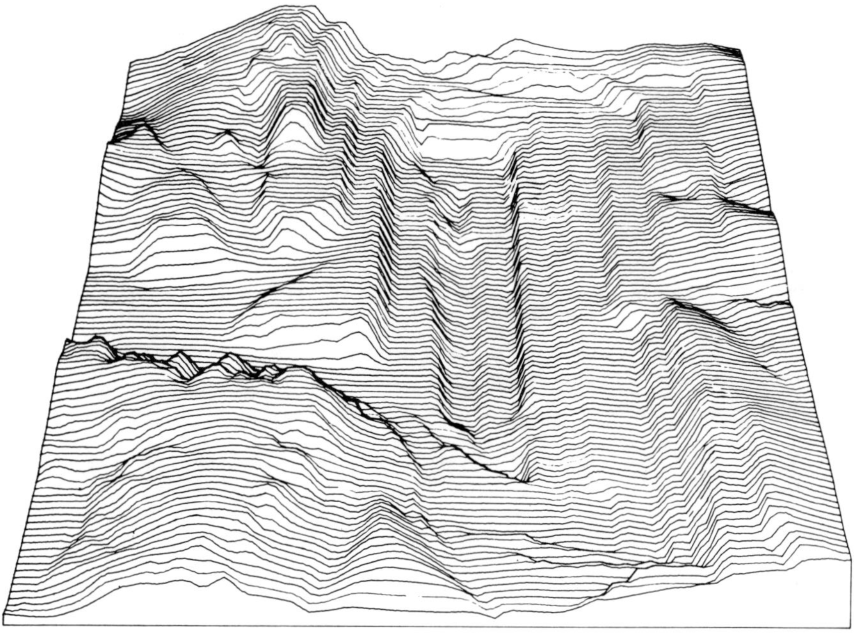

Figure 4. Typical three dimensional perspective view.

casting efforts, so a quick and convenient method is needed to re-evaluate projections.

Forecasting at the mine site is done on a fixed-block progression. Under this concept the mining area is divided up into strips or blocks representing typical advances. Figures 5 through 9 illustrate this procedure. Figure 5 shows the mining limits defined for this example. Figure 6 shows the coal progression. Each block is numbered sequentially so that the progression can be followed. The coal thickness grid is then integrated to determine the coal tonnage contained in each block.

Figure 7 shows the overburden progression. Each block is numbered sequentially so that the progression can be followed. The L-blocks represent material thin enough that removal will be done with front-end loaders. The SEU blocks represent material removed by the shovel on the upper bench. The SEL blocks and SW blocks represent material removed by the shovel on the lower bench immediately above the coal seam. The L blocks, SEL blocks, and SW blocks are integrated on an overburden thickness grid which was created incorporating no material above the upper-lower bench split to determine the overburden yardages contained in each block. The SEU blocks are integrated on an overburden thickness grid which was created incorporating no material below the upper-lower bench split.

Figure 8 shows the topsoil progression. Each block is numbered sequentially so that the progression can be followed. A topsoil thickness grid which was created from auger hole information ahead of the pit is integrated to determine the topsoil yardages contained in each block.

Figure 9 shows the backfill progression. Each block is numbered sequentially so that the progression can be followed. A backfill thickness grid which was created by subtracting the base of coal from the post-mining topography is integrated to determine the amount of backfill yardage which can be dumped into each block.

A material movement table is then set up such that to mine out any given block of coal, corresponding topsoil, front-end loader, shovel upper bench, and shovel lower bench blocks must also be excavated. A required in-pit inventory and operating set room distances are built into this table. From this information the material movement can then be easily sequenced given a coal production schedule. Backfill location can be determined by swelling the overburden removed during discreet time periods and sequencing these volumes through the backfill blocks.

The creation of updated topography and overburden thickness grids each month, utilizing the monthly photogrammetry results, then allows the user to update the quantities remaining in each fixed block polygon. This procedure is somewhat complex but is accomplished quickly and conveniently in three computer runs.

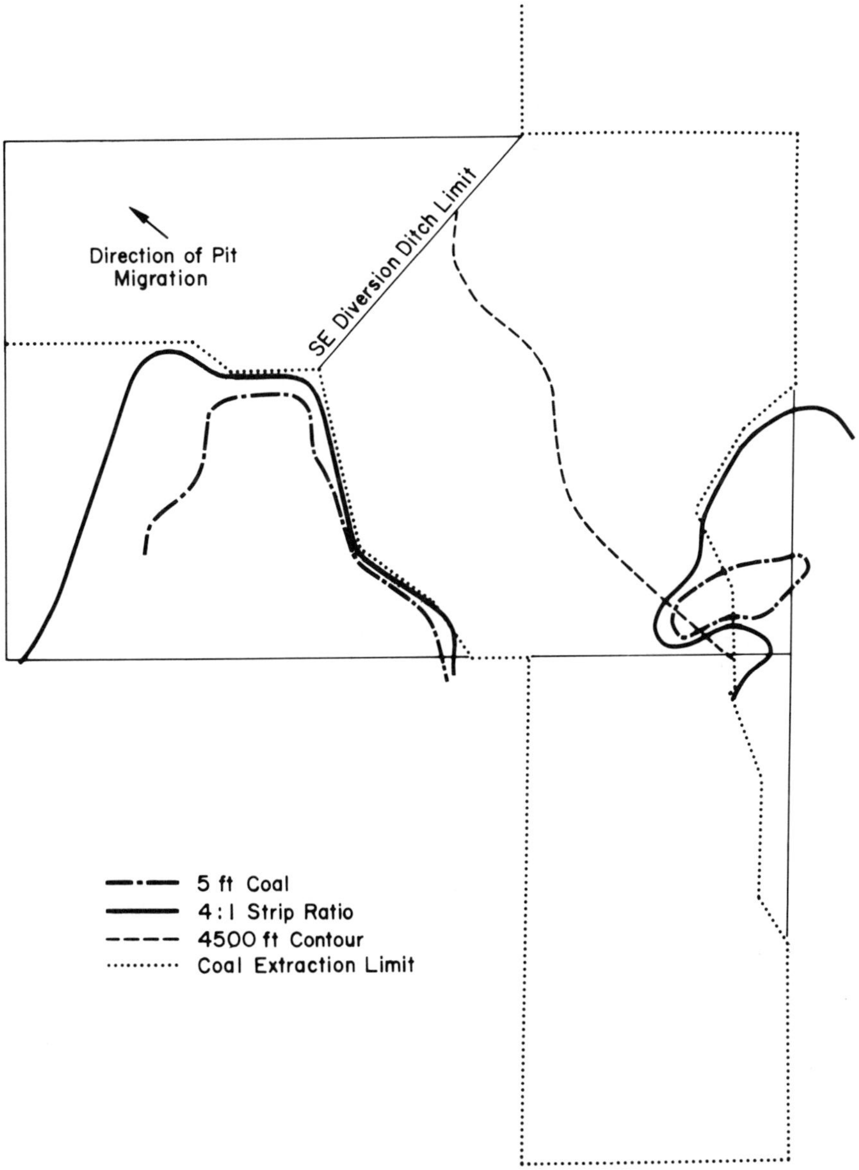

Figure 5. Pit cut-off limits.

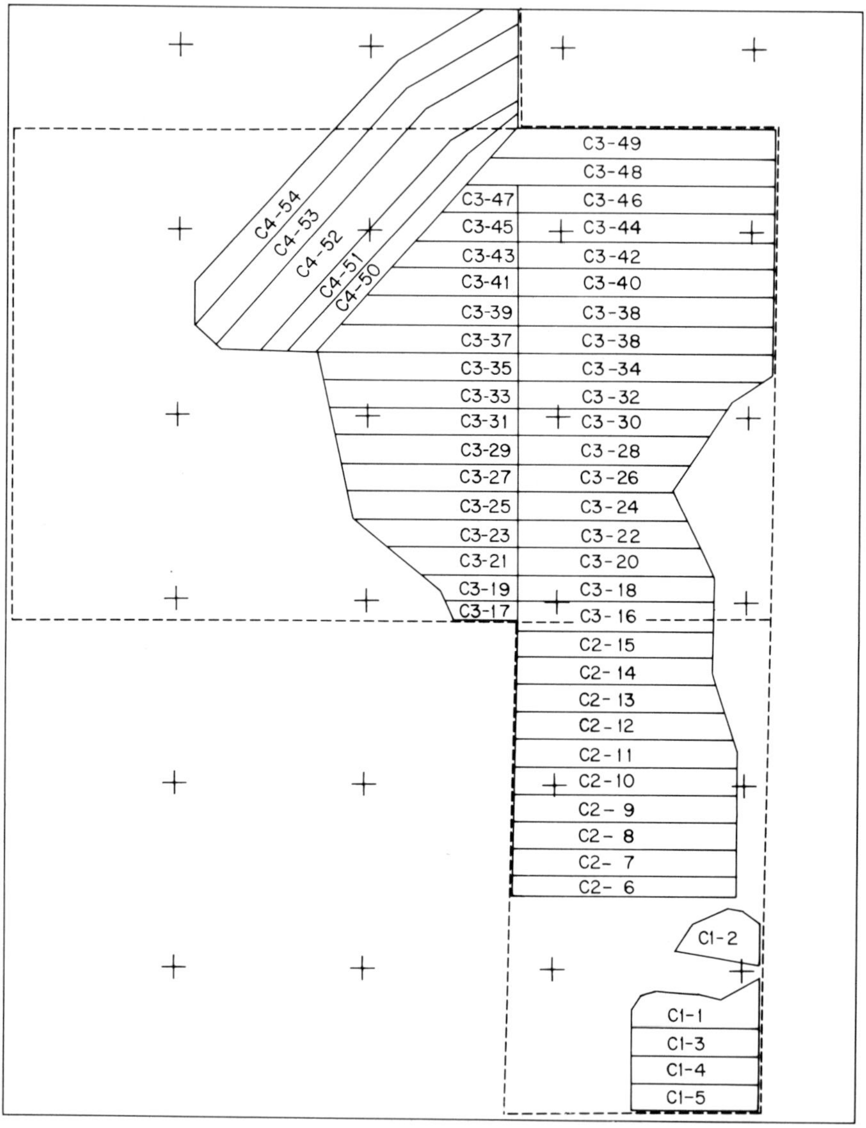

Figure 6. Planning polygons for coal.

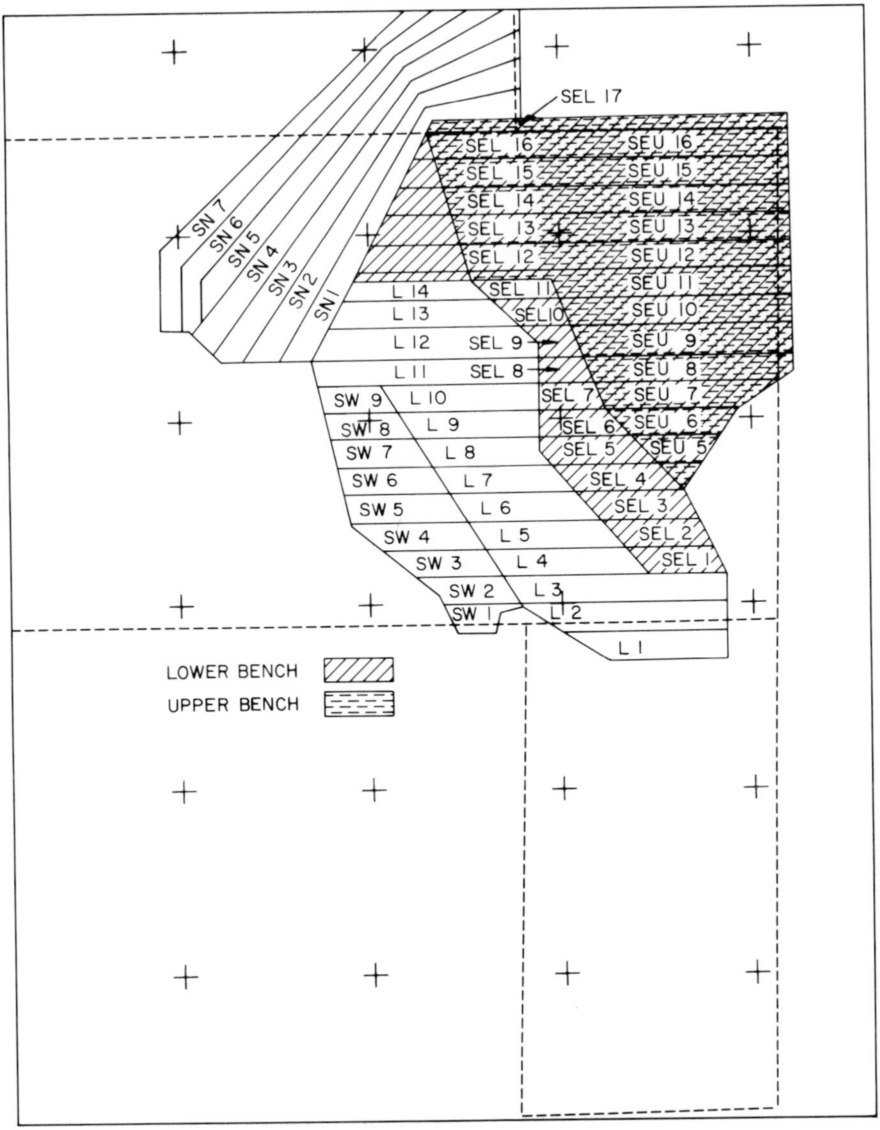

Figure 7. Planning polygons for overburden.

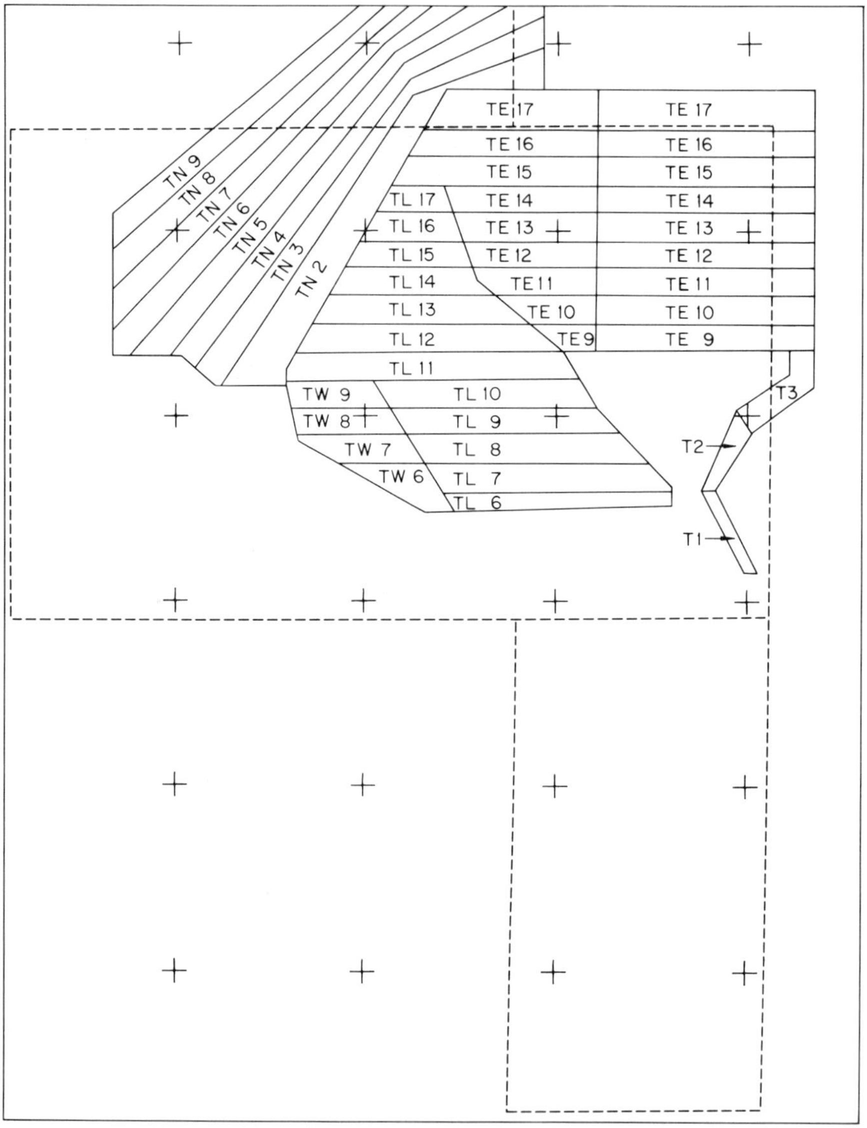

Figure 8. Planning polygons for topsoil.

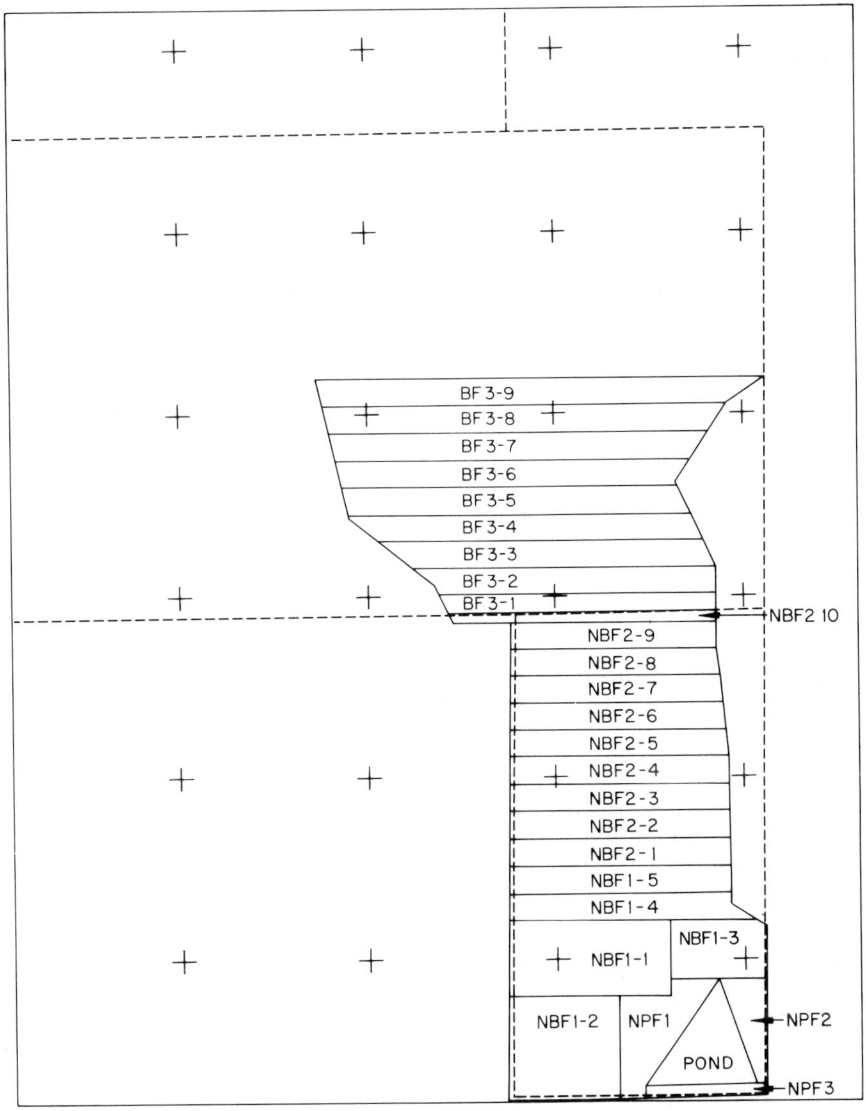

Figure 9. Planning polygons for backfill.

The first computer run updates topsoil and overburden volumes contained in their fixed block polygons. To update the topsoil quantities, the first step is to create an updated topsoil thickness grid. This is accomplished by subtracting the top of overburden grid from the updated monthly topography grid. This creates a topsoil thickness grid which incorporates areas from which topsoil has been stripped. The topsoil blocks are integrated on this updated topsoil thickness grid to yield updated quantities.

To update the overburden quantities requires a series of grids to be built. The first step is to create an updated overburden isopach. This is accomplished by subtracting the updated topsoil thickness grid from the overburden thickness grid created during the monthly procedure. This yields an updated overburden thickness grid which does not contain topsoil. The front-end loader and single shovel bench blocks (L and SW) are then integrated over this grid to yield updated quantities.

The next step is to create an updated top of overburden grid. This is accomplished by subtracting the updated topsoil thickness grid from the topography grid created during the monthly procedure. A special dummy grid established with values corresponding to the upper-lower bench split elevations is then subtracted from the top of overburden grid to yield an updated overburden thickness grid for the upper bench. The upper shovel bench blocks (SEU) are then integrated over this grid to yield updated quantities.

The next step involves subtracting the top of coal grid from the top of overburden grid below the elevation corresponding to the upper-lower bench split. This yields an updated lower bench overburden thickness grid. The lower shovel bench blocks (SEL) are then integrated over this grid to yield updated quantities.

A second computer run is required to update the coal quantities. The first step in this procedure is to lay out and digitize a polygon corresponding to the block of coal removed during the month. The coal thickness grid is then corrected within this polygon to account for coal already removed. The coal blocks are then integrated over this grid to yield updated quantities.

The third computer run is required to update the backfill room quantities. The first step in this procedure is to lay out and digitize a polygon corresponding to the backfill advance during the month. The backfill thickness grid is then corrected within this polygon to account for areas already backfilled. The backfill blocks are then integrated over this grid to yield updated quantities.

This entire procedure sounds somewhat complicated but can be accomplished in a matter of minutes. The procedure supplies the updated quantities contained within a fixed block progression which can be used to generate a revised material movement table. This can be used to sequence new production schedules or to re-evaluate if current production goals can be met.

SUMMARY

This discussion has focused on a system developed by The Carter Mining Company's engineering personnel which utilizes a computer modeling technique to create a baseline grid model, generate monthly quantities from photogrammetric information, and create a valuable forecasting tool using a fixed block progression concept. Since its inception, the system has undergone numerous changes in order to streamline the procedure and upgrade its usefulness. Future plans will be directed toward the development of a scheduling package to further enhance mine planning activities.

To date, engineering personnel at Caballo and Rawhide have been pleased with the results obtained from this system and have experienced a noticeable reduction in the time needed to generate these results. Consequently, more time is available to work the more detailed and important mine engineering problems at hand.

ACKNOWLEDGMENTS

Several Carter Mining Company employees aided in the preparation of this paper. Mr. T. D. Smothermon and Mr. G. W. Marker reviewed the manuscript and offered some helpful suggestions. Mr. Dan Duller drafted the figures, and Mrs. B. J. Wyninger typed the manuscript.

Chapter 21

OVERVIEW OF THE BELLE AYR MINE, THE LARGEST PRODUCING COAL MINE IN THE UNITED STATES

Gary J. Brown

Manager of Operations
AMAX Coal Company
Gillette, Wyoming

AMAX Coal Company, the nation's third largest coal producer operating in Indiana, Illinois and Wyoming, is a division of AMAX, INC., a United States based natural resources corporation that operates world wide. The operations in Wyoming (The Western Division) include Belle Ayr and Eagle Butte Mines.

Belle Ayr Mine, located 18 miles south of Gillette, Wyoming, presently employs 440 people. It was started into production in 1973 and has been the largest producing coal mine in the United States since 1977. Final figures will show that the mine was again the largest in 1980 with production levels of approximately 16.1 million tons, total production of coal in the United States being in excess of 800 million tons during this last year. As one of the first commercial producers of sub-bituminous coal in the Powder River Basin, Belle Ayr has served as an example for the many other mines which have since been constructed in that region.

The four coal storage silos are the first and most dominant features of the mine to be seen for miles across the Wyoming prairie. Each silo has the capacity to load a 110-car unit train in excess of 11,000 tons.

All service facilities at the mine are under one roof. Included in the service building are offices, warehousing, seven garage bays, and washhouse facilities for all employees.

Preparation of coal for shipping at Belle Ayr consists only of crushing from pit-run size to 2 x 0-in. Primary crushing is done under the truck hoppers to 6-8-in. product size. Secondary crushing is performed in the two large crusher buildings. The smallest plant can produce at the rate of 2,500 tph, while the larger can produce the finished product at the rate of 4,000 tph.

All coal handling facilities at Belle Ayr are equipped with dust collection and dust suppression equipment. The bag-type dust collectors are greater than 99% efficient in removing fugative dust. AMAX Coal is very concerned with protecting the environment as relates to surface damage, water, and air quality.

As both coal and overburden are hauled with diesel powered trucks, the mine has the ability to store in excess of 600,000 gallons of diesel fuel. Facilities are also included to allow small quantities of storage and blending of #1 diesel fuel for limited usage during the extreme cold weather periods.

To help keep the usage of #1 diesel fuel as low as possible and to keep the trucks easy to start in the winter, "Hot Starts" are used. These are heaters with recirculating pumps for fuel and engine coolant. Air and power are also supplied at these stations.

Belle Ayr has three drills to drill and shoot the majority of overburden and all of the coal. Most of the shooting at the mine is done with bulk ANFO. It is received in bulk and stored to allow lower cost. All shots are controlled to limit vibrations as regulated by the OSM rulings published in 1977.

Overburden is removed using five Bucyrus Erie 295B shovels with 23-cubic yard buckets and is hauled in 120 ton diesel/electric trucks. Coal is removed using two 295B's with 40-cubic yard buckets. The bench method is used for both overburden and coal removal. The 70-foot coal thickness requires only two benches for an efficient operation.

The coal is hauled up a central haulage road in 120-ton trucks to the truck hopper or receiving end of the preparation facilities. A 977 dozer (lump breaker) is used to traverse the truck hoppers to reduce the lump size and allow passage of the coal through the grizzly.

As part of the continued dedication of AMAX to protect the air quality and the safety of our employees, 20,000-gallon water trucks are used for dust suppression. These trucks are quickly filled and can easily cover the traffic of our haulage roads in one pass. Other items of support equipment include track dozers, graders, and RT dozers.

The reclamation of disturbed land advances with the mining activity which encompasses several acres of land. As mining is completed, the land is reclaimed for production use such as grazing. Prior to any activities at the mine, all top soil is removed and stored for later replacement on reclaimed topography. When an extremely thick deposit of top soil is encountered, it is saved for later activities where little or no top soil will be available for reclamation.

After the coal has been removed, the overburden is placed in the mined out areas as part of the first step of land reclamation. This material is replaced as it was removed using the bench method and with the shale on the bottom and the clay on top. The backfill benches are then shaped to conform to the desired final topography. This includes the development of necessary drainage to conform to pre-mining surface characteristics. The top soil is then replaced on the shaped overburden, and the area is seeded and mulched as required to attain the desired and state-approved ground cover.

The Belle Ayr reserve area is split by Caballo Creek. Prior to mining activities, this stream is relocated. After mining, the stream is reconstructed in the reclaimed land, and the special features are added to include piles of large rocks for small mammal habitats and shrubs and trees along the stream channel.

All types of wildlife including antelope and deer are abundant on the mine property and feed on the seeded areas.

Trains are loaded in one constant motion through the silos. They travel at approximately 1 mph and are loaded at a rate in excess of 5,000 tph. Trains are loaded 24 hours a day, 7 days a week, 365 days a year.

As with any operation the size of Belle Ayr mine and AMAX Coal's Western Division, people are what make things happen. AMAX recognized the need to assist the city of Gillette in developing housing for the influx of people and constructed the Prospector Village Mobile Home Community for the benefit of all citizens of the community.

Chapter 22

USE OF LASER GUIDANCE SYSTEMS FOR LARGE HAULAGE TRUCKS

S. Jess Larsen

Kerr-McGee Coal Corporation
Gillette, Wyoming

Modern technology has provided the world with many innovations to improve living standards, ease of job performance, efficiency, and managerial control. While not expected to compete with the computer revolution, Kerr-McGee Coal Corporation has found a way using laser technology to solve one of their haulage problems.

Operating successful and profitable surface coal mine facilities in Wyoming's Powder River Basin require economies of scale, flexibility of operation, conscientious cost control, and ever-present innovative effort. Because haulage costs are the major source of expenditure in surface mining, Kerr-McGee recognized early the need for maximum haulage productivity and the economies of scale provided by large haulage trucks.

Special conditions that exist in the Powder River Basin coal fields require consideration. The coal thickness (up to 100 feet and more) and varying depths of overburden dictate that a haulage unit with good gradability characteristics be used to avoid undesirable long ramps. Two and often three coal seams are encountered by most mines, which necessitates operating from several coal and parting faces simultaneously. This situation consumes much space, and to minimize the scope of problems, a coal haulage unit with good maneuverability and stability is needed. Finally, the low density of coal (± 1500 pounds/loose cubic yard) demands a large volume capacity to utilize truck load carrying capability effectively.

These conditions prompted Kerr-McGee to select unitized two-axle bottom dump trucks of the 160-170 ton size for primary coal haulage at its two Gillette-area mines. Use of 170-ton bottom dump trucks at the Clovis Point Mine handles most of the coal haulage, but the truck dump facility must also accommodate conventional end-dump trucks. Working with a finite dump hopper size, it was desirable to limit the dump bridge for the bottom dump coal haulers to its minimum size. This permits using the maximum remaining area for end-dump unloading.

Guiding a large haulage truck (22 feet wide) onto a dump bridge of minimum width required a review of available technology. At the cost of manpower rates, use of a "spotter" or "dumpman" was clearly not the proper choice. An accurate and reliable system which could endure the rigors of the mining environment was necessary.

A low voltage laser application was found best suited to our needs. Similar systems have been used for several years in underground mine surveying and in controlling grades within the construction industry. Such uses have proven accurate, reliable and durable in all weather conditions and require virtually no maintenance.

The guidance system installed at Clovis Point Mine effectively permits bottom dump coal haulers to enter the truck dump at speeds of approximately three miles per hour and be within 5/8-inch of the dump centerline. A clearance of only 5-inches per side has been provided between the truck and sidewalls with all other space being used for end-dumping capacity.

The laser unit is mounted on a concrete pedestal approximately 250 feet in front of the dump bridge centerline (Figure 1). The laser

Figure 1. Laser mounting pedestal and protective cover.

Figure 2. Laser receiver mounted under truck canopy.

is powered by a 12-volt electrical supply. A rotating 3/8-inch wide beam of light is emitted through a vertical plane to the center of the truck bridge, where it is sensed by a laser light sensor (Figure 2).

The sensor has a total receiving range of 15-inches or 7½-inches each side of the centerline. An indicator panel with amber-lighted directional arrows and a green-lighted center alignment panel is mounted in the cab of each bottom dump coal truck.

When a coal hauler approaches the dump bridge (Figure 3), the driver has a view like that in Figure 4. It is clear that the driving of such a large vehicle onto a bridge with only 5-inches clearance would be risky without assistance.

As the truck enters the receiver zone, the indicator lights activate in the truck cab and direct the driver to move left or right (Figures 5 and 6) and proceed onto the dump (Figure 7) after achieving an on-center green indicator light. In this way the truck is driven onto the dump accurately and efficiently and the load discarded.

The laser guidance system is proving to be a useful tool. Reliability is being tested and to date is impressive. While the experience is limited at this time, the confidence level is high. Long term success is expected, and all at a total cost of about 2/3 the annual outlay for one employee.

Figure 3. Truck approaching the drive-across bridge.

Figure 4. View of the truck dump from the cab of an approaching truck.

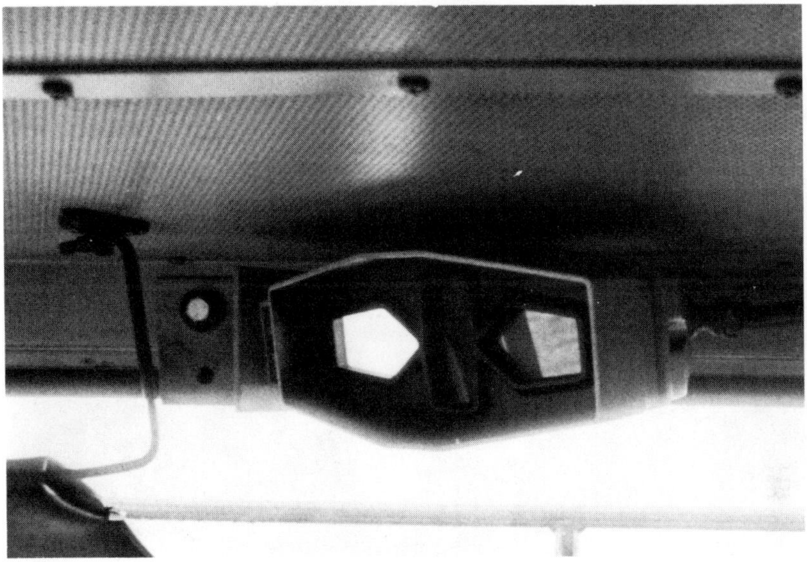

Figure 5. Amber signal displayed in the cab indicating that the truck should be steered to the right.

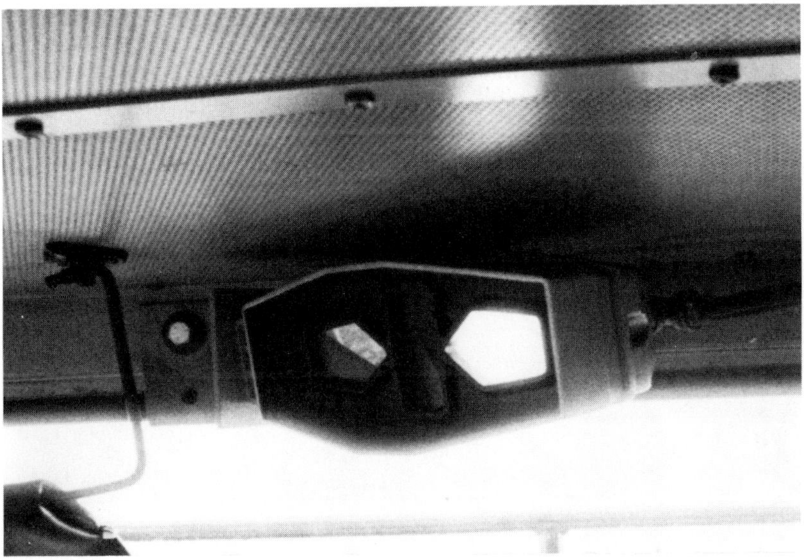

Figure 6. Amber signal displayed in the cab indicating that the truck should be steered to the left.

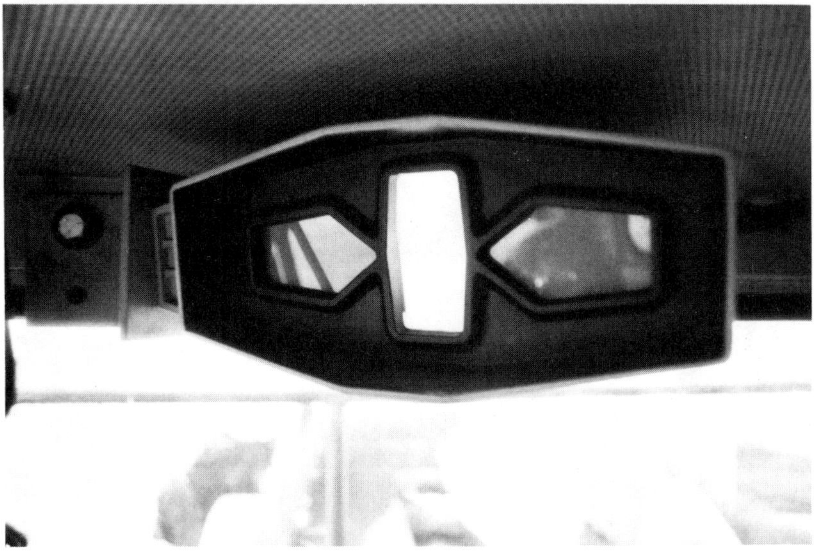

Figure 7. Green signal indicating that the truck is within 5/8" of the centerline.

Coal II

Session Chairman
Peter R. Oslund
Cordero Mining Company
Gillette, Wyoming

Chapter 23

COAL CHARACTERIZATION AND ITS APPLICATION TOWARD MINING AND UTILIZATION OF WYOMING SUB-BITUMINOUS COALS

Frederick J. Rich

Department of Geology and Geological Engineering
South Dakota School of Mines and Technology
Rapid City, South Dakota

INTRODUCTION

Coal characterization involves a variety of analyses which are designed to yield as much information as possible about the properties of coal. The types of analyses which are now regarded as necessary for any truly quantitative assessment of coal include the following:

1. proximate analysis

2. ultimate analysis

3. petrographic analysis

The two traditional forms of analyses are the proximate and ultimate types. A proximate determination reveals the amount of moisture, volatile matter, fixed carbon, and ash which are present in the coal. Nowadays, a sulfur determination is usually included in the proximate analysis. The ultimate analysis, in contrast, includes a major alkaline oxide determination, which tells how much sodium, calcium, and potassium are found in a sample. Sulfur forms, i.e. sulfide and sulfate vs organic sulfur, are also identified as well as carbon, hydrogen, oxygen, nitrogen, and trace element concentrations.

Proximate and ultimate analyses have been used for many years. The petrographic analysis, one which identifies and quantifies organic microcomponents of coal, is a comparatively new technique of coal characterization. Petrography adds great detail to a coal analysis and, to be sure, is a necessary component of any complete analysis.

Coal is not at all uniform in its composition. Material from different mines within one coal-producing region will display variation in all measurable parameters. Coals of different ages from different states or countries will show even greater variability. There is no typical coal, for, just as the plant communities which produced the world's coal deposits changed from time to time and place to place, so have the coals themselves come to be different. Insofar as this presentation is concerned with the characteristics of Wyoming sub-bituminous coals, I will restrict my comparisons to that type of deposit.

Most of Wyoming's coal deposits consist of beds of sub-bituminous rank. Sub-bituminous coals are considered to have between 8,000 and 13,000 BTU per pound, with an approximate range of 42 to 52% volatile matter and 70 to 75% carbon on a dry coal, ash-free basis.

Within the sub-bituminous classification, there are three subranks of coal: sub-bituminous C, B and A. As the rank or level of metamorphism increases within a seam, the coal proceeds from the C to A classification. Lignite is lower rank than sub-bituminous C coal, and has correspondingly more moisture and less carbon, while high-volatile bituminous coal is higher rank than sub-bituminous A coal. With higher rank, moisture levels decrease, and carbon contents increase.

Wyoming sub-bituminous coals range in age from Cretaceous to Tertiary and embrace a wide variety of deposits from deltaic to fluviolacustrine types. Along with the variety of depositional environments which produced the coals, there had to have been variation in the types of plants which led to coal formation. These variations not withstanding, one may use the analyses which are discussed in this paper to characterize any coal.

DISCUSSION

Proximate Analyses

The utility of the proximate analysis is plainly seen. This gives fundamental information which is necessary for any form of coal utilization, whether it be for coking, steam generation, or conversion to liquid and gaseous fuels. Coal rank classification is based for the most part on proximate values and, thus, for example, any coal with a range of 70-75% carbon on a dry coal, ash free basis will be of sub-bituminous rank. The amount of ash present in any coal is one of the most important items of information which the proximate analysis provides to coal consumers. High ash yields are simply undesirable, and this form of assay provides information which is necessary for decreasing ash levels. Other than providing certain rank and marketing information, however, the proximate analysis provides little description about the coal itself.

Ultimate Analyses

Alkaline Oxides

Ultimate analyses are of more critical value in coal characterization. The evaluation of major alkaline oxide concentrations, for example, reveals what sorts of problems will be encountered as far as boiler fouling is involved. Boiler fouling is a common problem with some Northern Great Plains coals, especially those from the Williston Basin in North Dakota. High levels of sodium are especially bad as that element reduces the fusion point of boiler ash to the extent that ash fuses and adheres as sticky globs to boiler tubes and walls. Sufficient amounts of such deposits not only cause a drop in the efficiency of heat transfer but also can physically damage the boiler. There have been cases, for example, where large chunks of fused ash have either deformed the boiler pipes or have fallen from the pipes and damaged the boiler wall. Obviously, then, one must know whether there will be unusually high concentrations of sodium in the boiler fuel in order to reduce the quantity either by sodium removal or by careful blending with low-sodium coal. Fortunately for Wyoming's sub-bituminous coal industry, those coals tend not to have the undesirably high levels of sodium which North Dakota lignites, for example, may have. This appears to be the result of Wyoming's coals having in large part been deposited in low-sodium freshwater swamps. Exceptions to this exist in the Green River and Ham's Fork coal fields in southwestern Wyoming where near-marine conditions affected some coal swamps One should not, however, allow oneself to become complacent toward alkaline oxide concentrations in even the most well-known freshwater coal deposits, such as those in the Powder River Basin. Coal seams whose sodium levels may be as high as those in North Dakota may eventually be found there.

Sulfur Concentrations

The sulfur content of coal has been included previously in the ultimate analysis, but only as ancillary and not of essential interest. With increasing concern for the quality of the environment, the recognition that high sulfur levels in our surroundings is undesirable has become widespread. Until Western coals had begun to be extensively used within the last decade, sulfur levels in most coals were considered to be too high for those coals to be used in some urban areas. Now, with Wyoming coals and their 0.5% or less of sulfur being used, it is much easier to meet environmental quality protection standards. Still, we find ourselves somewhat at the mercy of generalizations in coal quality. It is true that most Wyoming sub-bituminous coals have very low sulfur contents. This is in accord with levels of alkaline oxide concentrations because elements such as Na, Ca, and S are most concentrated in deposits which have experienced some kind of marine influence. Some of Wyoming's coal seams which presumably are derived from freshwater deposits do contain anomolously high levels of sulfur, however. The Dietz #3 and Monarch seams in the

Sheridan area have been shown by Glass (1) and Penn State University (2) to contain between 1.0 and 3.1% sulfur on a dry ash-free basis. There are also places within the Wyodak seam where sulfur is easily visible as pyrite and where sulfur values rise above the average for the seam. This is not to say that these seams are, as a whole, undesirable. Powder River Basin coals in particular are the best in the country as far as sulfur concentrations are involved. The fact that unusually high concentrations do exist there cannot be overlooked, and coal characterization plans must be designed to detect those anomolously high levels where they occur. If sulfur contents rise near unacceptable levels, as they do within the Adaville coals near Kemmerer, Wyoming, then selective mining and blending may be necessary. At Kemmerer, FMC Corporation's Skull Point Mine relies upon selective mining to insure that portions of seams with one level of sulfur concentration are used for one purpose, such as form-coke production, while the remainder of the coal is sent off for a different use. Wyodak Resources also practices selective mining of its seam near Gillette. Strata within the Wyodak seam which have somewhat higher sulfur contents are blended with lower sulfur coal so that the mine product conforms uniformly to buyers' specifications.

As long as it is realized that Wyoming coal is not uniformly low in sulfur, and as long as it is possible to locate areas within seams where sulfur is abundant, those coals can either be avoided, disposed of, or blended so that sulfur emmissions do not again become the problem that they were 20 years ago. The only way that this kind of success can be ensured is by careful monitoring of sulfur levels as a part of coal characterization.

Trace Elements

The environmental toxicity of coal use by-products is another potential problem which can be approached by comprehensive coal characterization. Steam plant waste such as fly ash, bottom ash, and flue gas, and converter waste such as char, ash, or tarry water may contain concentrations of toxic metals which are environmentally unacceptable. Certain reports in recent years have claimed Powder River Basin coals to have anomolously high concentrations of mercury and radionuclides. The paper which reported the high mercury levels has since been discredited, and our research has shown that radionuclide levels in Powder River Basin coals are among the lowest in the country. Wyoming coals, however, are still subject to hearsay and speculation regarding levels of toxic elements. These coals enjoy an excellent reputation as clean fuel as published analyses from Penn State University (2), the Wyoming Geological Survey (1), and the U.S. Geological Survey (3) will corroborate. It would be foolish to believe that existing accounts will suffice, though, and that we need no longer be concerned with the possibility of encountering coal in a certain area which may contain unexpectedly high concentrations of some toxic metal.

Certain of the critically toxic metals, such as Pb, Hg, Cd, Co, Ni, Cr, Be, and As are proven carcinogens in mammals, and the oxides of some of these elements are known carcinogens in humans. I advocate a program of regular analyses for these particular trace elements for several reasons:

1. We really do not know how abundant they are in all of Wyoming's major coal mining areas.

2. In most cases, we don't know what an acceptable environmental level of concentration of these elements even is, and geochemical surveys could advance our knowledge in this critical area.

3. We must know from the standpoint of personal health and that of the environment how much of these elements can be expected to be encountered as coal development in Wyoming accelerates.

4. We must be able to satisfy the public that coal development can proceed without undue risk to themselves, their livestock, etc.

Trace elements are, by their very nature, present in minute quantities. Their concentrations vary from place to place within a seam and cannot be expected to be constant among seams. In our search for these substances we have, in a sense, developed the capability to scare ourselves by improving detection techniques to the point where we find some level of toxic substances everywhere. As scientists, it is our duty now to define the difference between mere background occurrence and truly worrisome concentrations. Once again, this type of investigation must accompany accurate coal characterization.

Hydrogen–Carbon Ratios

Emphasis must also be placed on the determination of hydrogen and carbon relationships among sub-bituminous coals. The abundances of these two elements are usually expressed as the H/C ratio of the coal. While this ratio is not of particular interest where steam generation is involved, it can be shown that the ratio is significant if coal is to be used for liquefaction or gasification. Much of the coal in the Powder River Basin seems destined for conversion to other fuels. As conversion processes become economically and technically feasible, vast quantities of coal will be consumed in order to produce synthetic natural gas and other substances. Coals with greater quantities of hydrogen inherent in their structures behave better during conversion and will make the processes more economical. Thus, coals whose composition will favor conversion should be located, delineated and conserved for that use instead of simple steam generation.

Petrographic Analyses

If all sub-bituminous coals were alike, then ranges of volatile matter, carbon contents, etc. would suffice to describe the coals. Experience has shown, however, that even though two coals may have very similar proximate analyses, the volatile matter, fixed carbon, and ash will be variously derived from quite different coal constituents. There are a number of organic coal microcomponents, termed macerals, which produce the volatile matter, carbon and ash, and it is really the relative abundance of those coal macerals which indicates how coal will react rather than the abundance of the volatile matter, carbon, and ash which the macerals produce.

Coal petrography is used routinely by the steel industry in order to optimize coke production. The science has been used more recently for describing petroleum source rocks. It is now time to use petrographic techniques for describing Western coals for the purpose of locating the best conversion coals as well as coals which can be used for the production of form coke. Both the coal fuels and form coke industries are destined to grow increasingly important in Wyoming as present sources of petroleum and metallurgical coal are depleted. Insofar as the only presently reliable method of identifying the best coals for those purposes is to analyze them visually, coal petrography is also destined to grow in importance as an analytical technique.

Petrography has both advantages and disadvantages over other forms of analysis. It requires specially trained personnel, but can be performed on comparatively inexpensive equipment. Individual petrographic analyses are not performed as quickly as proximate analyses, but the data derived from petrography are really more informative.

In addition to providing base-line data concerning the physical and chemical characteristics of coal, petrographic compositions can be used for correlating coal seams and identifying environments of coal deposition. These data can then be used to predict coal characteristics in areas where actual analyses are lacking.

Petrography actually consists of several methods of investigation. One which we are presently emphasizing at the South Dakota School of Mines and Technology is fluorescence microscopy, a technique using near ultra-violet light. Fluorescence can be used for identifying lipid-like substances such as waxes, resins, pollen, etc. in the coal.

These particular macerals are enriched in hydrogen and comprise the best raw materials for gasification and liquefaction. Liptinites, as these macerals are termed, are only identifiable visually and are easily located in coal if one has the proper optical equipment, such as a fluorescence microscope. Any corporation which is seriously considering marketing its coal for conversion to other forms of fuel needs to seriously consider having petrographic analyses made so that its coal can be properly marketed.

SUMMARY

It will be inevitably concluded that the types of analyses which have been described will be necessary for rational development of Wyoming's sub-bituminous coal industry. The question which remains, though, is whether or not investigations will be initiated soon enough. Just as individuals in charge of oil refineries must know what the characteristics of their crude oils are, consumers of sub-bituminous coal must know how the coal they buy will perform. More important to coal producers is the knowledge that they must have to maintain product quality control as well as information which may improve coal marketability. Coal characterization, as a part of cooperative coal research between schools and industry will result in a greatly improved understanding of Wyoming's sub-bituminous coals, and will provide substantial benefits of many kinds to both coal producers and consumers.

REFERENCES

1. Glass, G. B., 1975, Analyses and Measured Sections of 54 Wyoming Coal Samples: Geol. Surv. Wyoming, R.I. No. 11, 219 p.

2. The Pennsylvania State University, Coal Data Base: Coal Research Section, The Pennsylvania State University, University Park, Pennsylvania.

3. Swanson, V. E., Medlin, J. H., Hatch, J. R., Coleman, S. L., Wood, G. H., Jr., Woodruff. S. D., Hildebrand, R. T., 1976, Collection, Chemical Analysis, and Evaluation of Coal Samples in 1975: U.S. Geol. Surv. Open-file Report 76-468, p. 336-380.

Chapter 24

ESTIMATING THE FLOATABILITY OF WESTERN COAL

F. F. Aplan

The Pennsylvania State University
University Park, Pennsylvania 16802

INTRODUCTION

Coal is a solid combustible mineral substance resulting from the alteration of vegetable matter largely in the absence of air. It occurs in nature with varying percentages of mineral matter associated with and incorporated into the coalified material. If the amount of this inorganic mineral matter is sufficiently small, and if selective mining can be practiced so as to avoid the incorporation of ash-bearing material from adjacent strata and/or parting seams, the coal can be used as-mined. Otherwise, it must be given a preparation treatment.

In 1980 the bulk of the approximately 525 coal preparation plants were located in the eastern half of the U.S., largely in the Appalachian and to a lesser extent in the midwest coal fields (1). Only 17 of these preparation plants were located in the west; these were in Colorado, New Mexico, Utah, Washington and Wyoming. In the past, however, a great many more preparation plants were operated in these and other western states. The upsurge in both the domestic and world-wide demand for quality coking and steam coal will certainly lead to an increase in the number of preparation plants in the west. In fact, there has been a massive development of coal mines and their associated preparation plants across our northern border in Alberta and British Columbia, largely to serve the export coking and steam coal market.

Most of the tremendous increase in newly-mined coal in the western U.S. has come from thick seams of inherently low-sulfur, low-ash coal, easy to mine selectively without significant contamination with ash-forming minerals and requiring no preparation. Because of this fact, there has been a tendency to ignore the role preparation may play in upgrading coals other than the Dakota lignites, the Wyoming-Montana sub-bituminous and other large deposits such as that at Black Mesa, Arizona. The reality is that there are many smaller deposits of coal, ranging in rank from sub-bituminous to semi-anthracite, that are widely scattered throughout the west. The quality of this coal, as mined, can often be significantly improved by preparation. These coals are often of reasonably high rank (HVC and above) and have a substantial potential for coking and steam use, either for the domestic or for foreign markets. Substantial deposits of these kinds of coal are located at such widely scattered places as Carbondale, Crested Butte, Somerset and Walsenburg, CO; Bear Creek, MT; Raton, NM; Salina and Sunnyside, UT; Bellingham, Centralia and Cle Elum, WA; Hanna and Rock Springs, WY, etc. Because of their place value (near local sources of population and use) and their proximity to west coast ports where they can serve the energy-short nations bordering the Pacific with a relatively inexpensive oil substitute, the development of these deposits may be expected to quicken in years to come. Indeed, Australia, Canada and South Africa have moved aggressively ahead to fill this demand. The U. S. has been a decided laggard in pursuing these markets even though we have the coal that can often meet the quality specifications. Nor have we aggressively gone after the domestic markets which a rational national energy policy could open up through substitution of the more economical coal for oil.

COAL PREPARATION METHODS

Coal preparation is usually done by wet methods (2,3), and the small percentage traditionally cleaned by dry methods has decreased significantly in the past few years, largely because of environmental problems. Coal preparation can be used for the following purposes:

1. For the reduction of ash.
2. For the production of quality coking coals.
3. For the production of relatively high BTU, low sulfur, steam coal.
4. For use as an integral part of the preparation plant water re-use, recycle system.
5. For special situations (e.g. separation of the maceral resinite).

Historically, preparation was used for the reduction of ash in run-of-mine coal, but in recent years the other purposes have become of increasing importance. Coarse and intermediate coal cleaning (say, +28 mesh, ∿0.5 mm) is normally done with jigs and heavy media processes and is efficient and effective. The fine coal cleaning methods using water-only cyclones, fine jigs, tables and flotation are not nearly so effective in separating coal from its associated refuse minerals (4). Of these, flotation offers the promise of providing the best separation and the highest recovery of clean coal. Furthermore, it is a proven, high capacity process. Its use has grown rapidly in the past two decades and may be expected to continue to grow as a fine cleaning process of merit.

Unfortunately, flotation has not always lived up to its promise. Not only has there been a dearth of research on coal flotation in general, but the flotation of western coals has been almost totally ignored. Many difficulties with coal flotation have been ennumerated (4) and principal among these, as applied to western coals, are: (1) difficulty in floating lower rank coals, lower rank macerals (e.g. fusinite) and those particles whose surface has become oxidized, (2) difficulty in floating relatively coarse particles (plus ∿0.5 mm), (3) difficulty in floating fine coal in the presence of a large amount of clay, and (4) difficulty in rejecting pyrite.

Very fortunately, this latter problem is not a very important one for the treatment most western coals as their sulfur content is usually low.

THE NATURE OF COAL

Coal is classified in the U.S. by ASTM (5,6) into the ranks given in Table 1. Perhaps a more meaningful presentation of coal properties with a change in rank is that of Given (7,8) shown in Table 2. Note that as the rank increases so does the percent carbon, the proportion of carbon in the aromatic form, and the calorific value, while the oxygen content, and especially the hydroxyl and carboxyl oxygen content, decrease. In most of the properties, there is a continuum from rank to rank except that the total oxygen content and the carboxylic oxygen content shows a sharp increase below the HVC bituminous rank.

SURFACE PROPERTIES OF COAL

The surface properties of coal, a solid usually considered to be naturally hydrophobic, are extremely relevant toward assessing its flotation potential.

The degree of hydrophobicity of a coal is related to its rank. Horsley and Smith (9) found that the captive bubble contact angle, θ, on coal could be correlated with the carbon content on a dry, ash-free basis (DAF). Their data, together with more recent data by Purcell (10), is shown in Figure 1. The data of Figure 1 indicate that coals of high carbon content would be quite hydrophobic, but below ∿80% C DAF, corresponding roughly to coals of HVB bituminous rank and below, the coals are not very hydrophobic (i.e. $\theta \cong 0$). Also shown in the same figure is the approximate rank of the coals used in this study based on their carbon content. It must be cautioned that such a procedure, while reasonably accurate, is but an approximation since only

Table 1. Classification of Coals by Rank
(after Yancey and Geer (6))

Class	Group	Fixed Carbon Limits, percent (Dry, Mineral-Matter-Free Basis)		Volatile Matter Limits, percent (Dry, Mineral-Matter-Free Basis)		Calorific Value Limits, Btu per pound (Moist[1] Mineral-Matter-Free Basis)		Agglomerating Character
		Equal or Greater Than	Less Than	Greater Than	Equal or Less Than	Equal or Greater Than	Less Than	
I. Anthracite	1. Meta-anthracite	98	--	--	2	--	--	--
	2. Anthracite	92	98	2	8	--	--	--
	3. Semianthracite	86	92	8	14	--	--	Nonagglomerating[2]
II. Bituminous	1. Low volatile bituminous coal	78	86	14	22	--	--	
	2. Medium volatile bituminous coal	69	78	22	31	--	--	
	3. High volatile A bituminous coal	--	69	31	--	14,000[3]	--	Commonly
	4. High volatile B bituminous coal	--	--	--	--	13,000[3]	14,000	agglomerating
	5. High volatile C bituminous coal	--	--	--	--	11,500	13,000	
						10,500	11,500	Agglomerating
III. Subbituminous	1. Subbituminous A coal	--	--	--	--	10,500	11,500	Nonagglomerating
	2. Subbituminous B coal	--	--	--	--	9,500	10,500	--
	3. Subbituminous C coal	--	--	--	--	8,300	9,500	--
IV. Lignite	1. Lignite A	--	--	--	--	6,300	8,300	--
	2. Lignite B	--	--	--	--	--	6,300	--

[1]Moist refers to coal containing its natural inherent moisture but not including visible water on the surface of the coal.

[2]If agglomerating, classify in low-volatile group of the bituminous class.

[3]Coals having 69 percent or more fixed carbon on the dry, mineral-free basis shall be classified according to fixed carbon, regardless of calorific value.

Table 2. Approximate Values of Some Coal Properties in Different Rank Ranges

	Lignite	Subbit.	High Vol. Bit. C	High Vol. Bit. B	High Vol. Bit. A	Bituminous Medium Vol.	Bituminous Low Vol.	Anthracite
% C (min. matter free)	65–72	72–76	76–78	78–80	80–87	89	90	93
% H	4.5	5	5.5	5.5	5.5	4.5	3.5	2.5
% O	30	18	13	10	10–4	3–4	3	2
% O as COOH	13–10	5–2	0	0	0	0	0	0
% O as OH	15–10	12–10	9	?	7–3	1–2	0–1	0
Aromatic C atoms % of total C	50	65	?	?	75	80–85	85–90	90–95
Av. no. benz. rings/layer	1–2	? ←		2–3	→		→ 5?	>25?
Volatile matter, %	40–50	35–50	35–45	?	31–40	31–20	20–10	< 10
Reflectance, %, Vitrinite	0.2–0.3	0.3–0.4	0.5	0.6	0.6–1.0	1.4	1.8	4
Density	→				→ increases	→		→
Total surface area	←				← minimum	→		→
Plasticity and coke formation			├←		← only	→	→┤	
Calorific value, moist, Min. matter free, Btu/lb.	7,000	10,000	12,000	13,500	14,500	15,000	15,800	15,200

Table prepared by Dr. Peter H. Given, Fuel Science Section, The Pennsylvania State University (7).

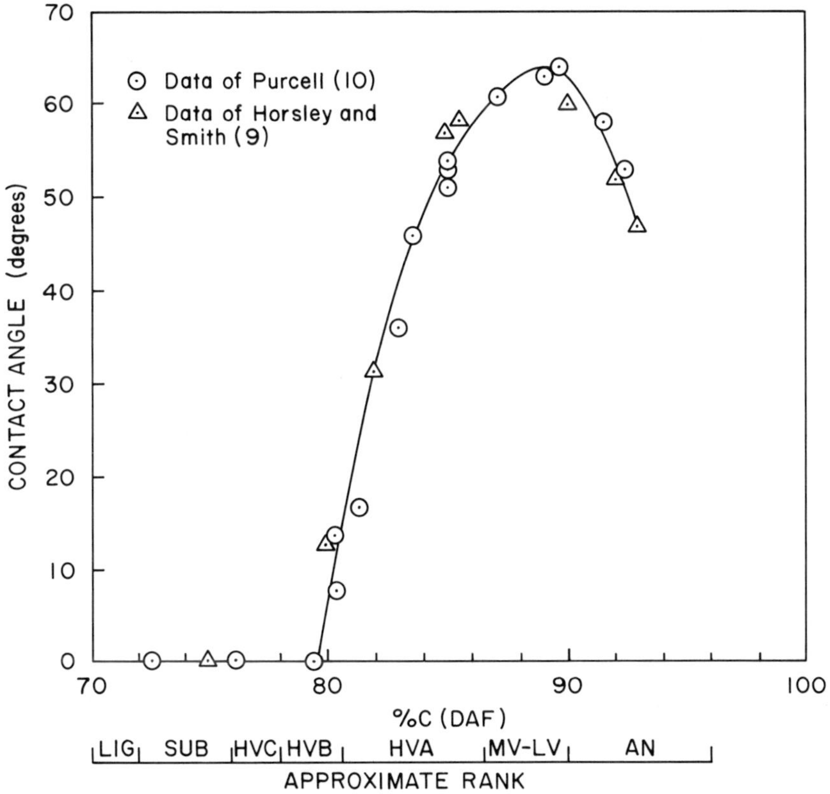

Figure 1. Variation of the captive bubble contact angle with carbon content of coal.

one of the four ASTM criteria (%C, % VM, BTU content, agglomeration characteristics) is used.

Unfortunately, many of the coals found in the western U.S. are of HVB rank and below, and conventional wisdom would indicate that with $\theta = \sim 0$ they would not be readily floatable without a strong collector. Other data by Parekh and Aplan (11) indicate that if the sessile drop contact angle technique is used instead of the captive bubble method, a smooth transition in the value of the contact angle of water on coal ranging from ∿50° for lignite to ∿75° for LV bituminous occurs. These data indicate that coal is a much more hydrophobic solid than would be assumed from the captive bubble data. This has been recently quantified for a broad range of coals by Gutierrez, Purcell and Aplan (12).

Parekh and Aplan (11) have found that the critical surface tension of wetting of coals of all ranks is about ∿45 ergs/cm^2 but that the slope of the Zisman (13) surface tension - cos θ plot, α, varies in a predictable way with rank, namely the steeper the slope, the greater the hydrophobicity. Lignite is shown to be the least hydrophobic material while semi-anthracite was found to be the most hydrophobic. The presence of pure MIBC and pine oil frothers did not significantly alter either γ_c or α while kerosene increased the value of the Zisman slope, α, showing that it serves to strengthen bubble-particle attachment (11).

Brown (14) has shown that for coals above 80% C, DAF, the quantity of creosote-cresylic acid reagent needed to float these coals closely parallels the captive bubble contact angle, ranging from nil at 89-92% C to 18 lb/T at 80% C. He presents no data for those coals containing less than 80% C which, unfortunately, are the coals of greatest interest here.

Another surface property that can sometimes be correlated with floatability is the zeta potential of the solid in water. Campbell and Sun (15,16) have studied in detail the zeta potential of coal and report iep values at ∿pH 4-5 for many coals though other of their data indicate iep values ranging from pH 2-10. These data, and that of other workers, are so contradictory that, given the sad state of our present knowledge, not many valuable conclusions can be drawn at this time except that coal will often be charged negatively in alkaline, neutral or weakly acid solutions.

ESTIMATING THE FLOATABILITY OF COAL

None of the surface properties discussed above are sufficient in themselves (especially in the range near 80% C and below) to allow a good estimation to be made of the floatability of a given coal. They do, however, provide many valuable clues. Quite obviously more data and much test work are needed.

In 1954 Sun (17) proposed a surface component hypothesis to explain differences in the floatabilities of various coals. From this premise he developed an ingenious equation to relate the floatability of a given coal to that of a standard, graphite, based on the ultimate (elemental) analysis of the coal. In essence, his equation gives a bonus for carbon and hydrogen and a penalty for oxygen, water and ash. His equation, unfortunately, is not always applicable because: (1) the detailed chemical analyses from several portions of the deposit may not be available, (2) the preparation plant may be treating coals from several different seams and/or sources, or containing a broad range of macerals, each with their own analytical and flotation properties, so that bulk analyses may be grossly misleading and (3) the calculated floatability based on bulk analyses may in no way correlate with the floatability of particles whose surfaces have become oxidized either in-situ or upon transit, storage or handling. The best available data (8,18) indicate that dimminution of floatability due to oxidation is a much more serious problem with lower rank coals than with higher rank ones. This aspect is being quantified in our laboratories and will be reported soon (19).

Another way to evaluate the floatability of coal is through a study of the data reported in the literature even though such data on the floatability of lower ranks of coal are largely missing. Sun (17) has one of the more extensive listings of such data, and his data have been condensed into the form shown in Table 3. While not agreeing with the data of Brown (14) in the specific details, the general trend is the same: coals require progressively greater quantities of oil for their flotation as the rank of coal is decreased. The amount of oil needed to float coals of HVA bituminous rank and above, is modest but increases sharply for ranks below this.

Data from the U. S. Bureau of Mines (20,21) on a MV bituminous coal from Washington state (Table 4) indicate that a somewhat greater quantity of oil is needed than that reported by Sun (17). The reason for the difference may lie in the special nature of western coal but is more likely related to flotation conditioning procedures and to differing size distribution of the crushed coal used. Rastogi and Aplan (22) have demonstrated clearly that coarse coal floats much more slowly than does fine coal and that the flotation of coarse coal can be massively improved by the addition of oil.

Because of these apparent inconsistencies, several flotation tests on western coals were undertaken. Table 5 clearly shows that for a HVC bituminous coal, oil can increase the overall yield very substantially and that a great deal of this increase in yield is due to the improved flotation of the coarser, +65 mesh coal. For the poorer ranks of coal, it is obvious that the flotation of the coarser particles massively influences the overall yield, and test data can only be compared for those samples having essentially the same size distribution.

Table 3. Floatability of Coal as a Function of Rank

(Data of S. C. Sun, Ref. 17)

-48 mesh coal

Rank	Location	Amount of combined light petroleum oil and pine oil for 90% yield*, lb/T
Low Vol Bit	Bedford, PA	0.32
Low Vol Bit	Windber, PA	0.47
Low Vol Bit	Cresson, PA	0.65
Med Vol Bit	Hastings, PA	0.32
Med Vol Bit	Sabinas, Mexico	0.75
High Vol A Bit	Boone Co., WV	0.68
High Vol A Bit	Richmond, WV	0.94
High Vol B Bit	Stockton, UT	0.78
High Vol B Bit	Clairton, PA	0.91 +
High Vol B Bit	Hopkins, KY	1.19
Sub-bit B	Sheridan, WY	3.6
Sub-bit C	Hanna, WY	6.0
Sub-bit C	Gillette, WY	10.6
Lignite	Velva, ND	16.0
Lignite	Minot, ND	25.0
Lignite	Noonan, ND	69.0
Lignite	Dickinson, ND	130.0

* These reagents added in the ratio of 1.5:1, light petroleum oil:pine oil.

Tests performed with older-style 500 g. Denver cell except that marked with (+) which was done with a 50 g. Denver cell.

Table 4. Flotation of Wilkeson Coal, Wilkeson-Carbonado Area, Pierce County, WA

Rank: Medium Volatile Bituminous

USBM Bul 337 (20)

Reagents: 1.4 #/T Cresylic acid
-48 mesh coal

	% Yield	% Ash
Float	88.3	14.4
Tails	11.7	83.4
Raw Coal Feed	100.0	22.5

USBM RI 6538 (21)

Reagents: 0.64 #/T Pine oil (added stagewise)
3.48 #/T Kerosene (added stagewise)

-20 mesh coal

	% Yield	% Ash
Float	74.8	15.0
Tails	15.2	46.4
Raw Coal Feed	100.0	22.9

Table 5. Flotation of Nominal -28 mesh 'D' Seam Coal

Yampa Field, Colorado

Rank: High Volatile 'C' Bituminous Coal

(Flotation tests by R. J. Purcell, Jr. and the author)

Test No.	Reagent Dosage, #/T		% Yield in 2 min.	% of Each Size Floated	
	MIBC	#2 Fuel Oil		-65M	+65M
T-3	0.18	0.5	22.0	48.5	21.8
T-4	0.18	2.0	44.7	89.5	51.4
T-5	0.18	6.0	95.7*	97.4	94.2

*Yield at 1 minute flotation time was 80.8%.

Operating Conditions:

Older WEMCO Fagergren lab cell
Collector conditioning: 2 min. at ∿40% solids
Frother conditioning: 2 min. at ∿16% solids
Flotation pH: pH 7
Aeration rate: 0.25 cfm
Impeller speed: 1750 rpm
Coal size: ∿10% 14x28 mesh, ∿45% 28x65 mesh, ∿45% -65 mesh

Another series of tests (Table 6) were run on a sub-bituminous 'C' coal from the Smith-Roland Seam. Note that the use of even 32 lb/T of pine oil did not produce a reasonable yield of coal nor did a comparable amount of fuel oil alone or as a 1:2 fuel oil:Barrett oil mixture. Only when 35 lb/T of Barrett Oil No. 4 was used did a reasonable yield result. Apparently, the cresol or similar compound in the Barrett oil is absorbed strongly enough to act as an effective collector. Our data clearly illustrate that lower rank coals may take massive amounts of oil for their flotation and, in general, the lower the rank, the greater the quantity of oil required. Any flotation engineer testing western coals should consider that much greater quantities of oil and oily collectors may be required than has traditionally been required for the flotation of eastern coals.

Since lower rank coals contain more oxygen than do higher rank ones, another collecting agent to be considered for use is an amine. Sun (23) has shown this to be an effective way of floating oxidized coal. While satisfactory for the flotation of oxidized higher rank coals, the author has not found this to be an especially good way to float extensively weathered or very low rank coals in the few instances he has used it. A further complication is that in the presence of fine clay, a circumstance not uncommonly encountered when floating fine coal, the use of the amine as collector may be counterproductive as it is also an excellent collector for clays.

PYRITE FLOATABILITY

If pyrite is present in a coal seam, its floatability is greatly increased when fuel oil is added. The data of Bonner and Aplan (24, 25) shown in Table 7 illustrate that for two high rank (LV and HVA) bituminous coals, the addition of only a modest amount of oil dramatically increases the floatability of the pyrite. Direct comparisons in the presence and absence of oil could be made in this case, since these high rank coals were so hydrophobic that they could be floated, in good yield, with a frother only.

Since recovery of the lower ranks of western coals will require relatively large amounts of oily collector, the pyrite contained therein may be expected to be readily floated. To test this premise, the tests shown in Table 8 were made. It will be seen that sulfur was not rejected during either the flotation of the 'as-received' coal or when a few percent of purified pyrite from the Pittsburgh Seam was added to the coal to be floated. In the former case, the sulfur was present largely as organic sulfur and in the latter case as pyritic sulfur. These data clearly illustrate that when massive quantities of Barrett oil are required to float the coal, the rejection of pyrite is essentially nil.

Very fortunately, most western coals do not contain much pyritic sulfur and so use of large amounts of oily collector should not

Table 6. Flotation of Nominal -28 mesh Coal from

Campbell County, WY

Smith-Roland Seam

Rank: Sub-Bituminous C

(Flotation tests by V. Choudhry and the author)

Reagents Added, #/T			
Fuel Oil #2	Barrett Oil #4	Pine Oil	% Yield
--	--	32	12
30	--	0.56	56
10	20	0.56	58
--	35	0.56	94

Operating Conditions:

Older WEMCO Fagergren lab cell
Weight of sample: 200 g.
Collector Conditioning: 10 minutes high solids, ∿40%
Frother Conditioning: 1 minute, ∿8% solids
Flotation Time: 1 minute
Air Rate: 0.25 cfm
Impeller Speed: 1750 rpm
Coal Size: ∿10% 14x28 mesh, ∿45% 28x65 mesh, ∿45% -65 mesh

Table 7. Effect of Fuel Oil on Pyrite Floatability During Coal Flotation (24, 25)

Coal	Rank	lb/T MIBC	lb/T #2 Fuel Oil	% Yield*	% of SLP** Floated
Lower Kittanning	LV	0.17	0	90	13
Lower Kittanning	LV	0.17	0.93	97	56
Pittsburgh Seam	HVA	0.34	0	84	16
Pittsburgh Seam	HVA	0.34	1.86	97	45

*Yield at 7.75 minutes; most of the floated within 1-2 minutes in each case.

**SLP = liberated pyritic sulfur

Table 8. Effect of Barrett Oil No. 4 on Pyrite Floatability During Coal Flotation

Coal: Smith-Roland Seam, Campbell County, WY
200 g., 100% -28 M, 90% -48 M

Rank: Sub-bituminous 'C'

Test No.	Product	Yield	Assay % S	Dist % S
8	Collector: 37 #/T Barrett Oil No. 5 Frother: 0.56 #/T Pine oil 10 min. collector cdng ∿40% solids; 1 min. addl. cdng with frother ∿8% solids			
	Float 1 min.	70.8	---	---
	Float 2 min., cum.	75.0	0.39	73.1
	Tails	25.0	0.43	26.9
	Calc. Head	100.0	0.40	100.0
9*	Collector: 35 #/T Barrett Oil No. 4 Frother: 0.56 #/T Pine oil 10 min. collector cdng ∿40% solids, 3 min. addl. cdng with frother ∿8% solids			
	Float 1 min.	85.8	---	---
	Float 2 min., cum.	93.8	2.03	93.5
	Tails	6.2	2.25	6.5
	Calc. Head	100.0	2.14	100.0

Operating conditions same as those given in Table 6 except for conditioning time.

*For this test 5% freshly crushed pyrite from a coal source (Pittsburgh seam) was purposely added.

ordinarily lead to any significant increase in the sulfur content of coal produced by flotation. When pyrite is present, however, special efforts will have to be made to facilitate its rejection in this case.

The flotation tests shown in Table 8 also show the importance of conditioning on the flotation of this low rank coal. A small increase in conditioning time gave a substantial increase in yield (compare tests 8 and 9). This factor needs much additional study.

SPECIAL SITUATIONS

Since coal is a highly variable natural product, certain special situations may exist in a given deposit. One such special circumstance is illustrated by the presence of resinite in some Utah coals. These coals have been found to contain the fossil resin largely in fracture seams in the coal and can be liberated by grinding to about 6 mesh. During World War II, Combined Metals Reduction Co. ran a large pilot plant to recover the resin using 0.4 lb/T of amyl alcohol (U.S. Patent 1,773,997, 1930, issued to W. D. Green) and possibly proprietary reagents (26). The resin was removed from the bulk of the -6 mesh coal by roughing, cleaning and recleaning in air lift type cells, and a concentrate containing 70-80% material soluble in benzol was obtained. The resin was used for printing inks, varnishes, water proofing agents, etc.

SUMMARY

1. Lower rank coals, lower rank macerals, coal particles with oxidized surfaces and coarse coal particles are difficult to float.

2. The floatability of coals decreases progressively with decreasing rank from semi-anthracite and low volatile bituminous to lignite coals. Such a decrease is floatability is related to a decrease in the carbon content and an increase in the oxygen content of the coal.

3. The floatability of coal can be increased by the use of an oily collector (fuel oil, Barrett oil, etc.) or possibly an amine.

4. The floatability of western coals may be simply estimated from their rank; the lower the rank, the greater the quantity of collector needed. For coals of especially low rank, rather massive quantities of oily collectors will be required.

5. The flotation of relatively coarse coal (approaching 0.5 mm) of low rank requires rather large quantities of an oily collector; minus 0.2 mm (65 mesh) fines require much less collector unless such complications as intense surface oxidation of the coal or the presence of much fine clay exists. In the latter case, the operator may be

well advised to deslime the -0.5 mm coal at, say, 325 mesh, throw away the slimes and float only the coarser fraction.

6. The overall yield of very low rank coal is likely related to the conditioning time of the coal with the flotation reagents, higher yields being related to longer conditioning times.

7. If pyrite is present in the coal to be floated, the use of large quantities of oily collector may be expected to result in the co-flotation of the pyrite.

8. Certain coals and/or their macerals may respond to flotation somewhat differently than do other coals of the same rank. This is illustrated by certain Utah coals which contain a fossil resin that may be removed from the bulk of the coal by a relatively simple procedure.

9. With our present state of knowledge, the flotation of lower rank western coals will require an individual development and testing program for each coal to determine those reagents and reagent quantities required and the best operating conditions. A good many of the problems may be solved by use of larger than normal amounts of oily collectors such as fuel oil and Barrett oil, by desliming in those situations where such a procedure is indicated, and by adopting special conditioning procedures.

ACKNOWLEDGEMENTS

The author wishes to acknowledge his present and former graduate students: C. M. Bonner, V. Choudhry and R. J. Purcell, Jr. for performing some of the flotation tests reported herein. Acknowledgement is also made to Mr. Purcell for measuring certain contact angle values reported in Figure 1.

REFERENCES

1. Nielsen, G. F. and Richardson, C. V., eds., 1980 Keystone Coal Industry Manual, McGraw-Hill, New York, 1980.

2. Leonard, J. W., ed., Coal Preparation, 4th Ed., AIME, New York, 1979.

3. Staff, U.S. Bureau of Mines, Minerals Yearbook, 1975, Vol. 1, U.S. Govt. Printing Office, p. 435, 1977.

4. Aplan, F. F., "Fine Coal Preparation - State of the Art, Problems and Predictions for the Future" in Somasundaran, P. and Arbiter, N., eds., Beneficiation of Mineral Fines - Problems and Research Needs, NSF Workshop, Sterling Forest, N.Y., Aug. 27-29, 1978,

AIME, New York, pp. 101-110, 1970.

5. ASTM, Annual Book of ASTM Standards, 1979, Part 26, Std. D 388-77 Spec. for Classification of Coals by Rank, ASTM, Philadelphia, 1979.

6. Yancey, H. F. and Geer, M. R., "Properties of Coal and Impurities in Relation to Preparation", Coal Preparation - 3rd Ed., Leonard, J. W. and Mitchell, D. R., eds., AIME, New York, pp. 1-3 to 1-56, 1968.

7. Given, P. H., Short Course on Coal Characteristics and Coal Conversion Processes, The Pennsylvania State University, May 19-23, 1975.

8. Aplan, F. F., "Coal Flotation", Chapt. 45 in A.M. Gaudin Memorial Flotation Symposium, Fuerstenau, M. C., ed., AIME, New York, pp. 1235-1264, 1976.

9. Horsley, R. M. and Smith, H. G., "Principles of Coal Flotation", Fuel, Vol. 30, pp. 54-63, 1951.

10. Purcell, R. J., Jr., Mineral Processing Section, The Pennsylvania State University, unpublished data.

11. Parekh, B. K. and Aplan, F. F., "The Critical Surface Tension of Wetting of Coal", Chapt. 9 in Recent Developments in Separation Science, Vol. IV, Li, N. N., Long, R. B., Stern, S. A. and Somasundaran, P., eds., CRC Press, West Palm Beach, Florida, pp. 107-113, 1978.

12. Gutierrez, J. A., Purcell, R. J., Jr., and Aplan, F. F., "Estimating the Hydrophobicity of Coals", submitted to Colloids and Surfaces.

13. Zisman, W. A., "Relation of the Equilibrium Contact Angle to Liquid and Solid Constitution", Chapter 1 in Contact Angle, Wettability and Adhesion, Adv. in Chem. Series 43, Fowkes, F. M. and Gould, R. F., eds., ACS, Washington, pp. 1-51, 1964.

14. Brown, D. J., "Coal Flotation", Chapt. 20 in Froth Flotation - 50th Anniversary Volume, Fuerstenau, D. W., ed., AIME, New York, pp. 518-538, 1962.

15. Campbell, J. A. L. and Sun. S. C., "Bituminous Coal Electrokinetics", Trans. AIME, Vol. 247, pp. 111-114, 1970.

16. Campbell, J. A. L. and Sun, S. C., "Anthracite Coal Electrokinetics, Trans. AIME, Vol. 247, pp. 120-122, 1970.

17. Sun, S. C., "Hypothesis for Different Floatabilities of Coals,

Carbons and Hydrocarbon Minerals", Trans. AIME, Vol. 199, pp. 67-75, 1954.

18. Gayle, J. B., Eddy, W. H. and Shotts, R. Q., "Laboratory Investigation of Oxidation on Coal Flotation", USBM RI 6620, 1965.

19. Gutierrez, J. A. and Aplan, F. F., "The effect of Oxidation on the Surface Properties and Floatability of Coals of Differing Rank", manuscript in preparation.

20. Bird, B. M. and Marshall, S. M., "Jigging, Classification, Tabling and Flotation of Coals Presenting Difficult Washing Problems", USBM Bul 337, 1931.

21. Geer, M. R., "Washability Studies and Cleaning Treatments on Coals from Pierce County, Wash.", USBM RI 6538, 1964.

22. Rastogi, R. C. and Aplan, F. F., "Coal Flotation as a Rate Process", submitted to Trans. AIME.

23. Sun, S. C., "Effects of Oxidation of Coals on Their Flotation Properties", Trans. AIME, Vol. 199, pp. 396-401, 1954.

24. Bonner, C. M., Mineral Processing Section, The Pennsylvania State University, unpublished data.

25. Aplan, F. F., "Use of the Flotation Process for Desulfurization of Coal", Chapt. 6 in Coal Desulfurization, Wheelock, T. D., ed., ACS Symposium Series, No. 64, ACS, Washington, pp. 70-82, 1977.

26. Klepetko, E., "Recovery of Resin from Western Coal", Trans. AIME, Vol. 173, pp. 410-424, 1947.

Chapter 25

INORGANIC CONSTITUENTS OF NORTHERN GREAT PLAINS LIGNITE AND THEIR MODIFICATION

L. E. Paulson[1], D. N. Baria[2] and W. R. Kube[3]

INTRODUCTION

The Northern Great Plains has 16 billion tons of strippable lignite (1). Presently, 15 million tons per year are mined; essentially all is used for generation of electrical power. In the future, additional large quantities will be consumed in the production of gaseous and liquid fuels.

A major consideration in design and operability of power stations and conversion plants is the nature of the ash or inorganic constituents of the lignite used. These inorganic components cause fireside fouling in boilers, exert a catalytic effect in liquefaction and gasification, and contribute to particulate and SO_2 emissions.

Ash fouling on fireside boiler tubes cause reduced heat transfer and gas flow which limits boiler performance. Severe fouling can result in shutdown of the boiler. A comprehensive test program in both pilot plant and field studies at the Grand Forks Energy Technology Center (GFETC) has shown that ash fouling increases with increased sodium content of low-rank coals (2). The sodium in the coal can be

1. Project Manager, Coal Preparation (Acting), Grand Forks Energy Technology Center, U. S. Department of Energy, Grand Forks, North Dakota.
2. Associate Professor of Chemical Engineering, University of North Dakota, Grand Forks, North Dakota.
3. Professor of Chemical Engineering, University of North Dakota and Chemical Engineering, Grand Forks Energy Technology Center.

removed by ion exchange techniques.

Some sulfur in lignite can be removed by gravity separation. Such separation methods reduce the pyritic sulfur content but have a very minor effect on reduction of organic sulfur. Any reduction of sulfur before combustion reduces stack gas clean-up requirements. Credit for such reduction can be taken in the computation of compliance as set by the Clean Air Act Amendments of 1977, which establishes emission limits and percent reduction requirements for SO_2 emissions.

Several methods which can be used to change the inorganic composition to improve utilization characteristics of lignite are discussed in this paper.

COAL ASH

All coals contain inorganic components which remain as ash after the organic material has been removed. Major inorganic components in lignite ash are sodium, calcium, magnesium, iron, aluminum, and silicon, and to a lesser extent potassium, titanium, and phosphorous. Comparison of ash composition of lignites and bituminous coals shows that the lignite ash has higher concentrations of Na_2O, CaO, and MgO and lower SiO_2 and Al_2O_3 contents. The lignite ash is "basic" rather than "acidic" as is the case for ash from high-rank coals. The range in ash content and composition for North Dakota lignites and for bituminous coals is illustrated in Table 1 (3). Inorganic constituents in lignites occur as organically bound cations, as finely dispersed clay and SiO_2, and as separable minerals (4). Figure 1 shows microprobe scans of a high-sodium (Beulah) and a low-sodium (Savage) lignite (4). The organically-bound atoms (calcium, magnesium, sodium, sulfur and some iron and aluminum) are shown to be rather uniformly distributed (humic acid salts). The silicon and some of the aluminum occur as discrete particles as indicated by peaks on the scan.

ASH REDUCTION BY DENSITY SEPARATION

The inorganic constitutents which occur as extraneous matter in lignite can be removed partially by float-sink methods. Separation is dependent upon the density of the medium, size of particles, moisture content of lignite, and time of contact. Float-sink tests utilizing carbon tetrachloride (sp.gr. 1.55) have been performed at GFETC on a number of samples from different lignite mines in the Northern Great Plains. The laboratory test procedure consisted of crushing the lignite sample to the size range of 50 and 150 mesh. The crushed material was divided into two portions; one portion was analyzed for ash content and composition of the ash; the other was subjected to float-sink separation tests.

Figure 2 illustrates the distribution of lignite between the

Table 1. Range of Ash Analyses (3).

	North Dakota Lignite (338 Samples)			Bituminous Coal (135 Samples)		
	Low	High	Average	Low	High	Average
Ash, pct mf coal:	5.8	17.9	9.5	4.0	18.5	8.6
Analyses, pct of ash:						
SiO_2	6.3	48.1	20.0	9.3	69.6	48.2
Al_2O_3	4.2	26.1	11.8	5.6	41.7	24.9
Fe_2O_3	1.2	28.5	8.0	3.8	84.2	14.9
TiO_2	.1	.7	.3	.1	2.5	1.1
P_2O_5	.0	1.6	.3	NA	NA	NA
CaO	11.7	52.0	26.3	.2	36.7	6.6
MgO	2.8	13.6	7.2	.4	6.3	1.7
Na_2O	.2	27.8	6.9	.0	7.6	1.2
K_2O	.1	1.7	.3	.1	6.0	1.5
SO_3	5.2	32.0	17.2	NA	NA	NA
ASTM Softening Temp, $^{\circ}F$:	1,970	2,730	2,322	1,940	2,930	2,355

NA - Not Available

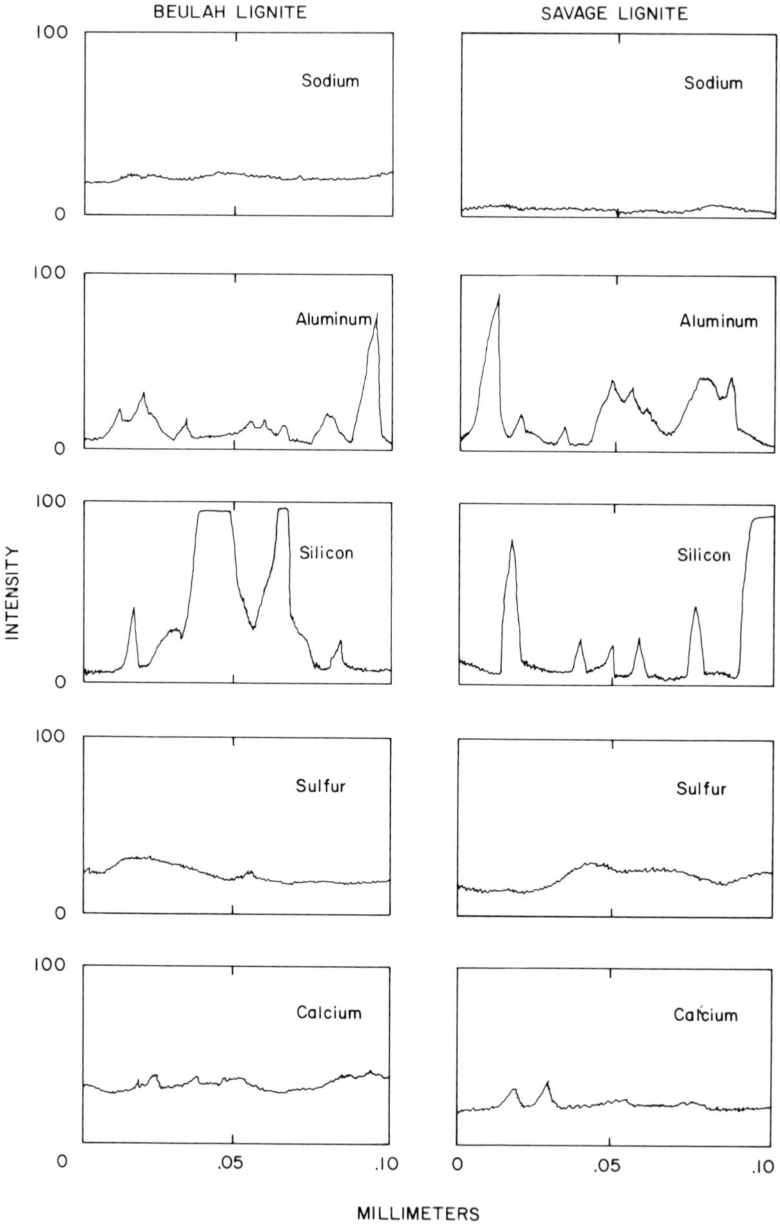

Figure 1. Microprobe scans of lignite showing the distribution of the inorganic constituents of the coal structure.

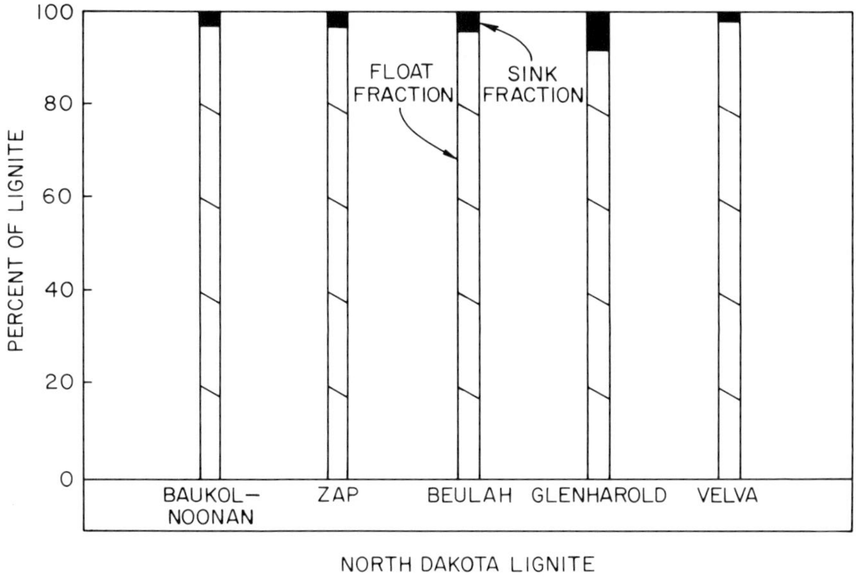

Figure 2. Distribution of float-sink fraction (s.g. 1.55) for various lignites.

float and sink fractions of samples taken from five mines in the Northern Great Plains region. In all cases, more than 90 percent of the original lignite was recovered as float material.

Table 2 summarizes the ash and sulfur contents for the startling lignite and for the float and sink fractions. Sulfur is reported on a moisture-free basis and ash on a moisture- and sulfur-free basis. These results show that the sink fraction in all cases contained 50 percent or more ash. Concentration of pyritic sulfur in the float portions was reduced by more than 50 percent. Reduction of organic sulfur was not significant.

Ash analyses of the original sample and of the float and sink fractions are presented in Table 3 on a sulfur-free basis. Minerals containing silica, aluminum, and iron tend to concentrate in the sink fraction. Other inorganics (e.g., sodium, magnesium, and calcium) are retained in the lignitic material.

Individual minerals in the sink fraction have been identified by X-ray diffraction and microprobe examination (5). Table 4 lists the minerals in descending order of concentration. None of the minerals

Table 2a. Ash and Sulfur Concentrations in Initial Coal and in Float and Sink Fraction Separation Made at a Specific Gravity of 1.55.

	BAUKOL-NOONAN			BEULAH			GLENHAROLD		
	Starting Material	Float	Sink	Starting Material	Float	Sink	Starting Material	Float	Sink
Distribution of coal, gms	100	96	4	100	95	5	100	92	8
Ash, % (Moisture, sulfur free)	7.5	6.0	49.2	6.6	5.4	47.2	11.0	6.8	65.7
Pyritic sulfur, % (Moisture free)	0.02	0.01	2.05	0.38	0.08	14.21	0.14	0.08	1.37
Organic sulfur, % (Moisture free)	0.40	0.34	1.76	0.99	0.78	0.00	0.87	0.88	0.69

Table 2b. Ash and Sulfur Concentrations in Initial Coal and in Float and Sink Fraction Separation Made at a Specific Gravity of 1.55.

	VELVA			ZAP		
	Starting Material	Float	Sink	Starting Material	Float	Sink
Distribution of coal, gms	100	97	3	100	96	4
Ash, % (Moisture, sulfur free)	6.3	5.2	65.1	6.0	4.8	48.6
Pyritic sulfur, % (Moisture free)	0.02	0.01	0.57	0.48	0.05	19.56
Organic sulfur, % (Moisture free)	0.36	0.32	0.57	0.32	0.43	0.74

Table 3a. Ash Analysis of Initial Coal and Float and Sink Portions Separation Made at a Specific Gravity of 1.55.

Ash Constituent	BAUKOL-NOONAN			BEULAH			GLENHAROLD		
	Starting Material	Float	Sink	Starting Material	Float	Sink	Starting Material	Float	Sink
SiO_2, %	30.4	26.6	46.1	17.5	16.2	19.4	45.8	27.5	65.2
Al_2O_3, %	19.2	19.2	21.1	15.4	16.0	13.2	15.0	14.3	17.6
Fe_2O_3, %	10.3	6.8	25.4	14.3	4.8	57.9	8.2	7.1	9.2
TiO_2, %	1.0	0.8	0.6	0.6	0.7	0.5	1.0	0.8	0.9
P_2O_5, %	0.6	0.9	0.1	1.3	1.4	0.2	0.6	0.8	0.1
CaO, %	20.7	25.3	3.9	32.6	37.8	7.2	14.6	28.4	1.9
MgO, %	4.8	5.7	0.8	9.6	11.4	1.3	4.3	6.7	1.5
Na_2O, %	12.3	13.3	1.2	8.3	10.9	0.1	8.9	13.0	1.5
K_2O, %	0.7	0.8	0.8	0.4	0.8	0.2	1.6	1.4	2.1

Table 3b. Ash Analysis of Initial Coal and Float and Sink Portions Separation Made at a Specific Gravity of 1.55.

Ash Constituent	VELVA			ZAP		
	Starting Material	Float	Sink	Starting Material	Float	Sink
SiO_2, %	16.3	13.1	40.5	21.1	19.8	22.6
Al_2O_3, %	10.2	10.0	11.7	13.4	15.0	8.2
Fe_2O_3, %	11.0	6.7	40.7	22.0	10.3	62.7
TiO_2, %	0.5	0.5	0.3	0.5	0.5	0.5
P_2O_5, %	1.1	1.1	0.1	0.5	0.6	0.1
CaO, %	39.3	44.1	4.3	23.2	29.3	3.9
MgO, %	9.6	10.8	1.1	8.5	11.6	1.1
Na_2O, %	11.6	13.0	0.5	10.4	12.1	0.5
K_2O, %	0.4	0.7	0.8	0.4	0.8	0.3

Table 4. Major Minerals Identified in Sink Fractions of Lignite (5). (Listed in descending order of prevalence).

Beulah		Baukol-Noonan		Glenharold		Savage	Velva	Gascoyne
High sodium	Low sodium	High sodium	Medium sodium	Top seam	Bottom seam			
Nacrite	Nacrite	Calcite	Nacrite	Quartz + Nacrite	Pyrite	Quartz + Nacrite	Quartz + Nacrite	Quartz + Nacrite
Quartz	Quartz	Nacrite	Quartz + Nacrite	Gypsum	Nacrite	Pyrite	Hematite	Nacrite
Pyrite	Pyrite	-	Quartz	Quartz	Calcite	Quartz	Nacrite	Pyrite
Hematite	Hematite	-	Calcite	-	Quartz	-	Quartz	Quartz
Barite	Barite	-	-	-	-	-	Barite	Barite

listed contain sodium as a component.

ION EXCHANGE IN LIGNITE

Replaceable atoms in lignite exist as salts of carboxylic acids and hydroxyl groups of humic acids. Almost all the sodium and potassium occur in this form as does most of the calcium and magnesium and some of the aluminum and iron (2,4,6). In general, any cation in solution can exchange with cations that are organically bound, with the lower valence ions being more easily replaced by ions of higher valence. Lignite is a weakly acidic cation exchanger. Ions which can be replaced from this exchange structure, starting with the easiest to replace, are: $Na^{+} > K^{+} > Mg^{++} > Ca^{++} > Al^{+++} > Fe^{+++} > H^{+}$ (7).

Hydrogen ion forms the most stable structure because of bonding with the oxygen atom on the carboxylic group. The other cations form strong electrolytes with the carboxylic group and are more readily exchanged (8). Replacement also depends upon the ionic strength of the exchanging ion in solution. The ionic strength is dependent on the valence and concentration of the ionic species.

ION EXCHANGE MECHANISM

The ion exchange process studied occurs in an aqueous medium and is postulated to occur in five steps:

1. Diffusion of the exchange ion from the bulk liquid phase through a boundary layer to the surface of the lignite.

2. Diffusion of the exchange ion through the pores and capillaries in the lignite.

3. Ion exchange in the interior of the lignite where the bound cation is replaced by the exchange ion.

4. Diffusion of the replaced ion through the pores and capillaries to the exterior.

5. Diffusion of the replaced ion from the exterior lignite surface to the bulk liquid phase through a boundary layer.

PARAMETERS AFFECTING ION EXCHANGE

The removal of sodium from lignite by ion exchange has been investigated in batch (9-11) and continuous processes (12), and was found to be a function of particle size, contact time between solids and the exchanging solution, moisture content of the lignite, ionic strength of the exchange solution (exchange ion), and the lignite-to-

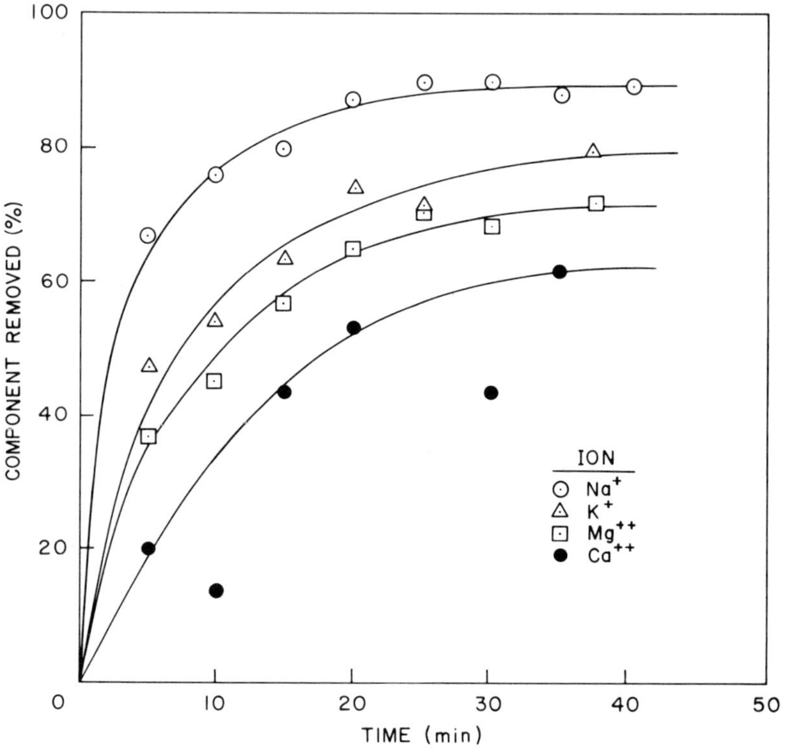

Figure 3. Removal of several ions from lignite with sulfuric acid as a function of time. The acid concentration, liquid/solid ratio, mean particle size, and initial sodium concentration were 0.1 N, 0.822 mm, 2.0 min, and 8.5 percent Na_2O in the ash, respectively.

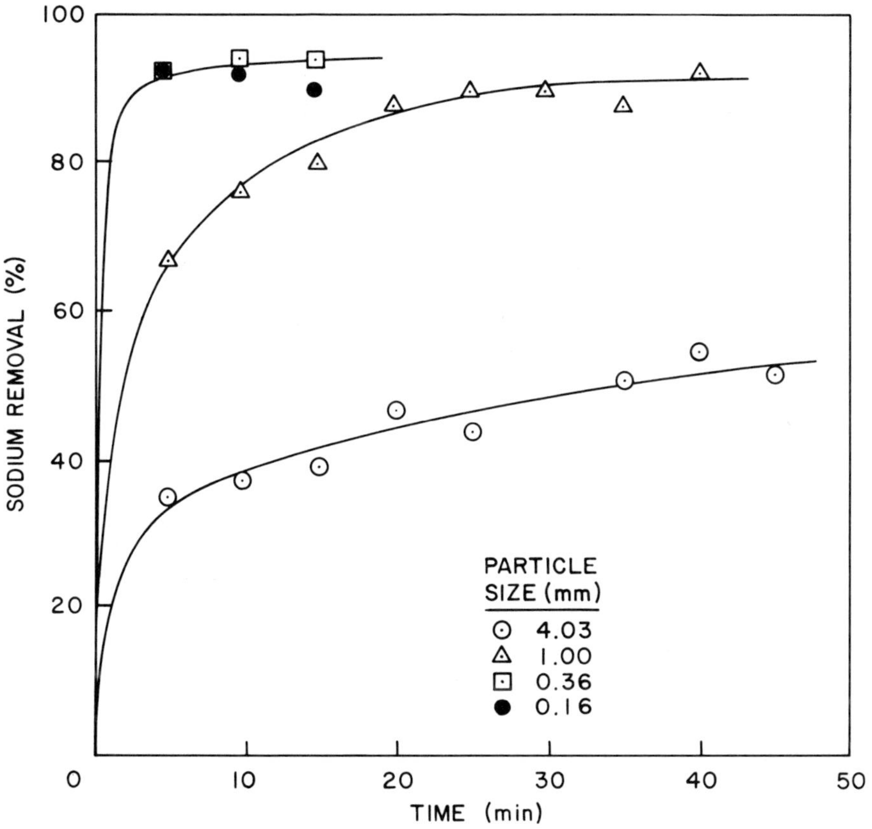

Figure 4. Sodium removal from lignite with sulfuric acid as a function of mean particle size. The acid concentration, liquid/solid ratio, and initial sodium content were 0.1 N, 10, and 8.5 percent Na_2O in the ash, respectively.

solution mass ratio. Results from the batch and continuous tests were similar.

The amount of sodium removed increases as contact time between the lignite and the solution increases, as shown in Figure 3, where the percent sodium removed from lignite in batch tests is plotted against contact time in a 0.1N H_2SO_4 solution. Removal of other replaceable ions is also shown. Sodium is indicated to be easiest to replace, followed by potassium, magnesium, and calcium. This agrees with the theory that the ease of replacement in lignite is $Na^+ > K^+ > Mg^{++} > Ca^{++}$ (7).

Removal of sodium in batch tests is presented as a function of contact time for various particle sizes of lignite in a 0.1N H_2SO_4 solution in Figure 4. These data indicate that as particle size decreases, the amount of sodium removed increases. Figure 5 is a similar plot using deionized-distilled water as an exchange medium and illustrates that sodium is removed to a lesser extent using deionized water. Removal is again size dependent, reaffirming the postulate that diffusion through the pores of the lignite may be controlling, especially for larger particles where there is a longer diffusion path.

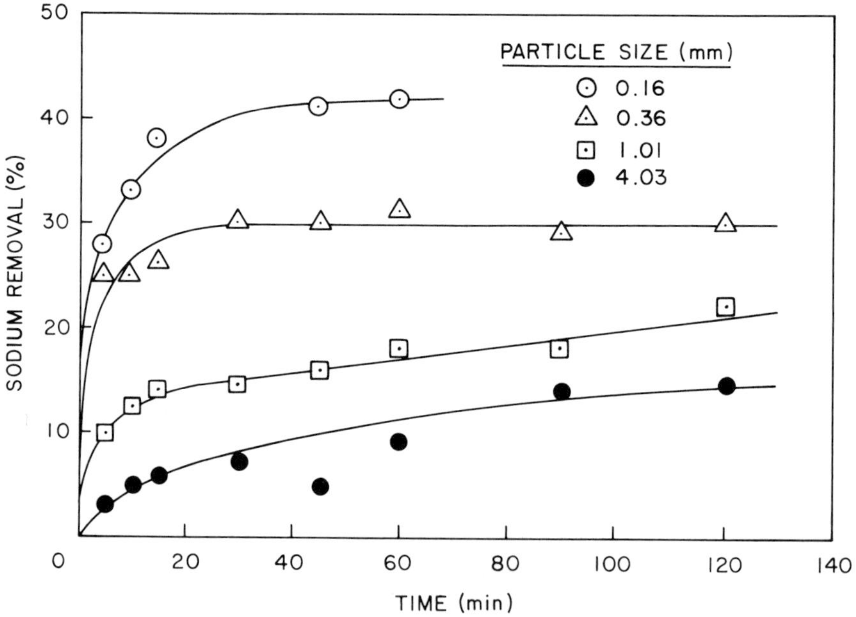

Figure 5. Sodium removal from lignite with water as a function of particle size. The liquid/solid ratio and initial sodium content were 10 and 8.5 percent Na_2O in the ash, respectively.

In Figure 6, sodium removal using $CaCl_2$ as the exchange solution in batch tests is presented as a function of the moisture content of the lignite. As moisture is removed from lignite, there is a decrease in both the rate and extent of ion exchange, which may be the result of shrinkage of the lignite particle and collapse of the pores and capillaries. Wetting of the lignite does not reopen the restricted pores (10).

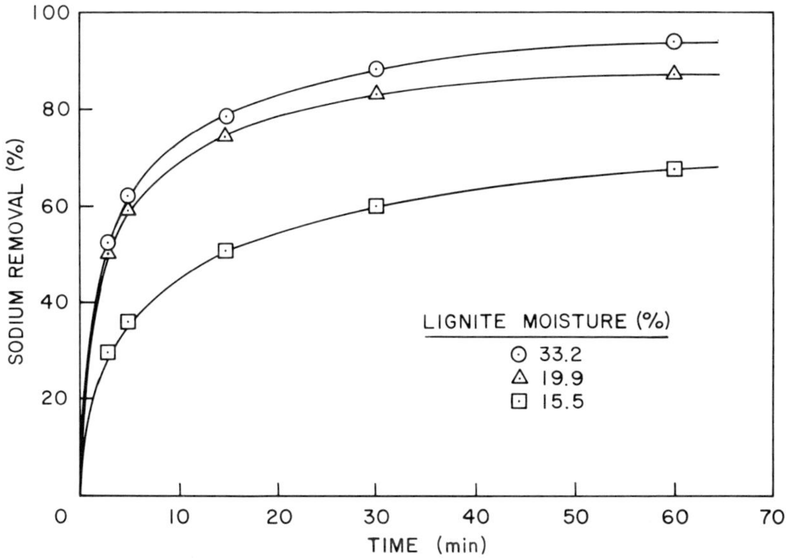

Figure 6. Sodium removal from lignite with calcium chloride as a function of moisture content. The $CaCl_2$ concentration (expressed as CaO), particle size, and initial sodium content were 162 ppm, 0.0492 mm, and 7.1 percent Na_2O in the ash, respectively (9).

Figure 7 shows the effect of $CaCl_2$ (exchange solution) on the amount of sodium removed in a continuous countercurrent exchanger (12). The amount of sodium removed increases as the concentration of the exchange solution increases. This is also shown in Figure 8 which is a plot of the amount of Na_2O remaining in lignite on a moisture-free basis against the ionic strength of various exchange solutions (batch tests).

The amount of metal ions removed as a function of the exchange

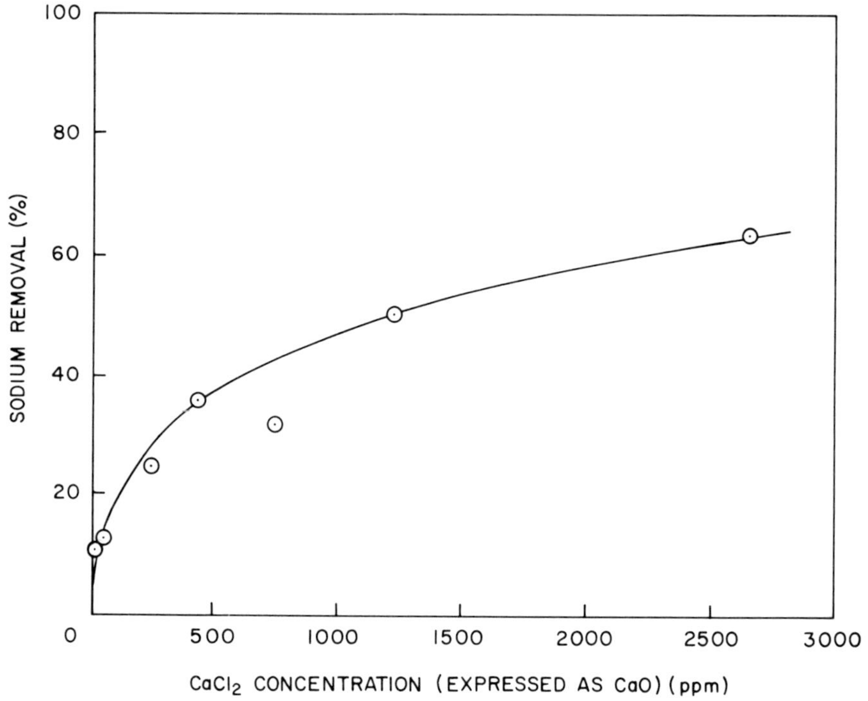

Figure 7. Sodium removal from lignite as a function of calcium chloride concentration. The particle size, contact time, liquid/solid ratio, and initial sodium content were 7.14 mm, 1.9 hours, 7.0, and 7.4 percent Na_2O in the ash, respectively. (Computed from Ref. 11).

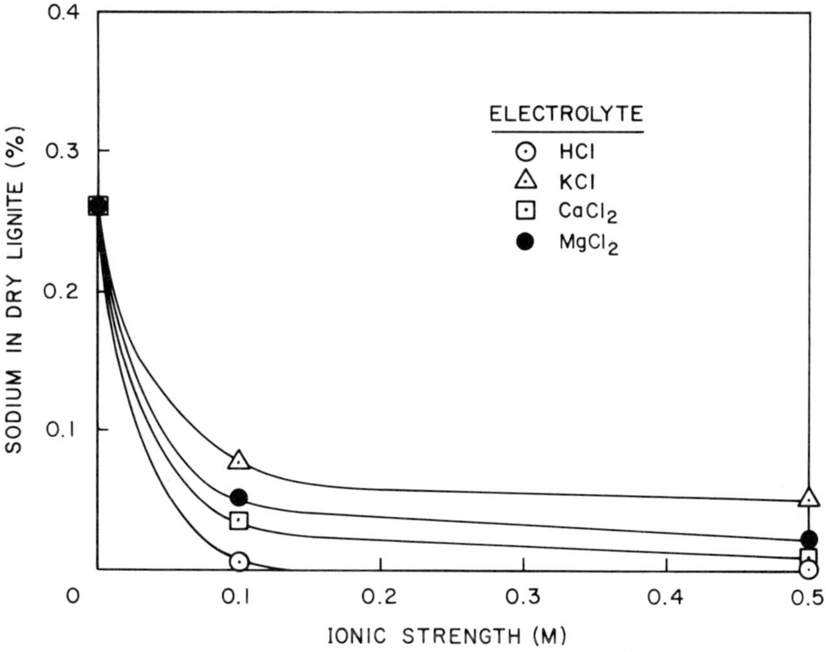

Figure 8. Sodium removal from lignite as a function of ionic strength. The mean particle size, contact time, liquid/solid ratio, and initial sodium concentration were 0.0821 mm, 2.5 min, 6.0, and 0.4 percent in the moisture free lignite(10).

solution-to-lignite ratio is plotted in Figure 9 using 0.1N H_2SO_4 in continuous tests. Percentage removal increases as the liquid-to-solid ratio increases up to 2; for ratios above 2, there is no significant increase. The hierarchy of removal of the ions again is similar to that predicted by theory.

It can be concluded that sodium can be easily replaced from lignite using ion exchange techniques. The most effective exchange ion for replacement of Na^+ is H^+. Decreasing the particle size, increasing the concentration of the exchange ion, and increasing the liquid-to-solid ratio all increase the rate and extent of sodium removal. Drying the original lignite decreases the rate of sodium removal.

CONCEPTUAL DESIGN OF A PLANT FOR SODIUM REMOVAL FROM LIGNITE

A conceptual design has been made of a facility for removing sodium from 1,584,000 tpy of lignite. Lignite from Mercer County, near Beulah, North Dakota, containing 8.5 percent Na_2O in ash was selected as the test lignite. The sodium content was to be reduced to 4 percent as Na_2O in ash to produce a low-fouling fuel. Since many costs are site specific, a mine-mouth operation in west-central North Dakota was selected because of the concentration of large power generating complexes in the area. The ion exchange solution selected was sulfuric acid.

Figure 10 presents a simplified flow diagram illustrative of major steps for the ion exchange and process water clean-up steps.

The nominal capacity of the plant is 200 tph of lignite as received with essentially full moisture content. The feed lignite is crushed to pass through 6 mesh. Sixty percent of the crushed lignite is fed to the countercurrent ion exchange unit with 40 percent by-passed to be mixed with the treated lignite to give a product containing 4 percent sodium oxide in the ash. In the ion exchange unit, the lignite is contacted countercurrently with 0.1N H_2SO_4. Residence time for the lignite is five minutes. The extracted lignite containing approximately 1.35 percent Na_2O in the ash is size separated on a vibrating screen. The plus 28-mesh fraction is dewatered using a vibrating screen centrifuge with the minus 28-mesh fraction blended in a mix tank with the filtrate from the centrifuge and the underflow from the first set of hydrocyclones. A horizontal belt filter is used to separate and dewater the fine material. Both dewatered fractions are combined and blended with the by-passed lignite. Concentrated H_2SO_4 is diluted to the required concentration in a mix tank prior to feeding into the ion exchange unit. The filtrate from the horizontal belt filter provides part of the water to dilute the acid.

The spent solution from the ion exchange unit is processed through a series of hydrocyclones and a backwash filter to separate the carry-over lignite fines. The high pressure filtrate from the backwash filter is sent through a power recovery unit before treatment

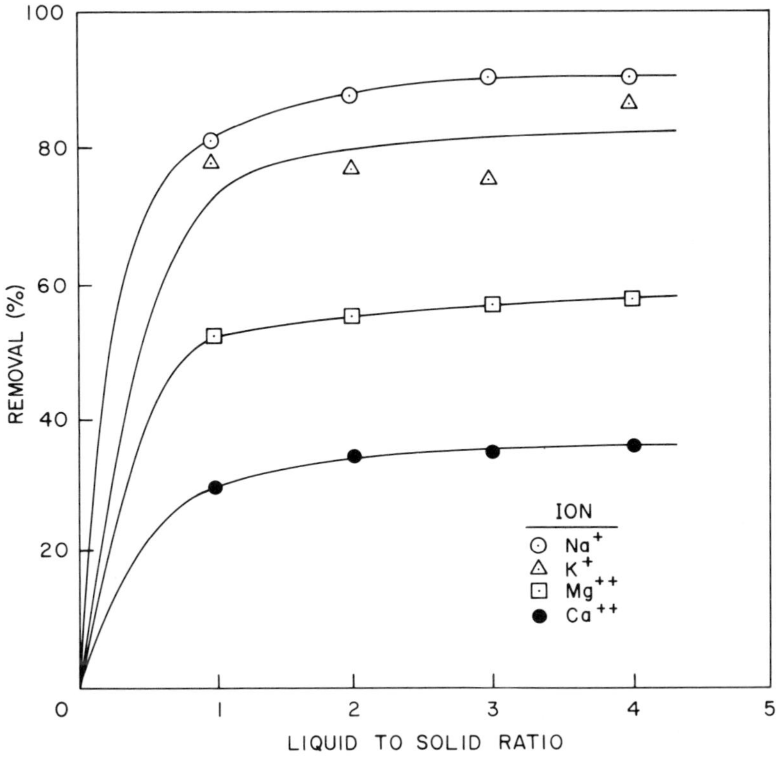

Figure 9. Removal of several organically bound metals from lignite with sulfuric acid as a function of the liquid/solid ratio. The acid concentration, mean particle size, contact time, and initial sodium content were 0.1 N, 0.822 mm, 2.0 min, and 8.5 percent Na_2O in the ash, respectively.

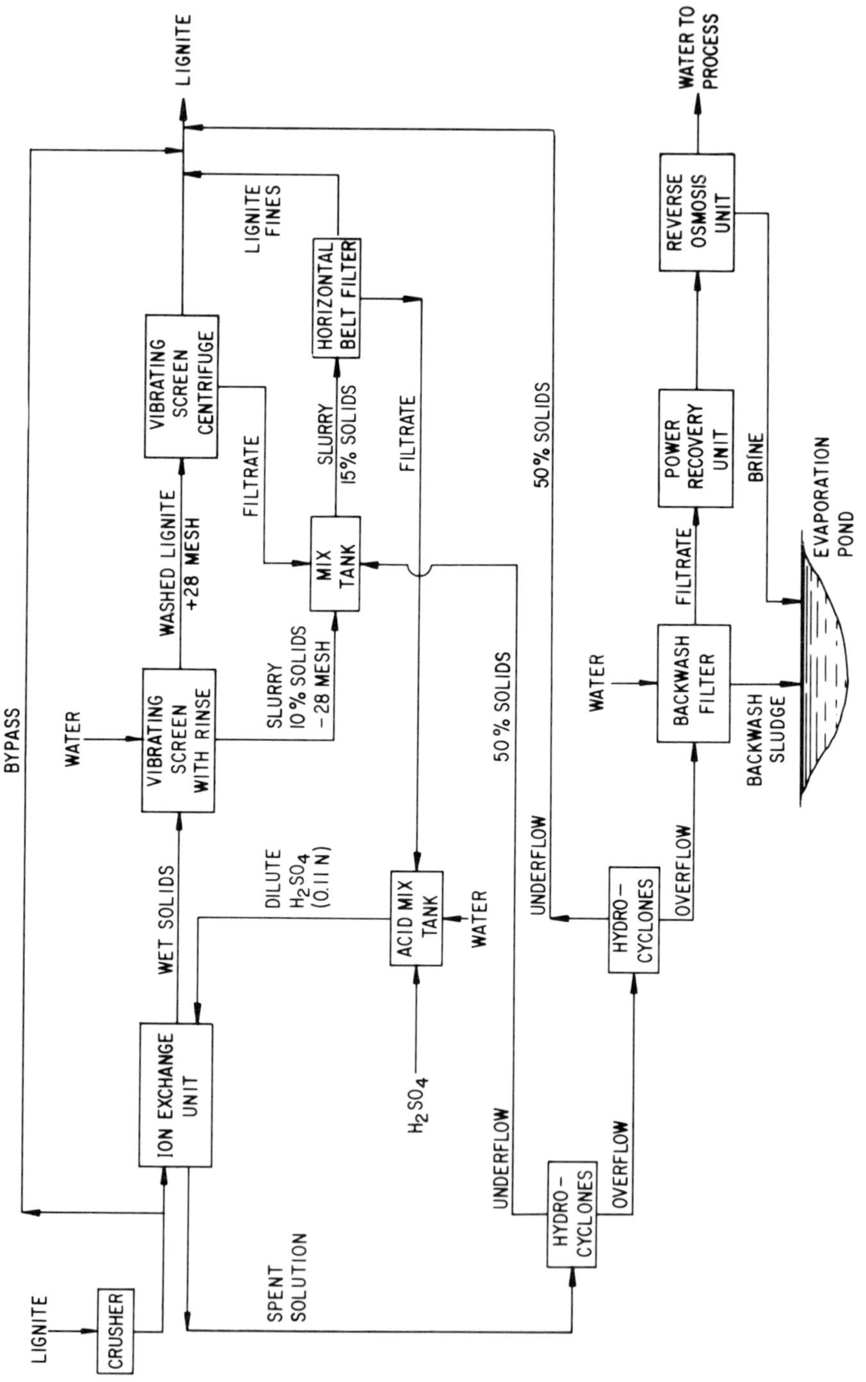

Figure 10. Process for the removal of sodium from lignite.

by reverse osmosis. The cleaned stream is reused as process water.

Underflow from the first series of hydrocyclones is passed to the mix tank before the horizontal belt filter, and that from the second series is sent directly to product blending. Sludge from the backwash filter and the brine from the reverse osmosis unit are discharged to an evaporation pond for disposal.

Reuse of process water reduces water requirements. Environmental impact of the process is low as the only waste streams are those from the backwash filter and reverse osmosis unit which are not discharged into the general surroundings.

FUTURE WORK

The ion exchange mechanism in lignite needs to be studied in greater detail. Work is underway using both batch and continuous systems to obtain data for construction of a model equation which will predict the ion exchange behavior of lignite. Plans also call for building and operating a 5-tpHr pilot plant to remove sodium from lignite by ion exchange.

Further work in separation of extraneous minerals from lignite is anticipated.

REFERENCES

1. Wiebner, J. D., "Lignite and North Dakota," Mining Engineering, 29, No. 8, 12 pp., August 1977.

2. Sondreal, E. A., Tufte, R. H. and Beckering, W., "Ash Fouling in the Combustion of Low-Rank Western U.S. Coals," Combustion Science and Technology, 16, pp. 95-110, 1977.

3. Sondreal, E. A. and Ellman, R. C., "Fusibility of Ash from Lignite and its Correlation with Ash Composition," U.S. Energy Research and Development Administration GFERC/RI-75/1, 121 pp., 1975; available from National Technical Information Service, Springfield, Virginia 22161.

4. Sondreal, E. A., Gronhovd, G. H. and Kube, W. R., "Research and Development Relating to Lignite Use in Power Production," Paper in Gulf Coast Lignite Conference: Geology, Utilization, and Environmental Aspects, University of Texas at Austin, Austin, Texas, June 2-4, 1976, comp. by W. R. Kaiser. U.S. Department of Energy CONF-7606131, pp. 100-124, 1978; available from National Technical Information Service, Springfield, Virginia 22161.

5. Paulson, L. E., Beckering, W. and Fowkes, W. W., "Separation and Identification of Minerals from Northern Great Plains Province Lignite," Fuel, 51, pp. 224-227, 1972.

6. Beckering, W., Haight, H. O. and Fowkes, W. W., Examination of Coal and Coal Ash by X-Ray Techniques. Paper in Technology and Use of Lignite, Proceedings: Bureau of Mines - University of North Dakota Symposium, Grand Forks, North Dakota. May 1-2, 1969, comp. by J. L. Elder and W. R. Kube. BuMines IC 8471, pp. 89-102, 1970.

7. Kunin, R. and McGarvey, F. X., "Principles of Deionization with Ion Exchange Resins," Chapter in Ion Exchange Technology, ed. by F. C. Nachod and J. Schubert, Academic Press Inc., New York, p. 97, 1956.

8. Kunin, R. and Berry, R. E., "Carboxylic Weak Acid Type, Cation Exchange Resin," Ind. Eng. Chem., 41, pp. 1269-1272, 1949.

9. Gronhovd, G. H., Harak, A. E. and Paulson, L. E., "Ash Fouling Studies of North Dakota Lignite," Paper in Technology and Use of Lignite. Proceedings: Bureau of Mines - University of North Dakota Symposium, Grand Forks, North Dakota, April 27-28, 1967, comp. by W. R. Kube and J. L. Elder. BuMines IC 8376, pp. 76-94, 1968.

10. Crystal, J. T., "The Removal of Sodium from Lignite by Ion Exchange," M.S. Thesis, University of North Dakota, 24 pp., August, 1970.

11. Paulson, L. E. and Fowkes, W. W., "Changes in Ash Composition of North Dakota Lignite Treated by Ion Exchange," BuMines RI 7176, 19 pp., 1968.

12. Paulson, L. E. and Ellman, R. C., "Reduction of Sodium in Lignite by Ion Exchange: A Pilot Plant Study," U.S. Department of Energy GFETC/RI-79/1, 50 pp., 1979; available from National Technical Information Service, Springfield, Virginia 22161.

Chapter 26

THERMAL DRYING OF WESTERN COAL - A REVIEW

Larry G. Bauer

Department of Chemical Engineering
South Dakota School of Mines and Technology
Rapid City, South Dakota

The vast coal reserves in the Wyoming, Montana, North and South Dakota region are sufficient to supply the total energy needs of the United States for several hundred years. Not only is there an abundance of this sub-bituminous and lignite coal available, but it is easily obtainable by strip-mining. The overburden varies from several feet to several hundred feet.

This western coal has yet another asset, namely, a low sulfur content. Whereas eastern and midwestern coal has a sulfur content of 2.5%, the upper Great Plains coal contains sulfur in the 0.5% range. The importance of the sulfur content of the coal is a direct result of federal government emission standards for sulfur dioxide (SO_2). Hand (1) points out that "sub-bituminous coals and lignites have a propensity for retaining this sulfur in the ash under certain conditions which is not evident in higher rank coals." This would tend to reduce further the SO_2 concentration in the combustion gases of boilers.

Western coal is no panacea, however, in that it is far removed from most population and industrial centers. Furthermore, the moisture content of the coal varies from 20 to 40%. This moisture is almost exclusively inherent, that is, it is chemically and physically bound to the organic matter of the coal. It is believed that the water forms a part of the coal structure perhaps as water of hydration of organic salts or organic acids, perhaps by hydrogen bonding, or perhaps by simple adsorption (2).

The primary market for this coal is the electric power industry. Lesser markets will be as coke for the steel industry and possible as a raw material for coal gasification and coal liquefaction plants.

When one considers shipping western coal to eastern and mid-western markets, an obvious question arises. Why should one pay the freight charges to haul water one third to one half the distance across the country. Consider, for example, a 100-car unit train carrying sub-bituminous coal having an inherent moisture content of 25%. A hopper car will generally carry 100 tons of coal. Consequently, a unit train of sub-bituminous coal would contain

$$(100 \text{ cars})\left(\frac{100 \text{ tons coal}}{\text{car}}\right)\left(\frac{25 \text{ tons } H_2O}{100 \text{ tons as-mined coal}}\right) = 2{,}500 \text{ tons } H_2O$$

The transportation charges will be on the order of 1 to 2 cents per ton-mile; so if the coal is being shipped 1,500 miles at 1.5¢ per ton-mile, one is shipping

$$(2{,}500 \text{ tons } H_2O)(1{,}500 \text{ miles})\left(\frac{\$0.015}{\text{ton mile}}\right) = \$56{,}250$$

for shipping water halfway across the nation. Certainly a reduction in transportation charges of approximately $50,000 per unit train would appear to be sufficient justification for drying coal prior to shipping.

The literature (1,3-8) reveals additional reasons for drying western sub-bituminous and lignite coal prior to shipping. Some of these are:

1. Transportation savings. (Added for sake of completeness.)

2. Increase in boiler efficiency. The rule of thumb generally accepted by the industry is that there is an increase in boiler efficiency of 1/8% for each percentage point reduction in moisture.

3. Reduced coal handling problems associated with freezing.

4. Reduction in severance taxes and royalties in some states if these were paid on a per ton of coal shipped basis.

5. Improved quality of coal for special-purpose uses such as the production of coke, briquettes and chemicals.

6. Reduced operating and maintenance costs of boilers:

    a. Moisture content of the dried coal does not limit the induced draft fans but actually results in a reduced operating pressure of about 8%. This results in a lower fan horsepower consumption.

    b. Carbon carryover due to incomplete combustion is reduced, and fouling in areas of the boiler not covered by the original soot blower is minimized.

c. Combustion occurs in the proper part of the furnace. When using high-moisture coal, burning can occur in a different section of the unit than the one in which the boiler designer intended.

d. Lower gas exit temperature is achieved without danger of corrosive sulfuric acid condensate forming in the air heater.

7. Higher flame temperature in the boiler since large volumes of water aren't being evaporated.

8. Reduced potential bottlenecks in crushing, grinding and conveying capacity as well as in the capacity of the induced draft fans.

9. Less coal loss during transportation in open hopper cars if dried coal has been "oiled."

Even though there are many advantages to be gained by drying western coal, drying does magnify conventional coal handling problems (1,3-8). For example;

1. Dried coal is more susceptible to spontaneous combustion than as-mined coal.

2. Dried coal generates more dust, and

3. Dried coal is more friable; consequently degradation in particle size distribution due to handling is commonplace.

Kindig and Phillips (3) mention that "spontaneous ignition involves three important variables: availability of oxygen, adsorption of moisture, and temperature."

Cooling of the coal, after drying, is advisable prior to loading in hopper cars or stockpiling. Ellman et al. (4) state "that stockpiling and handling requirements do not appear to be significantly different from as-mined coal. The need for thorough compaction (to minimize oxygen and moisture penetration) is somewhat greater, but normal coal handling equipment can be used." Paulson et al. (6) in their discussion on transportation of dried coal also mention that the oxygen of the pile gases (within the standard open top, bottom dump hopper car) is rapidly consumed by the dried coal to provide an inert atmosphere. If no additional air is provided, additional temperature increases can not occur. They had experienced ignition at the openings of the dump doors, but this was eliminated when the doors were filled with fiber insulation and secured with plastic sheeting prior to loading.

In regard to the problem of dust generation, Paulson et al. (6) concluded that dried, low-rank coal must be treated to reduce dust problems in subsequent handling operations unless a completely closed system is provided. The use of 1.5 to 2 gallons of number 6 oil per ton of coal has been shown to be an effective dust abatement procedure. It is common to spray the oil on the coal, after cooling, on leaving the dryer. The cost of the oil is partially offset by its heating value contribution on combustion. An added benefit of the oil spray is that it significantly reduces the coal loss due to windage (during transportation) compared to as-mined coal.

Although the dried coal is more friable than as-mined coal and handling of it will result in a larger fines content, the combination of oil spray and a properly designed enclosed system can overcome this liability.

Regardless of the relative merits of shipping dried coal as compared to shipping as-mined coal, the decision to do so will be made after considering "the bottom line." Kindig and Phillips (3) have recently published an economic study on the transportation of dried western coal. The basis of their study was the amount of coal required to operate a 1,000 MW electric power generating plant. Two different coal sources were considered. The first was a Denver Basin lignite and the second was a Powder River Basin sub-bituminous. For the economic analysis, these coals were considered to be dried from 35 to 5% and 28 to 5% moisture, respectively. Kindig and Phillips point out that in their study only the transportation savings are compared to drying costs; other benefits constitute an added advantage. Table 1 shows some of the relevant physical properties of these two types of coal as well as the annual coal requirements for operation of a 1,000 MW power generating plant. Tables 1 through 4 and Figures 1 through 5 were taken from Kindig and Phillips (3).

In order to determine the economic benefits, if any, of drying coal prior to shipping, it will be necessary to know the cost associated with drying. An estimate of the capital required for the drying facilities can be obtained after considering both the conventional means of preparing coal for shipping and the proposed means of shipping dried coal.

Figure 1 shows the current processing steps used in shipping as-mined coal, while Figure 2 depicts the processing steps that would be needed to ship dried coal. Inspection of these two figures reveals that drying of as-mined coal requires the addition of equipment for (1) crushing coal from 2" x 0 x 3/4" x 0, (2) a fluidized bed dryer, and (3) additional storage facilities.

The incremental capital and operating costs for drying of low-rank coal is shown in Table 2 of Kindig's and Phillip's paper. One notes that the drying costs are in the neighborhood of $2.60 per ton of as mined coal.

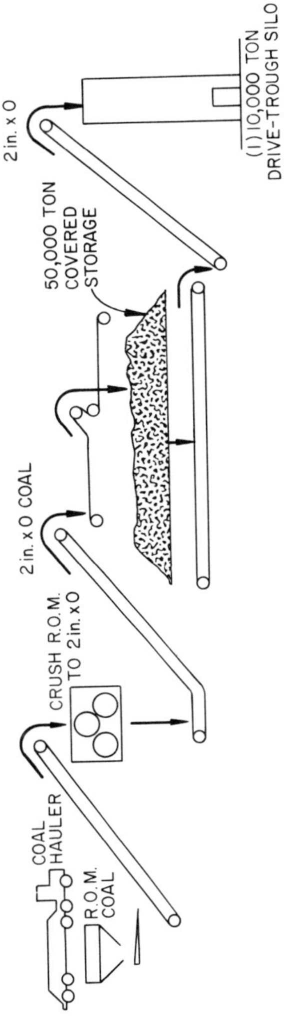

Figure 1. Coal Crushing, Handling and Load-out of Undried Low-rank Coal

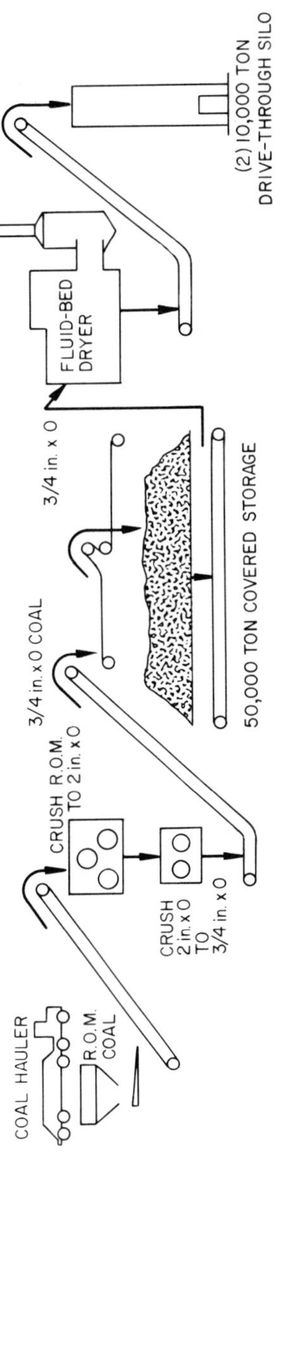

Figure 2. Coal, Crushing, Handling, Thermal Drying and Load-out of Dried Low Rank Coal

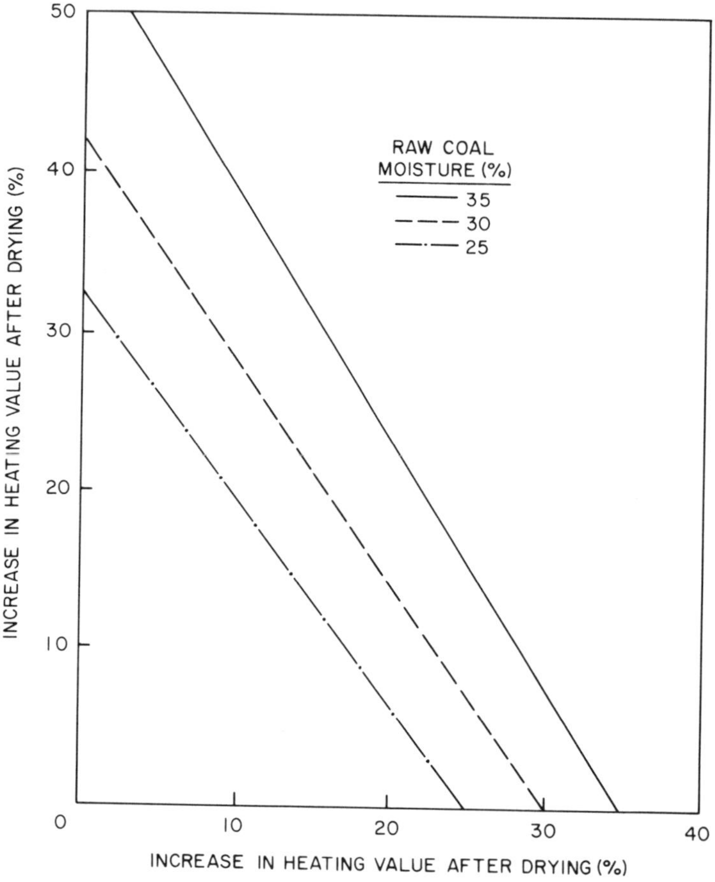

Figure 3. Increase in heating value after drying for various moisture contents.

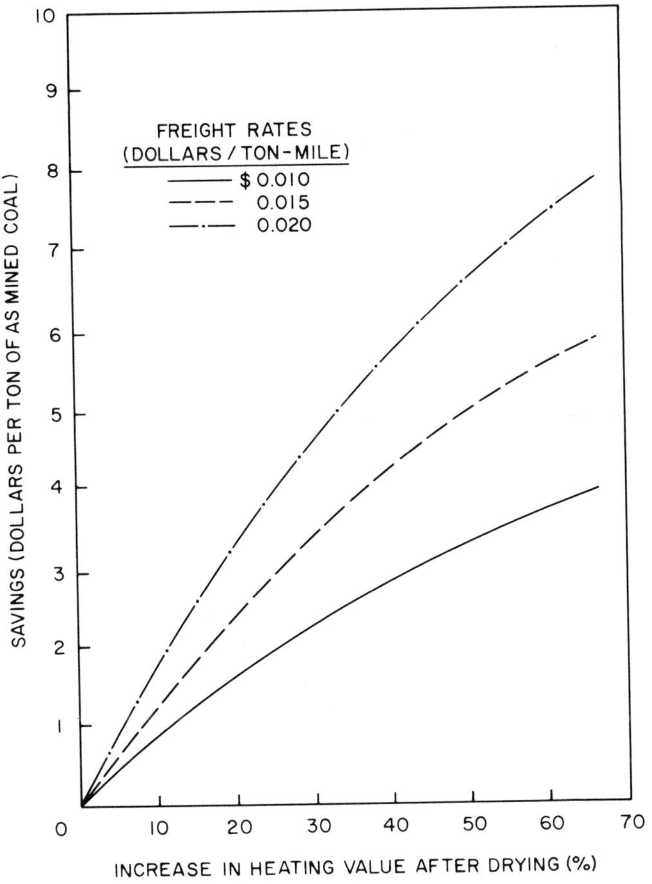

Figure 4. Transportation savings for various increases in heating value after drying for a transportation distance of of 1,000 miles.

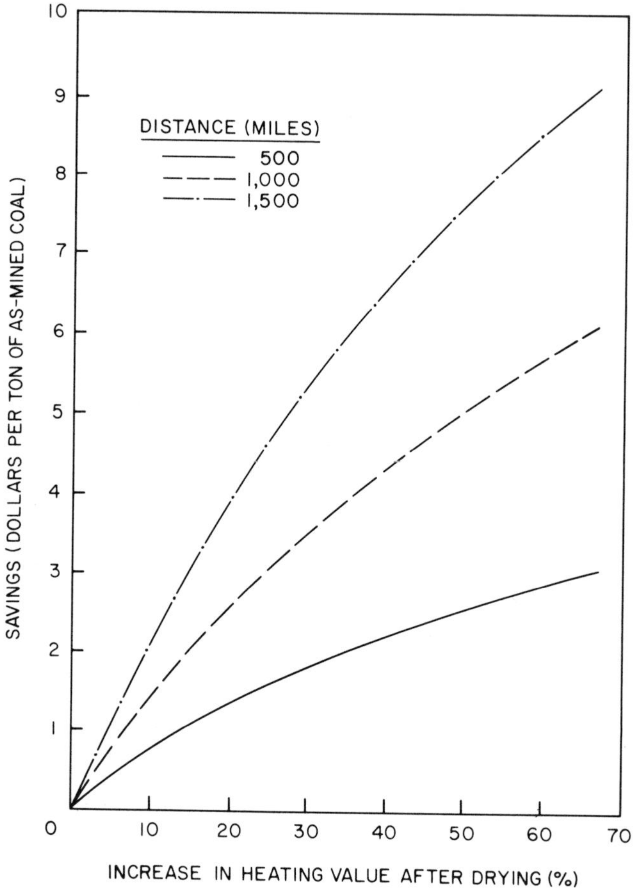

Figure 5. Transportation savings for various increases in heating value after drying for a freight rate of $0.15/ton-mile.

Table 1 - Tonnage of Various Types of Coal to Satisfy Thermal Requirement of a 1,000-Mw Generating Station

Coal Rank	Moisture Percent and Type	Heat Content Btu/lb	Annual Coal Requirement for a 1,000-Mw Generator* Millions of Tons
Lignite			
As mined	35 inherent	6,300	4.25
Dried	5 inherent	9,208	2.91
Subbituminous			
As mined	28 inherent	8,700	3.08
Dried	5 inherent	11,479	2.33
Bituminous			
As received	6 surface	12,500	2.14

*Assumed 9,400 Btu per net kwh and a plant capacity factor of 0.65. This calculates to a requirement for 5.352×10^{13} Btu per year.

In order to compare the cost of drying coal to the transportation savings that would result, it is helpful to consider Figure 3 which indicates the increase in heating value associated with drying the coal to various moisture levels. After the increase in heating value due to drying has been established, one can determine the transportation savings for various freight rates and shipping distances from Figures 4 and 5.

The break-even distance (where drying costs just equal transportation savings) are shown for various freight rates in Table 3. These break-even distances do not reflect any expenses associated with stabilizing or "oiling" the coal to reduce the problems of spontaneous combustion or dust generation. If one assumes a cost of $0.75 per ton of as-mined coal for stabilization, the break-even distances shown in Table 3 will be increased by about 30%. The break-even distances for shipping stabilized dried coal are shown in Table 4. Inspection of information presented in Table 4 suggests that the economics of shipping dried stabilized coal distances of one thousand miles or more appears very favorable. Recall that these savings are only savings in transportation. They do not include any of the other benefits mentioned previously.

Kindig and Phillips indicated the use of a fluidized bed dryer for removing moisture from low-rank western coals. This is one technique that is currently being used; there are others, however.

Table 2 - Incremental Capital and Operating Costs for Thermally Drying Low-rank Coals.

	Powder River Basin Sub-bituminous	Denver Basin Lignite
Basis (Btu for 1,000-Mw station, million tpy)		
As-mined	3.08	4.25
Dried to 5% moisture*	2.33	2.91
Incremental capital requirement (millions of dollars)		
Crushing to minus 3/4-in.	$ 3.87	$ 4.69
Drying	9.75	13.00
Silo storage, 10,000 tons	3.50	3.50
Total	$17.12	$21.19
Capital recovery factor, 15% for 15 years	$ 2.92	$ 3.62
Incremental operating costs (millions of dollars)		
Fuel, 4 x 16^6 Btu/ton H_2O evaporated $1/$10^6$ Btu^2	$ 3.00	$ 5.36
Power	0.19	0.34
Maintenance, taxes and insurance	1.00	1.30
Labor, supervision	0.70	0.70
Total	$ 4.89	$ 7.70
Operating costs (per ton of as-mined coal)		
Direct, fuel, power, maintenance, labor	$ 1.59	$ 1.81
Indirect, amortizing capital	0.95	0.85
Total	$ 2.54	$ 2.66
Total operating cost per ton of water evaporated	$10.41	$ 8.44

*Increase in Btu values, 32% for subbituminous, 46% for lignite

Table 3 - Break-even Distance Where Drying Costs* Equal Transportation Savings (Various Freight Rates).

Freight Rate, $ per ton-mile	Distance, miles Subbituminous	Lignite
0.010	1,049	842
0.015	699	562
0.020	525	421

*Drying cost $2.54 per ton of as mined subbituminous coal, $2.66 per ton of as-mined lignite.

Table 4 - Break-even Distance Where Drying and Stabilizing* Costs Equal Transportation Savings (Various Freight Rates).

Freight Rate, $ per ton-mile	Distance, miles Subbituminous	Lignite
0.010	1,359	1,080
0.015	906	720
0.020	679	540

*Stabilizing cost assumed at $0.75 per ton. Total drying and stabilizing costs: $3.29 per ton of as-mined subbituminous coal, $3.41 per ton of as-mined lignite.

Dryers can be classified as being either convective or conductive in nature. Convective dryers are characterized by the fact that the hot, drying gases are in intimate physical contact with the coal particles being dried, such as in a fluidized bed. Conduction dryers, on the other hand, use a solid metal wall to separate the hot, drying gases from the coal particles being dried. In this situation heat is conducted through the metal wall to the coal particles. There is generally some mechanical means of "turning over" the coal particles in contact with the metal wall to facilitate a more even drying of the particles. Equipment used in coal drying is discussed in References 1,5, 8-14.

Thermal drying of coal in the U.S. dates back to at least the 1940's. Coal was dried convectively and cleaned mechanically for use primarily in the steel industry. At that time and into the early 1950's, moving screen, cascade, multi-louvred and flash dryers were essentially the only type of dryers being used. Today, very little coal is dried by these three techniques. Another dryer type, the rotary dryer is in wide spread use in other industries, but it has never made much of an impact in coal drying. This has been due to considerations of space requirements, capital and operating costs, and generally low through-put capacities.

The flash dryer, or suspended bed dryer or entrained bed dryer as it is sometimes called, is still in use in the U.S. and in Europe. However, the fluidized bed dryer, introduced in 1955, has dominated the U.S. market. The Europeans still seem to prefer the entrained bed dryer to the fluidized bed dryer, though.

In the operation of the flash dryer, Jones (13) reports that hot gases supplied by a furnace are drawn into the system by a vent fan which is connected to the outlet of a cyclone dust collector. Coal to be dried is fed into the hot gas stream where instantaneous or flash drying occurs. The coal is in contact with the hottest gases, about 1200°F, for a fraction of a second. The hot combustion gases carry or entrain the coal particles to the cyclone. The dried coal and moisture laden combustion gases (now about 200°F) are drawn into the cyclone where they are separated. The coal is removed through a rotary air-lock at the bottom of the cyclone, while the gases are exhausted through the vent fan.

The fluidized bed dryer has dominated the thermal drying of coal in the U.S. in recent years for several reasons. First of all it is able to provide a final product over which one has good control of the moisture content and, also, the particle size distribution. Secondly, high through-put capacities are possible. Thirdly, these are possible to do in a relatively safe manner. Figures 6-9 illustrate typical fluidized bed dryers.

Fluidization is achieved by passing a gas(es) upward through a perforated horizontal distribution plate which also serves to support the solid particles during startup. When proper gas velocities have been attained, the finely-ground coal particles are suspended in this gas stream. This forms a dense, turbulent mixture of solids and gas. The characteristics of this agitated mass are similar to that of a boiling liquid in that the material will overflow weirs, seek a constant level and exert hydrostatic pressure. These characteristics suggested the name fluidized bed.

Turner and Donohue (7) reported some of the design parameters used in the fluidized bed dryer at Mid-Continent Resources' coal operations on Colorado's Western slope. The coal fed to the dryer contained approximately 20% water and was dried to 4%. The feed was about 75%, 28 x 0 mesh. About 50,000 SCFM of air at 1,100 to 1,400°F entered the fluidized bed and left it at about 150°F. Through-put was 150 tons per hour of dried coal.

The advent of the conductive dryer has been primarily in response to the federal emissions laws. Because of the physical sepa-

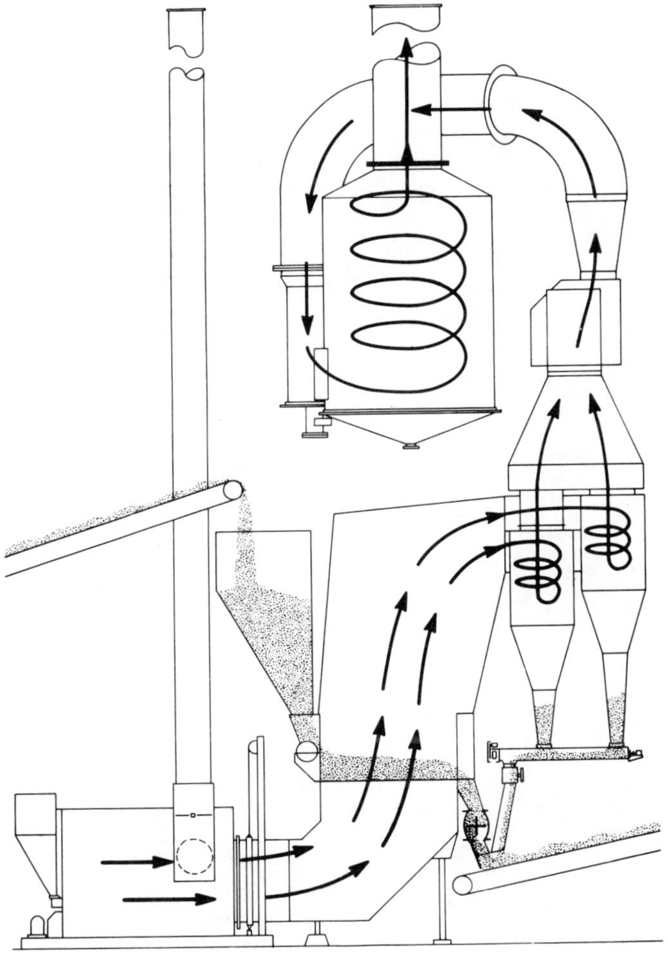

Figure 6. Cross section of the FMC Fluid-Flo coal dryer (Courtesy: FMC Materials Handling System Division).

Figure 7. Fluid-Flo coal dryer which evaporates 38 tph water from minus 3/8 in. coal (Courtesy: FMC Materials Handling System Division).

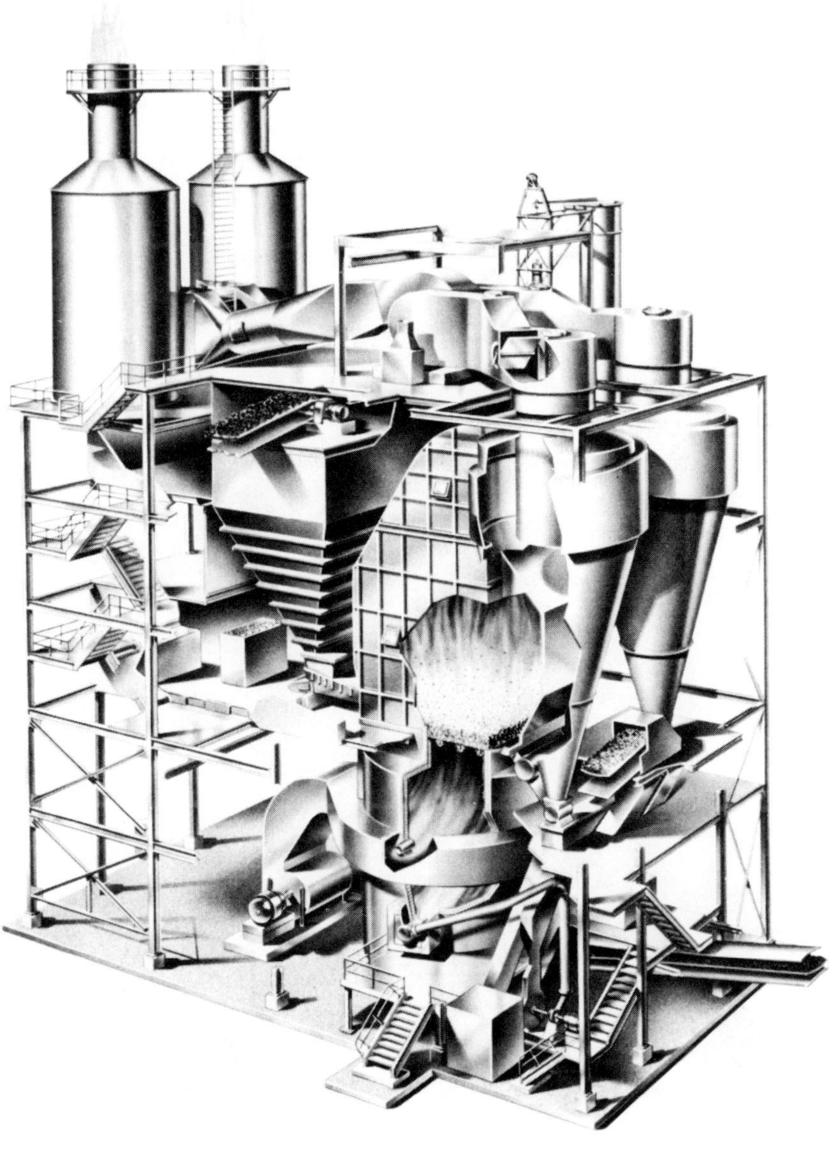

Figure 8. Cutaway view of the McNally Flowdryer (Courtesy: McNally Pittsburg, Inc.).

Figure 9. Flowdryer installed at the Mapco, Inc. Mettiki Mine. (Courtesy: McNally Pittsburg, Inc.).

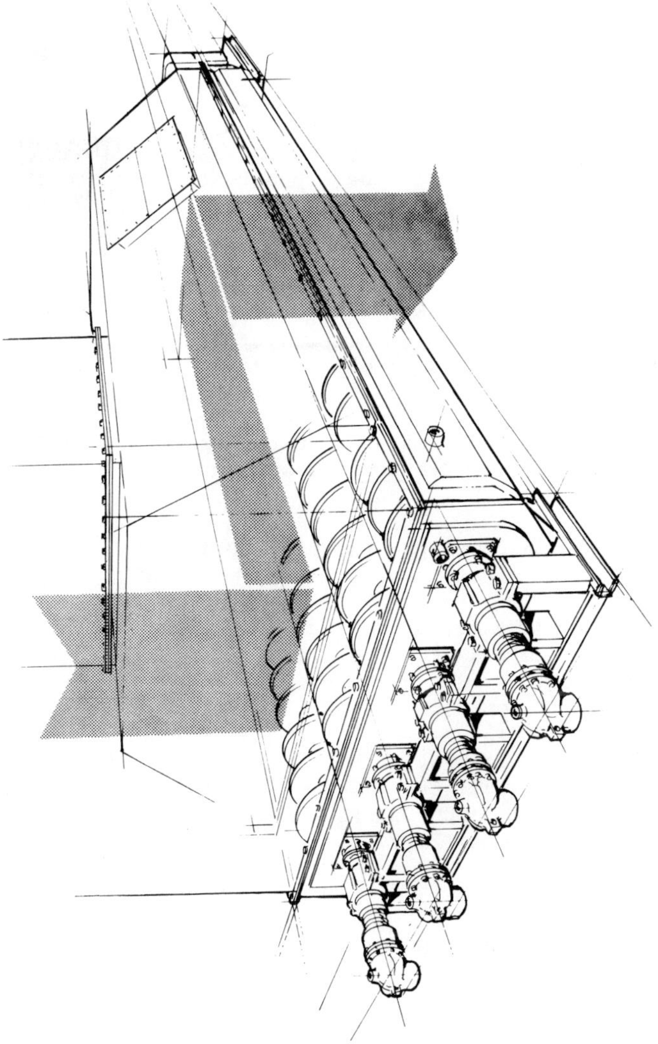

Figure 10. Holo-Flite dryer. (Courtesy: Joy Industrial Equipment Co.).

ration of the hot, drying gases from the finely-ground coal, coal dust isn't carried out of the dryer with the combustion gases. This eliminates the need for expensive dust collection and scrubbing devices. It also reduces the danger of fire and dust explosions. See Figure 10. While the literature doesn't report the use of conduction type dryer for western coal, conceptually they would appear to work satisfactorily under these conditions.

In summary, then, it can be said that the technology for drying western coal is available and proven. The problems of dust production and spontaneous combustion are real, but these can be minimized by design of proper handling facilities. Any decision on drying will probably be made based simply on economics, and the economics are such that the shipping of dried coal appears to be economically feasible for markets of about one thousand or more miles from the coal source.

REFERENCES

1. Hand, John W., "Drying of Western Coal," Mining Congress Journal, pp. 30-32, May 1976.

2. Ode, W. H., "Coal Analysis and Mineral Matter," Section of Chemistry of Coal Utilization, supplementary volume, H. H. Lowry, Editor, John Wiley and Sons, N.Y., 1963.

3. Kindig, James and Phillips, Patrick, "Dried Coal is a Better Buy Than Wet," Coal Age, pp. 149-158, May 1980.

4. Ellman, R. C., Paulson, L. E. and Cooley, S. A., "Commercial-Scale Drying of Low Rank Western Coals: Part I - Rail Shipment Test Observations," ERDA-University of North Dakota Lignite Symposium, Grand Forks, N.D., pp. 312-340, May 1975.

5. Wegert, C. J. and Jensen, H. M., "Commercial Scale Drying of Low Rank Western Coal: Part II - Utilization Feasibility," ERDA-University of North Dakota Lignite Symposium, Grand Forks, N.D., pp. 341-344, May 1975.

6. Paulson, L. E., Cooley, S. A., Wegert, C. J. and Ellman, R. C., "Experiences in Transportation of Dried Low Rank Western Coals," Trans. AIME, 260, pp. 300-305, Dec. 1976.

7. Turner, John H. and Donohue, William J., Jr., "Heat Drying of Coal at Mid-Continent," Mining Congress Journal, pp. 17-19, Sept. 1979.

8. Ziesing, G. F., Forgaard, K., LeFever, J., Rhodes, B. and Griffiths, V., "Drying of Low-Rank Coal for MHD Application," Paper presented at the 10th Biennial Lignite Symposium at Grand Forks, N.D., May 31, 1979.

9. Castanoli, A. F., "A New Concept in Coal Dryers," Coal Mining and Processing, pp. 36-37, Oct. 1974.

10. Charmbury, H. B., "The Bearce Dryer Goes Commercial," Coal Mining and Processing, pp. 72-74, Oct. 1977.

11. Venkatasan, Stan, "Research Imperative to Improve Thermal Drying," Coal Mining and Processing, pp. 80-82, Sept. 1975.

12. Chambers, R. H. and Alderman, L., "Coal Preheating as Applied to Coking Installations," Iron and Steel Institute. Publication No. 127, 1970. Coke in Iron Making, pp. 63-67.

13. Jones, Donald C., "So You Are Considering Thermal Drying," Coal Mining and Processing, pp. 36-43 and p. 74, Oct. 1969.

14. Luckie, Peter T. and Draeger, Ernie A., "The Very Special Considerations Involved in Thermal Drying of Western Region Coals," Coal Age, pp. 106-109, Jan. 1976.

Environmental I

Session Chairman
C. Dudley Blancke
Union Carbide Corporation
Edgemont, South Dakota

Chapter 27

LOW LEVEL EFFECTS FROM MILL TAILINGS

Jamie Deuel

Deuel and Associates
Albuquerque, NM

The "Radon Connection"

For the purpose of this paper, we will concentrate on the ubquitous radon gas as the principle, low-level radiation effect from uranium tailings piles. Standards are also proposed for gamma radiations but, by most accounts, this is much less of a low level problem than is radon. I am also aware of some just completed research at Oak Ridge National Laboratories that suggest polonium may be a major concern in tailings, but those data are preliminary at this time.

What is radon gas? In the first place it is not a gas in the accepted sense in that it has no driving pressure head. It might prove a more convenient description to call it radon vapor. Radon vapor, then, is 7-1/2 times heavier than air. Radiologically, it is an alpha emitter which means it spontaneously gives off high ionizing potential but low penetrating power particles. In fact, the alpha particle has such low penetrating power that this sheet of paper will completely shield a person. The problem, though, is breathing the radon decay products, called radon daughters. If radon daughter products are ingested into the lungs, their alpha emissions may cause cancer. The principle carcinoma associated with breathing radon, or more basically, associated with radiological effects is oat cell carcinoma of the undifferentiated small cell type. Radon has a 3.8 day half life which means in less than four days half of its initial radioactivity has decayed away. In less than a month it essentially becomes inert or non-radioactive.

The Tailings Situation

The NRC, in its April 1979 Generic Environmental Impact Statement

(GEIS) on uranium milling, estimates that by the year 2000 we will have almost 2900 x 10^6 MT of radon emitting tailings in the U.S. However, since this estimate was developed, a significantly lower (by about 30%) national electrical demand is predicted, and none of the new mills projected for start-up in 1980 are on-line as of the Fall of that year. There are basically two categories, chronologically, of tailings - active and inactive. The later group are those WWII, Manhatten Project related tails, that have been abandoned following accepted procedures at the time. In 1978 the government adopted the legal position of "compassionate responsibility" for those crisis generated wastes and passed the Uranium Mill Tailings Radiation Control Act (UMTRCA). The Department of Energy has seven years, starting in 1981, and over $500 M to clean up 25 designated inactive sites across the country, most of which (22) are in the western states. The Environmental Protection Agency (EPA) is responsible for writing the cleanup standards for this program. In a Federal Register issue (April 22, 1980) when their interim standards were published, the EPA stated,

> "Tailings usually only pose a significant risk to people who are exposed for a long time."

and

> "The greatest hazard from tailings on open lands is increased levels of radon decay products in buildings."

Tailings Reclamation Regulations

At the moment these depend on whether one wishes to use the Nuclear Regulatory Commission's (NRC's) or the EPA's. Since the EPA is still going through the initial rule making process to promulgate theirs, the existing NRC's will be used as guidance. Last year, an amount of soil cover that had grown from two feet in 1977 to seven feet in 1978, including a compacted two foot clay cap, to, finally, a minimum of three meters or about 10', was used. At first the standard was designed to reduce the radon emission level to no more than twice background (and the gamma ray levels to equal background). The revised regulations proposed in the GEIS specify an absolute amount of radioactive flux - a measure of radioactivity per unit of area in a given time increment - without regard for how high or how low the natural background conditions actually are. The NRC proposes 2 pCi/m^2-sec.

Field Experience with Tailings Reclamation

To date there has been only one full scale test of a modern tailings reclamation action. My firm designed and executed the post-operation reclamation evaluation plan for this facility when it ceased operations. In a report to the responsible state regulatory agency (Deuel & Associates letter of December 21, 1979 to Mr. C. Mattson, Colorado Department of Health), the following results for a tightly engineered, compacted clay cap of 2' - 3' were provided:

Background (17 stations, 43 measurements) - 2.45 pCi/m^2-sec

Tailings site (8 stations, 24 measurements) - 1.81 pCi/m^2-sec

When discussing these actual results with less than three feet of clay cover versus the requirements for 10' (3 m), regulatory officials state that they are not surprised. The extra volume is needed, however, to protect the site from erosion for 1,000-year time periods or longer. But, doesn't it seem likely that over a long enough time period, erosion will expose the uranium in nature, and then who will reclaim it? In my professional life I have been fortunate enough to have worked closely with three brilliant, eminently successful business leaders. They all agree that even 5-year plans are mostly crystal ball projections after the second year. Isn't it more than a little presumptious by our regulatory agencies to suggest a solution today for an event 1,000 or more years in the future? Wouldn't it be more prudent, and less environmentally damaging, to have some or all of the money companies would have spent on getting the extra seven feet of cover material placed in an interest earning trust account. These funds could be used by agencies to maintain the sites and adopt new, less impacting reclamation techniques as science develops them. Several interesting, man-made stabilization schemes are currently under study that could greatly reduce the countryside scar from finding so much (3 m) cover material.

Linear Hypothesis

To discuss the low level health effects related to tailings, one must keep in mind this assumption. It can best be described as saying if a lot of something is bad for you, then a little of it is also bad in a proportionate way. Until recently, there was extensive data on radiological effects at high levels but almost a total void of information in the low level regime. Most agencies responsible for radiological health have adopted the linear, no threshold hypothesis as a conservative way to proceed. However, some others feel that the body does have a tolerance level for radiation just like it does for aspirin. Below a certain level aspirin can be very therapeutic; well above that level it can cause serious physical effects. Those who challenge the linear hypothesis as being unrealistically restrictive, point out cases like Denver which has about twice as much natural radiation (because of its altitude) as New Haven but also has a lower cancer incidence rate. Recently, a small number of researchers have suggested that successively lower levels of radiation have greater and greater impacts on the individual. This would seem to suggest that the maximum impact would occur with no radiation present. The linear, no threshold hypothesis is a major bone of contention in many low level effects assessments.

Low Level Health Effects

From Mining

We hope that most people would accept the assumption that uranium miners should be, as a group, potentially at greater risk from low level (and high level) radiation effects than the general public. After three years of searching through death records, the Uranium Epidemiology Study of the UNM Cancer Research and Treatment Center in Albuquerque has found no cases of lung cancer deaths among New Mexico miners who started work in or after 1961. Their study consisted of almost 4,100 underground miners with one year, or more, experience. Although this study has many more years to run these initial data are a dramatic early statistic. In Grand Junction, CO, widely known pathologist Dr. Geno Saccamenno has been following lung cancer deaths in the Colorado Plateau for over 25 years. The cases of oat cell carcinoma (radiation related) in 292 lung cancer deaths through 1979 have dropped steadily in every five year recording period measured. From 1954 - 1959, over 76% of miner lung cancer deaths were of the oat cell type. In his most recent 5-year period (1974 - 79), that percentage has dropped to under 23%, close to the frequency of occurrence in the general public. In 1979 there were no oat cell carcinoma deaths.

From Tailings

World famous health physicist with extensive credentials in academia and major government agency contracting, Dr. Keith Schiager, testified (Albuquerque - October 19, 1979) against the NRC low level effects predicted in the GEIS. Their estimate of 10 deaths per year through the year 3,000 should only be "one-fourth of one premature cancer death per year" and for completely uncovered tailings. "The proposed radon emission criterion of 2 pCi/m^2-sec is not based upon either the necessary protection of individuals or on considerations of public health risks," he pointed out. The GEIS assumed outdoor, background radon concentrations (not flux) in the west of 240 pCi/m^3. A 1969 government study by Shearer and Sill reports levels in Grand Junction (800), Durango (500), Salt Lake City (380) and Monticello (340) all above the NRC stated average. The use of averages instead of site specific conditions was decried. Schiager added that,

> "No mining company would be in business if the only ore available contained the <u>average</u> mineral concentration found in the earth's crust."

Even for a worst case scenario of (1) a residence built directly on a reclaimed tailings pile with (2) a radon emission rate of at least

10 times the NRC limit it was shown that the indoor rate "would still fall within the range of natural background."

Risk Comparisons

At the same hearing in Albuquerque when Dr. Schiager testified in 1979, chemical engineering PhD, Dr. Harrison Rhodes with the Union Carbide Corporation, presented an American Mining Congress (AMC) developed set of curves showing some comparative risks. (See Figure 1). Because of the wide range of risks included, it was necessary to represent them with a logarithmic vertical axis.

Consider first the risks taken directly from the draft GEIS. The far-field and continental risks shown in the bottom two curves are lower than those for the regional population by factors of about 200 and 250, respectively. On a far-field basis, the risk from uncovered tailings is nearly an order of magnitude below the lower end of a suggested acceptable risk range of 0.1 to 1 health effects per million persons per year, and is about the same as an FDA recommendation of an acceptable level of risk for carcinogenic residues in meat products.

The draft GEIS risks are shown by the upper curve in the figure. At the base case level of no control (flux of 450 pCi/m^2-sec), the risk is about 4. This is about 1/20 of the risk of natural background radiation, about 1/3 the risk of being in a room with a smoker or the risk from common household accidents. The 2 pCi/m^2-sec requirement suggested by the GEIS is clearly not supported by the risks.

The Question in Perspective

Do you remember the overly generous NRC prediction, in the GEIS, that there would be 2900 x 10^{6}MT of uranium mill tailings by the year 2000? Well, that huge sounding volume of tails would occupy only 0.0013% of the land mass of the contiguous 48 states. Let's, for the sake of perspective, assume that the city limits of Rapid City, South Dakota, represent our continental national boundaries. That land mass would be the same as the two end zones of any football field in this city! Does this compare with the horror stories of vast acreages of poisonous tailings painted by the media and nuclear power opponents?

Whether you accept fully the arguments for the lack of any real evidence for a proven low level radiation effects problem, surely the point is made that it will not have a major geographic impact on our nation. This should be more obvious when one considers the normal remote, semi-arrid location of most uranium mills.

University of Pittsburgh physicist and researcher, Dr. Bernard Cohen, has viewed the perspective in another way. He has taken the

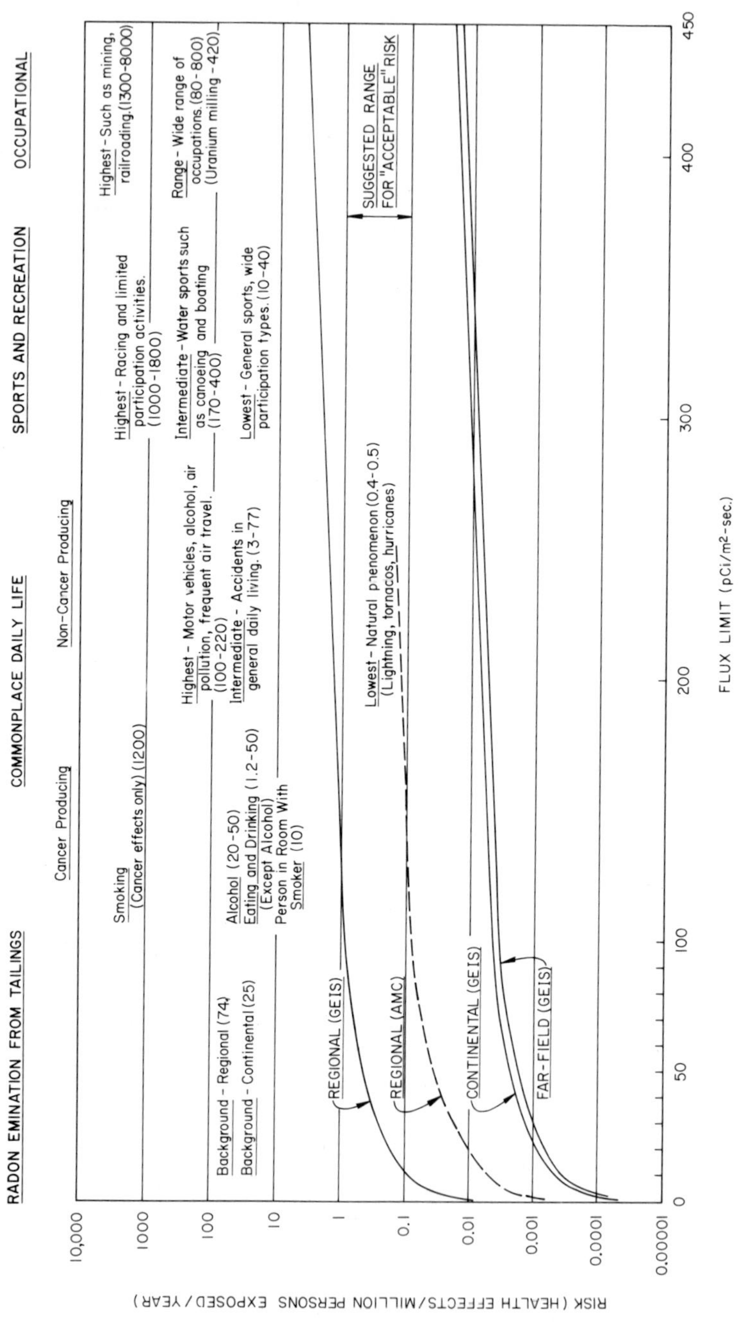

Figure 1. Risks of occupational and daily living.

GEIS fatality predictions of 10/year (which Dr. Schiager says should be only 1/4 person/year) and compared them, unadulterated, with other known risks. The GEIS mill tailings risks, using NRC's figures, equate to 4 fatalities per year with today's population. Compare this with -

All accidents	-	100,000 vs 4 !
Suicides	-	28,000 vs 4 !
Drownings	-	8,000 vs 4 !
Choking on food	-	3,000 vs 4 !

At this point I hope the audience can evaluate the low level radiation effects from uranium mill tailings.

The Price Tag

During a graduate school course of economics, my professor told a story about Alexander the Great crying because he had no more worlds to conquer. As a substitute he set out to learn all there was to know about economics. The professor developed the story at some length with Alexander's wise men failing each time to be concise enough. Finally, on their last try to satisfy Alexander, the chief vizier reported "Sire I have it all distilled down to one sentence." When Alexander unrolled the wise man's scroll it read simply,

THERE IS NO SUCH THING AS A FREE LUNCH!

Uranium has only one peaceful use - the generation of electricity. The petroleum fuels that are currently being burned up at an alarming rate represent irreplaceable sources of key elements of pharmaceuticals, plastics, fertilizers and other important products. If we do not use our uranium resources, they will gradually decay away. Our energy plight is so severe that we must also use coal for the time being, but it has a far greater impact on our environment than nuclear ever will have. If we are to survive and not use nuclear power, an alternative must be provided. None exist today or in the forseeable future that will meet the expected demand. Solar, for the large central power generating requirements of schools, hospitals, theaters, etc. is still 20-50 years away. Geothermal, wind, ocean currents and others can help but only for limited portions of our energy needs. Even assuming the more gradual energy growth curve of 1980, we will face crippling shortages in a few years if we don't get going very soon. We need nuclear power today. Future generations will thank us for not letting our society regress because of an unnecessary loss of essential electricity services.

Opposition to Uranium and Nuclear Power

We do not suggest that what has been presented is the only viewpoint, the only perspective. We have tried to be as factual, practical and as honest as we can. Others will differ from our perspective, some for philosophical reasons, some for personal bias and even some for political considerations. What we would like to avoid, however, is the phobia approach. Dr. Robert L. DuPont, a practicing psychiatrist and President of the Non-profit Institute for Behavior and Health, describes it this way:

> "A phobia is a malignant disease of 'what if's'. The phobic thinking process is a spiraling chain reaction, to use an atomic energy analogy, of 'what if's', and each 'what if' leads to another. 'What if this happens and then what if that happens, and then what if the other thing happens?' Phobic thinking always travels down the worst possible branching of each of the 'what if's' until the person is absolutely overwhelmed with the potentials for disaster."

Let us work very hard to avoid phobic thinking about uranium and nuclear power. The other real dangers today are too great, far greater than the 'what if' potentials for disaster from the nuclear fuel cycle.

Chapter 28

RADON DAUGHTER CONTROL IN THE URAVAN MINERAL BELT

Roger W. Swindle

Union Carbide Corporation
Uravan, Colorado

INTRODUCTION

The Uravan Mineral Belt in southwestern Colorado and southeastern Utah is a uranium-vanadium mining district with a unique set of radon daughter control problems. This paper describes these problems and the methods that have been developed to deal with them. The greatest emphasis is placed on the problems caused by air stagnation because these are the most frequently encountered ventilation problems, and this is an aspect of uranium mine ventilation that is often misunderstood.

VENTILATION PROBLEMS

Most of the mine ventilation problems in the Uravan Mineral Belt are caused by one or more of the following conditions: (1) the highly irregular ore deposits are difficult to locate and delineate, (2) air stagnation occurs in many areas of the extremely irregular mine workings, and (3) it is often necessary to reactivate old workings with small airways that are often caved.

Planning Problems

Orebodies in the Uravan Mineral Belt are so discontinuous and unpredictable that they are difficult to locate and delineate. Data supplied by surface drilling and underground test holing are often misleading; very little ore may be found in the vicinity of an ore hole, and, conversely, an unmineralized hole may be adjacent to a high-grade ore deposit. Also, ore guides are often ambiguous, and their interpretation is usually a matter of opinion.

As a result of the difficulties encountered in locating and delineating the orebodies, insufficient information is available to permit detailed, long-range planning of ventilation systems. Mine development and ventilation can be planned only in a very general way, and it must be done in such a manner that frequent revisions can be made as mining progresses.

It is usually impossible to determine the farthest limits of ore in a mine. This precludes the use of a retreating system of mining that would permit contaminated air to be exhausted through mined-out workings, thereby reducing the amount of ventilation required. Instead, mining activities are usually widely dispersed. Consequently, in order to maintain haulageways and ventilation airways to serve the areas being mined, it is usually necessary to ventilate extensive, otherwise inactive workings that are the principal source of radon daughter contamination.

Inactive workings often cannot be sealed off from active areas because the openings between active and inactive areas are too large or too numerous. Also, it has been demonstrated repeatedly that the indiscriminate construction of stoppings that cannot be installed and maintained airtight can cause more radon daughter contamination than existed before they were installed. This effect will be described in detail in the following section.

Air Stagnation Problems

Ore in the Uravan Mineral Belt is mined by a random room and pillar method that produces workings with highly irregular configurations as is shown in Figure 1. The workings contain many peripheral stope areas that are difficult to ventilate because they are partly or completely by-passed by the primary air currents. In order to provide sufficient ventilation to work these by-passed areas, it is necessary to install numerous primary booster fans, auxiliary fans and air diversion stoppings.

The slow movement of air through a stope area can also cause an increase in the radon daughter contamination farther downstream in the ventilation circuit. The slow air movement is usually caused by the complex geometry of the workings as described in the preceding paragraph, but it can also be caused by the improper installation of stoppings that are intended to seal off an area but actually only impede the flow of air.

The manner in which radon daughter contamination builds up in poorly ventilated areas will be demonstrated by an example, illustrated in Figure 2, in which the air flow through a stope area is impeded by a regulator at the upstream end and a stopping with a fixed orifice at the downstream end. The regulator and stopping duplicate the effect of two poorly constructed stoppings that impede the flow of air through the area. Adjustment of the regulator

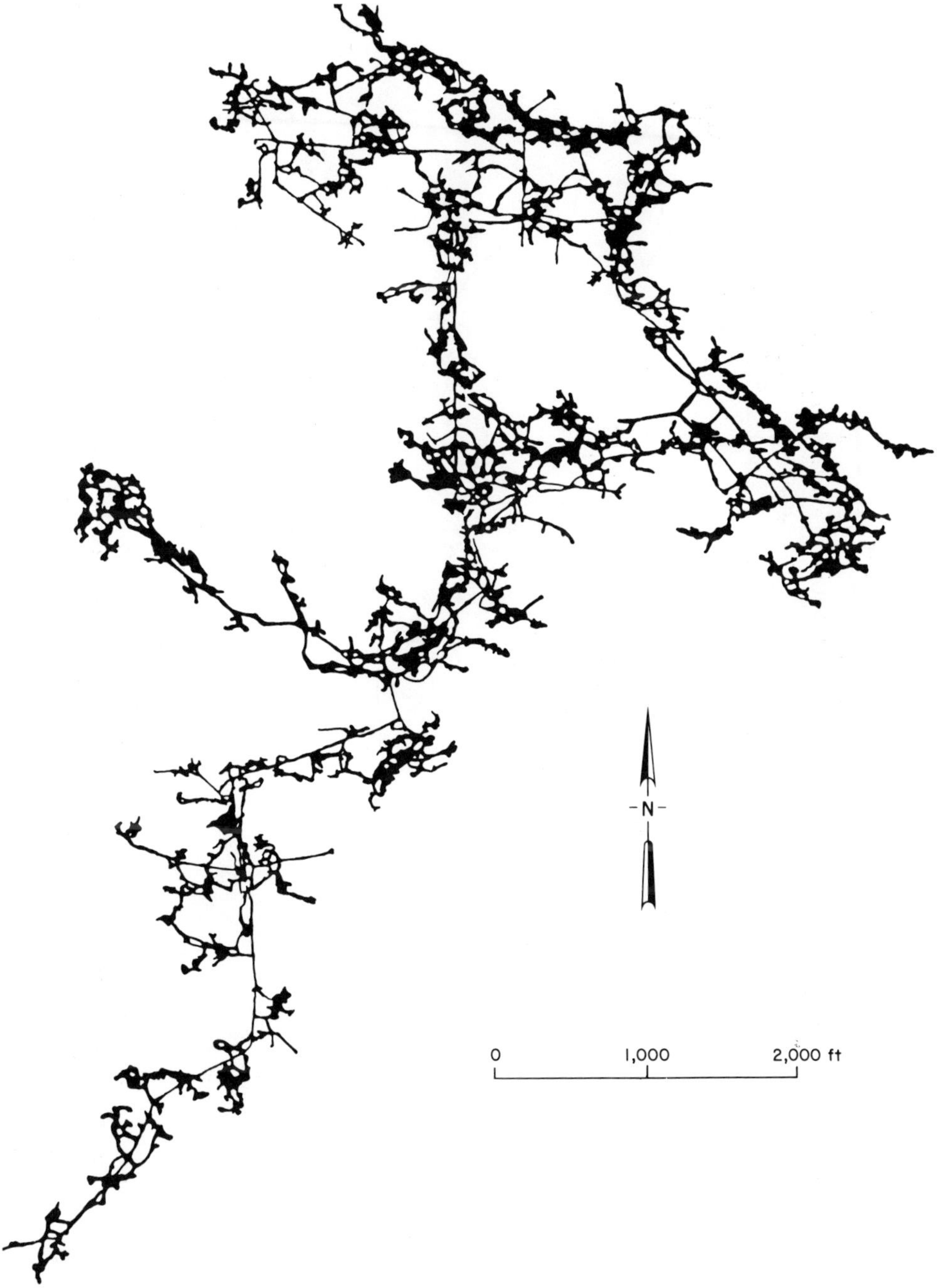

Figure 1. Deremo Mine. This map shows the complexity of typical workings in the Uravan Mineral Belt.

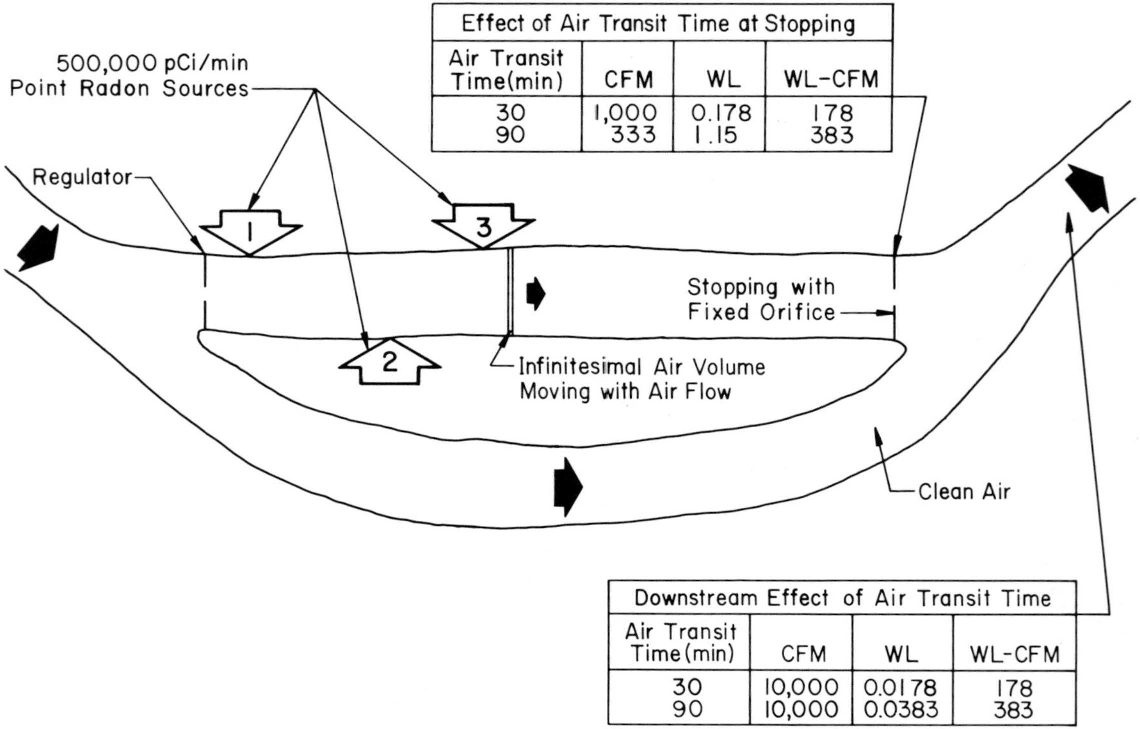

Effect of Air Transit Time at Stopping			
Air Transit Time(min)	CFM	WL	WL-CFM
30 90	1,000 333	0.178 1.15	178 383

Downstream Effect of Air Transit Time			
Air Transit Time(min)	CFM	WL	WL-CFM
30 90	10,000 10,000	0.0178 0.0383	178 383

Figure 2. Stope area with simulated, leaking stoppings to show the effect of the air transit time (from the regulator to the stopping) on the downstream radon daughter contamination. The effects of adjusting the regulator for air transit times of 30 and 90 min. are shown. The growth of the working levels in the infinitesimal volume as it moves with the air is shown in Figure 4.

permits the air flow rate to be adjusted to demonstrate its effect on the downstream radon daughter contamination.

The simulated stope area in Figure 2 has three 500,000 picocurie (pCi) per minute radon sources. The movement of an infinitesimal volume of air will be traced as it moves through the area, and the growth of the working level* (WL) in this volume will be calculated. The following assumptions are made for mathematical simplicity: (1) the indicated radon sources are point sources, (2) the air velocity is constant over any cross section of the stope area, (3) no radon daughters are removed from the air by "plate out" (the attachment of daughters to fixed surfaces such as rock and timber), and (4) no transfer of radon or daughters occurs between the infinitesimal volume and adjacent air volumes.

As the infinitesimal volume moves through the stope area in Figure 2, the contained working level from each radon source grows in the manner shown in Figure 3. The working level builds up at a rapid but decreasing rate until it reaches radioactive equilibrium with its decaying radon source after a period of approximately four hours. It then decreases at a rate nearly equal to the exponential decay rate of the radon activity.

Figure 4a shows the growth of the working level in the infinitesimal volume as it moves from the upstream regulator to the downstream stopping with a transit time of 30 minutes. The working level contribution from each of the radon sources is indicated by a separately crosshatched area, and the uppermost boundary of the combined crosshatched areas is the total working level growth curve. The value of this curve at 30 minutes is the working level in the air discharged through the downstream stopping.

Figure 4b shows the effect of increasing the air transit time in Figure 2 to 90 minutes. The large increase in the working level at the downstream stopping is caused by two factors. These are: (1) the decrease in the rate of radon dilution, which is directly proportional to the ventilation rate (CFM), and (2) the increase in the time available for the working level to grow.

In order to calculate the extent of the downstream radon daughter contamination from an area such as that in Figure 2, it is necessary

---

*The expression "working level" has a dual meaning in conventional usage. The working level is defined as a unit of potential radon daughter alpha energy in a liter of air (130 billion electron volts per liter). It is also used to indicate the degree of radon daughter contamination (e.g. the working level is high) in a manner that conflicts with the definition because the working level unit can obviously have only one value.

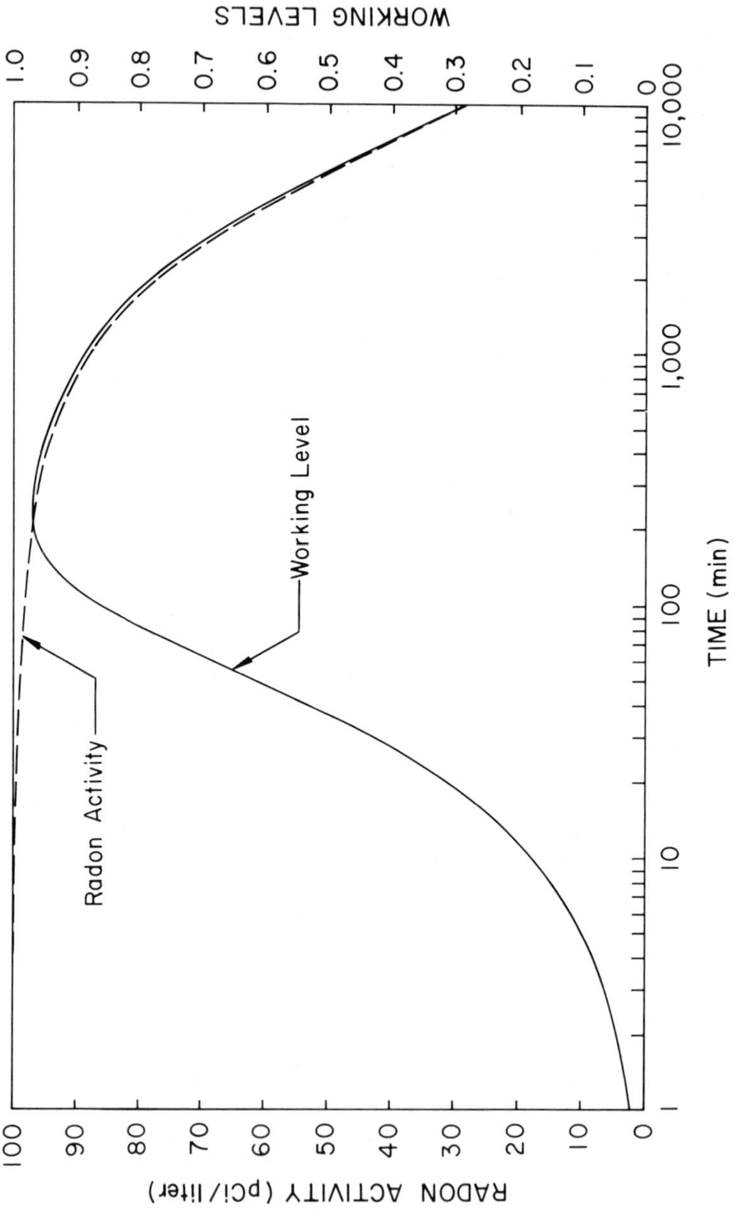

Figure 3. The growth and decay of the working level derived from a decaying radon source with an initial activity of 100 picocuries (pCi) per liter.

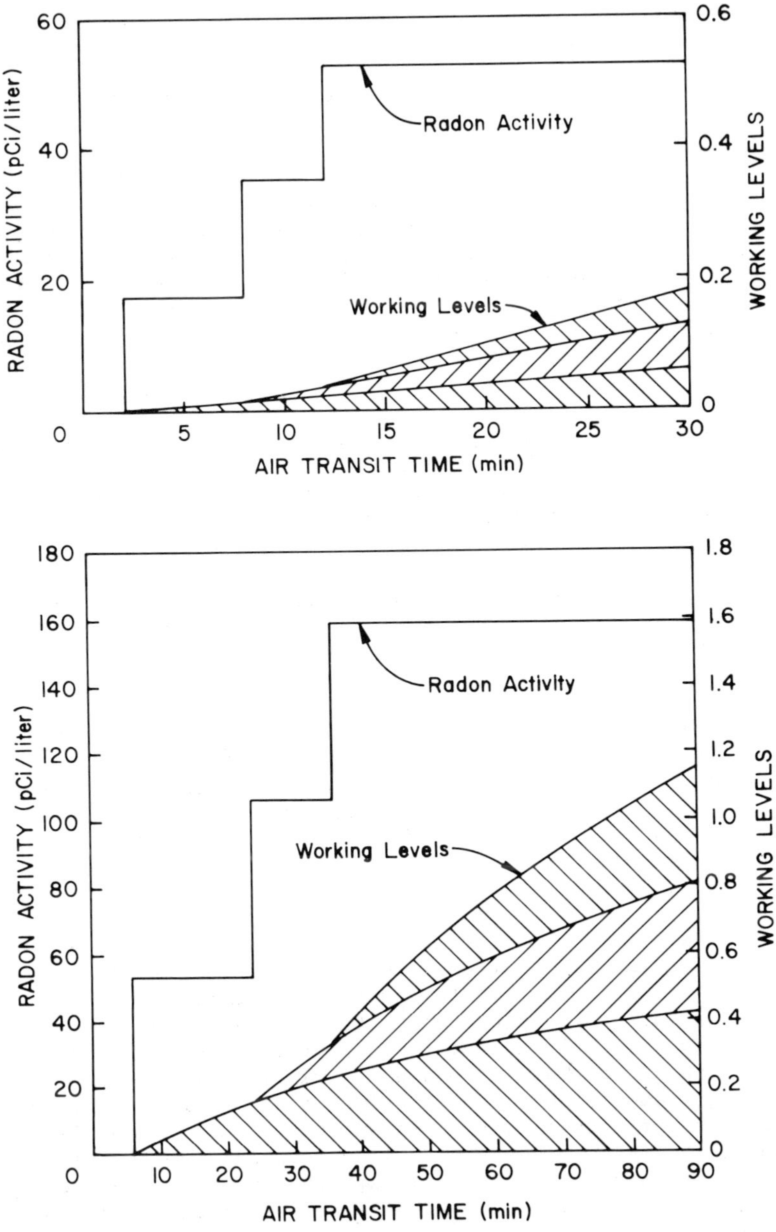

Figure 4. The growth of working levels due to the three point sources shown in Figure 2 as a infinitesimal volume of air moves through the area indicated. Transit times involved are 30 minutes (top) and 90 minutes (bottom).

to employ a new unit, the WL-CFM, which is the product of the working level and the ventilation rate in CFM. The derivation of the WL-CFM unit is illustrated graphically in Figure 5 which shows the WL, CFM and WL-CFM discharged through the downstream stopping in Figure 2 for air transit times of 10 to 10,000 minutes.

The WL-CFM unit is useful because it can be used to calculate the downstream WL component derived from a radon source. The calculation of the downstream working level is illustrated in Figure 2 for transit times of 30 and 90 minutes. The working level in the main airway immediately downstream from the downstream stopping is determined by dividing the appropriate WL-CFM value in Figure 4c by the CFM in the airway. It should be noted that the calculated working level applies only to that component of the working level in the airway that was derived from the radon sources under discussion.

In actual practice, the WL-CFM discharged from an area will usually be determined by measuring the WL and the CFM of the discharged air. It can also be determined by measuring the WL and the CFM upstream and downstream from the discharge point, calculating the WL-CFM at each location and taking the difference.

Figure 5 shows that the WL-CFM remains high for extremely low air flow rates. This demonstrates the futility of attempting to seal off any such areas unless the stoppings can be constructed and maintained airtight.

The slow movement of air through some areas of highly complex workings causes the same sort of radon daughter build-up as was demonstrated in the preceding example, but the mechanism is far more complex. The airflow patterns through such workings consist of many individual air currents with widely varying velocities, and each current contributes its own unique radon and daughter mixture to the primary air stream.

The irregularity of the air flow pattern in a typical stope area is compounded further by temperature differences between the primary air stream and the rock. A temperature difference of just a few degrees can cause pronounced convective flows in dead-end workings that can significantly affect the downstream radon daughter contamination. The convection in dead-end workings varies with the season of the year, and it depends on the orientation of the workings above or below the primary airways; in summer the convection is greatest in dead-end workings that are inclined upward from the airways, and in winter it is greatest in dead-end workings that are inclined downward below the airways. The effects of temperature changes are often unpredictable; a change of air temperature can either increase or decrease the overall radon daughter contamination depending on the geometry of the workings.

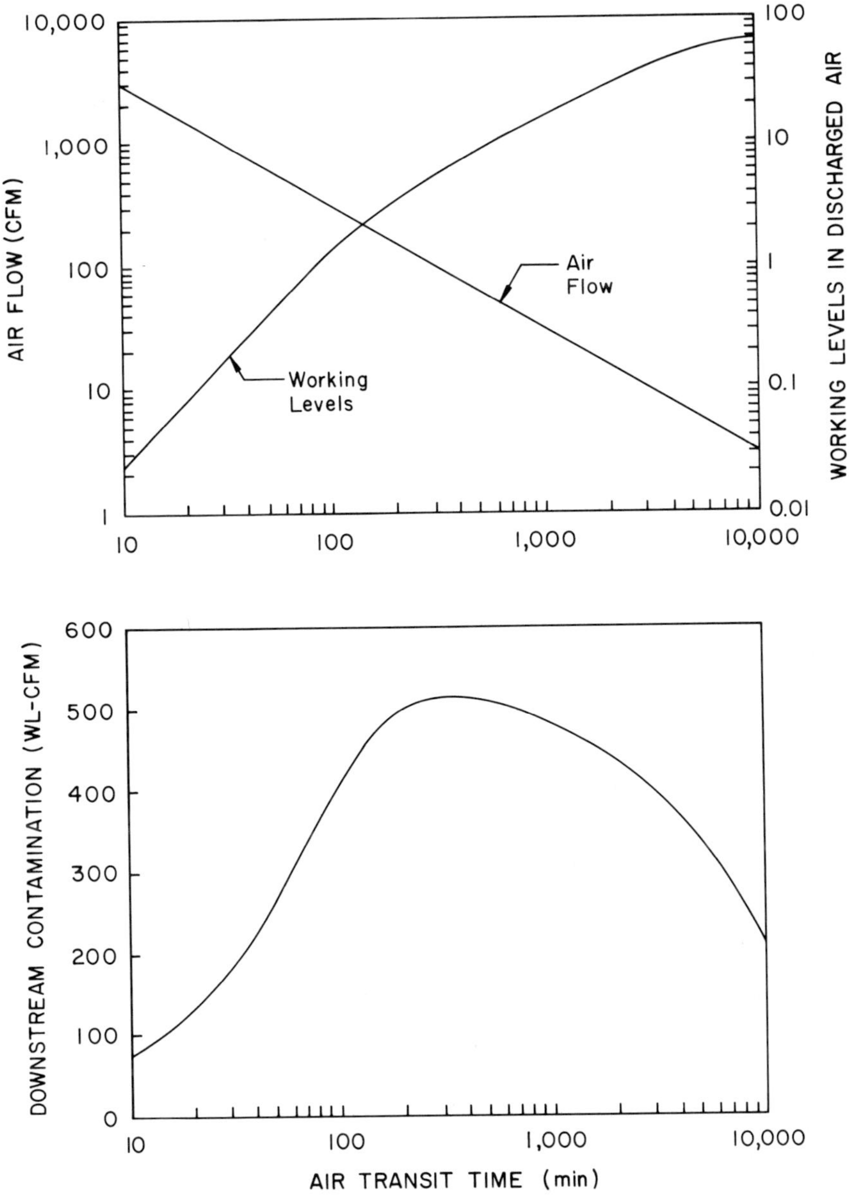

Figure 5. The effect of air transit time through the stope area in Figure 2 on the downstream radon daughter contamination. The WL-CFM curve, the product of the WL and CFM curves, shows the extent of downstream radon daughter contamination.

Reactivation Problems

Additional ventilation problems are caused by the frequent necessity of reactivating old workings. Many of these workings were previously worked many years ago when ventilation requirements were far less stringent than they are today. Consequently, surface ventilation holes, drifts and other airways are often too small to be of much use for ventilation purposes. Also, haulageways were often driven in mudstone beneath the ore horizons and they are frequently caved.

CURRENT VENTILATION TRENDS

The two principal methods that have been developed to improve the ventilation of mines in the Uravan Mineral Belt are: (1) the drilling of large-diameter ventilation raises from the surface to provide more ventilation at lower cost, and (2) the use of electronic air cleaners to remove radon daughters from mine air.

Large-Diameter Ventilation Raises

The ventilation problems described in the previous sections and the increasingly strict state and federal ventilation requirements for the control of radon daughter products and diesel exhaust have provided the impetus for drilling increasingly larger surface ventilation holes over the last two decades. The diameters of ventilation holes have increased from a range of 12 to 20 inches in the late 1950's to the 5 to 8 foot diameter raises currently being drilled.

Union Carbide has had an Ingersoll-Rand Model RBH-5E7 raise boring machine in operation since March, 1976. This machine is equipped to upream 5, 6, 7 and 8 foot diameter raises.

The increased diameters of the raises have aggravated ground control problems, especially in the thick mudstone horizons that occur in most areas of the Uravan Mineral Belt. The influx of groundwater has also increased, and the obstruction of ventilation raises by ice during cold weather is a serious problem. The original solution to the icing problem consisted of heating the air with direct-fired propane heaters, but the high cost of propane soon made this approach impractical.

Shotcreting is the most commonly used method for controlling unstable ground and for reducing the influx of water. The raises are lined by a Union Carbide crew with 3 inches of shotcrete with a compressive strength of 3,000 to 5,000 psi.

In some raises the ground is so unstable that steel casing with wall thicknesses of 3/8 to 1/2 inch must be installed. Grout is placed outside the casing in water-influx zones.

Electronic Air Cleaning

Ionizing type electronic air cleaners have been used by Union Carbide to remove radon daughters from mine air for several years. Approximately 300 air cleaner cells were in operation in the early 1970's, but the number has declined in recent years because so many large ventilation raises have been drilled. At the present time nine cells are in operation.

This type of air cleaner is a two-stage electrostatic device that is widely used to clean air in office buildings, residences, etc. The model that has been used almost exclusively is the Westinghouse Model PM Precipitron, which has a solid-state, high-voltage power pack mounted on each cell. The 3 ft. wide by 2 ft. high cells are individually mounted in specially built steel housings that can be stacked and bolted together in any configuration. The cells are usually installed in banks to clean all of the air flowing through an airway, but they have also been installed in ducts at the inlets of auxiliary fans.

The cells are washed with a detergent and water spray at one to four week intervals as required. Originally, the cells were sprayed with a water soluble adhesive after washing, but this was found to be unnecessary.

The cleaners are operated at a face velocity of approximately 300 feet per minute to achieve a working level reduction efficiency of about 90%. The efficiency at any face velocity can be expected to vary somewhat with variations in the size distribution of the particles to which the radon daughters are attached.

The extent to which air cleaners can be used to control radon daughter contamination is limited because they do not remove radon from the air. The rate of regrowth of the working level in cleaned air (assuming 100% daughter removal) is shown by the working level growth curve in Figure 3, and it is proportional to the radon activity. Electronic air cleaners can be a valuable supplement to ventilation, but they should be used only when the radon activity in the air is low, or the cleaned air is to be reused quickly.

The utility of the electronic air cleaners is also limited by their fragility. The collectors are constructed of thin aluminum plates that are easily bent, and the 0.006 inch diameter tungsten ionizer wires are easily broken. Corrosion of the aluminum collector plates is also a problem if the humidity is high. More durable equipment is needed to make this method more suitable for mining applications.

Chapter 29

TECHNICAL UNCERTAINTIES IN MINED GEOLOGIC DISPOSAL OF RADIOACTIVE WASTES

Paul F. Gnirk

RE/SPEC Inc.
Rapid City, South Dakota

INTRODUCTION

The notion of permanent disposal of high-level radioactive waste in natural salt formations was formally proposed in 1957 by the Committee on Waste Disposal of the National Academy of Sciences-National Research Council. This proposal was the result of the Committee's deliberations in response to a request from the U.S. Atomic Energy Commission (AEC) in 1955 to assess the viability of geologic formations for such disposal. The research investigations for waste disposal in salt were initiated at the Oak Ridge National Laboratory (ORNL) in the late 1950's and continued through the 1960's and early 1970's at a relatively modest level of funding. During the period of transfer of Federal Administrative authority from the AEC to the Energy Research and Development Administration (ERDA) to the Department of Energy (DOE) during the latter part of the 1970's, financial support for the research program was enhanced substantially. The program at ORNL was regrouped under the Office of Waste Isolation (OWI) and later transferred to the Office of Nuclear Waste Isolation (ONWI) at Battelle Memorial Institute. With this enhancement in the research program, consideration of basalt, granite, shale, and tuff as repository host media was initiated. In February, 1980 the President of the United States established officially a comprehensive waste management program which called for the selection of the first repository site in 1985 from a suite of four or five potentially suitable sites, and initiation of a repository operation at that site in the mid-1990's.

The current disposal system concept involves the emplacement of radioactive waste packages in mined disposal rooms at depth in suitable geologic formations, with closure of the facility effected by a combination of backfilling and sealing of the various facility excavations. The waste package, backfill and sealing materials, and the host rock mass comprise a system of multiple barriers to loss of containment and isolation of the wastes. The intent of this paper is to discuss technical uncertainties in the mined geologic disposal system and their potential impacts on containment and isolation, in conjunction with the minimization of uncertainties through the use of technical conservatism in repository siting and design.

MINED GEOLOGIC DISPOSAL SYSTEM

A mined geologic disposal system for radioactive wastes is defined as the combination of the rock mass which contains the repository facility, the facility excavations and attendant backfill and sealing materials, and the radioactive wastes and their associated containers (NWTS Program Information Meeting, 1979). The volume of the system is described by a vertical dimension which extends from the ground surface to several hundred meters below the waste emplacements, and by lateral dimensions which extend several kilometers away from the outer boundaries of the facility. A conceptual repository facility consists of a systematic collection of disposal rooms, haulageways, and ventilation drifts, commonly situated on a single level at a depth of 500 to 1000 meters and connected to surface by several vertical shafts. Current conceptual designs envision the emplacement of containers of radioactive waste in drillholes in the floors of the disposal rooms. To aid in the retrievability of the waste, if required at some time prior to closure and decommissioning of the repository facility, the emplacement drillholes are to be lined with steel sleeves. The container-sleeve and sleeve-drillhole wall annuli will be filled with engineered overpack and backfill materials. The collected aggregate of the waste form, container, sleeve, overpack, and backfill materials is commonly referred to as the waste package. Closure of the repository facility involves backfilling of the disposal rooms, haulageways, and ventilation drifts, and sealing of the shafts with "natural" materials.

The purpose of the disposal system is to contain and isolate the radioactive wastes to the extent that releases of radionuclides to the biosphere in the future will not endanger humans, i.e. in excess of those quantities permitted by radiation standards. The time period of concern is ten thousand years because that is the duration of time required for the radiotoxicity of spent fuel to decay to levels below that of naturally occurring uranium ore bodies (Voss, 1980). The mined geologic disposal system may be thought of as a system of multiple barriers to the loss of containment and isolation. Specifically, the waste package consists of elements which are resistant to the corrosive action of circulating groundwater and to the transport of radionuclides if the container is breached, and the waste form is subjected to leaching. The rock mass forms a second barrier in that radionuclides must be transported through chemically-retarding environments over distances which are both lengthy and time consuming.

LOSS OF CONTAINMENT AND ISOLATION

Loss of containment and isolation implies that the waste emplacement is breached by some type of disruptive phenomena, followed by the release of radionuclides from the waste package, and transport of these radionuclides along a pathway to the surface. The disruptive phenomena can generally be categorized as repository induced and externally induced. Repository-induced disruptive phenomena are those associated with the repository facility per se and the emplaced waste. Examples commonly sited are those due to heat generation from the emplaced wastes. In particular, the notion has been advanced that the heat will give rise to degradation of the rock around the waste emplacement, followed by a reduction in the thermal conductivity of the rubblized rock mass, and the significant rise in the temperature within the waste form. This temperature rise is supposed to be sufficient to cause degradation of the waste form, which then permits enhanced surface area for leaching by circulating groundwater. Another scenario for salt envisions the migration of brine towards the waste package, leading to enhanced corrosion of the waste container and exposure of the waste form to brine solutions. Externally-induced disruptive phenomena are visualized as sabotage, nuclear warfare, drilling, meteorite impact, volcanism, faulting, and erosion. In all instances, apart from some form of human intrusion, leaching of the waste form is required, together with transport to the biosphere by groundwater along some pathway. Pathways may be visualized as ruptured shaft seals, enhanced fracture systems due to faulting, or the natural fracture system of the host rock. Entry into the food chain or directly into humans must be accomplished by means of the surface water system.

TECHNICAL UNCERTAINTY AND TECHNICAL CONSERVATISM

Technical uncertainty is a phrase which is intended to mean in the context of this paper that a given characteristic of a mined geologic disposal system cannot be clearly identified or precisely quantified. A characteristic may be a physical property of the rock mass, such as thermal conductivity or coefficient of thermal expansion, or of the waste form, such as leaching rate for a given thermal/ chemical environment. A characteristic may also be phenomenological in nature, such as the thermomechanical or thermochemical responses of a rock mass to radiogenic heating. Finally, a characteristic may be some form of natural disruptive phenomenon, such as frequency and magnitudes of earthquakes. Thus, we may categorize several characteristics of a mined geologic disposal system to which technical uncertainty may be attached as:

1. Physical properties of the rock mass, including such items as thermal/chemical/mechanical/hydrological properties, states of in situ stress and geothermal temperature, spacing and orientation of joints, fractures, and structural discontinuities, etc., and of the waste form and backfill and sealing materials;

2. Phenomenological responses of the rock mass to excavation, backfilling and sealing, radiogenic heating, and radiation, and of the waste form and backfilling and sealing materials;

3. Frequency and magnitudes of natural disruptive phenomena, such as earthquakes, volcanism, meteorite impact, erosion, dissolution, etc.

Disruptive phenomena due to the actions of mankind, such as sabotage, drilling into the repository, etc., are detached from these deliberations as they are not truly intrinsic characteristics of the disposal system.

An obvious deduction from the proceeding discourse is that uncertainty associated with any characteristic of the disposal system will give rise to uncertainty in the containment and isolation of the radioactive wastes. This is certainly true if an intrinsic characteristic cannot be identified and quantified a priori. Identification of characteristics is a product of observations from geologic history and from analogues due to nature and mankind, and from in situ tests in geologic media. In situ tests with electrical heaters (to simulate the heat production from radioactive wastes) and spent fuel are capable of subjecting the geologic medium in its natural setting to conditions of waste emplacement (Gnirk, 1980). Although the time durations of the tests may be relatively short, of the order of a few years, the rock in the near vicinity to the emplacements is subjected to temperature levels that both are comparable and exceed anticipated design conditions. A particular purpose of the in situ tests is to identify the characteristics of the phenomenological response of the particular geologic medium to simulated or actual waste emplacement.

Identification is one aspect of the definition of technical uncertainty. The other is the quantification of that characteristic. Geologic materials rarely possess absolute homogeneity and isotropy of physical properties over distances of even a few meters. However, the range of variation in properties can generally be quantified by observation and testing.

In part to account for those characteristics which cannot be readily identified or quantified, and in part to account for those characteristics which are quantifiable only over a range, use is made of the notion of technical conservatism in site selection, repository design, and repository performance assessment. Technical conservatism is an approach wherein extremes in characteristics are employed to establish acceptance bounds. In essence, adverse characteristics are overestimated and favorable characteristics are underestimated. Site selection, for example, requires that the geologic medium has a history of tectonic quiescence, i.e. a very low frequency of earthquakes (Brunton and McClain, 1977). For the design of a repository system, the lower extreme of the thermal conductivity of the rock mass can be selected, which implies that the temperature in the waste form will be maximized (Charlwood, et al., 1979; Gnirk, et al., 1980). In repository performance assessment, faulting through the repository is assumed to occur, with attendant breaching of waste canisters (Shipler, et al., 1978). Site selection and repository design criteria are selected to minimize the potential for loss of containment and isolation due to repository-induced and externally-induced disruptive phenomena. Conversely, repository performance assessment assumes that disruptive phenomena will occur, whereby the consequences of radionuclide releases along pathways to the biosphere must be assessed. In effect, this approach tends to minimize and bound the range of technical uncertainty in mined geologic disposal systems.

ACKNOWLEDGMENTS

The author wishes to acknowledge Ms. Julie A. Annicchiarico of RE/SPEC Inc. for her careful attention in typing the manuscript.

REFERENCES

1. Brunton, G. D. and McClain, W. C., "Geological Criteria for Radioactive Waste Repositories," Office of Waste Isolation Document Y/OWI/TM-47, 16 pp., 1977.

2. Charlwood, R. G., Mahtab, M. A., Burgess, A. S., McCreath, D. R., Gnirk, R. F. and Ratigan, J. L., "Geological Engineering Factors in the Design of a Radioactive Waste Repository in Hard Crystalline Rock," Geological Survey of Canada (Paper 79-10), pp. 45-54, 1979.

3. Gnirk, P. F., "Synopsis of In Situ Testing for Mined Geologic Disposal of Radioactive Wastes," Proc. Workshop on Thermo-mechanical-Hydrochemical Modeling for a Hardrock Waste Repository, ONWI/LBL, Berkeley, 1980.

4. Gnirk, P. F., Callahan, G. D. and Ratigan, J. L., "Numerical Modeling of Radioactive Waste Repositories," ASCE Preprint 80-164, ASCE Convention and Symposium, Portland, 1980.

5. NAS-NRC Committee on Waste Disposal, Division of Earth Sciences, Disposal of Radioactive Wastes on Land, National Academy of Sciences--National Research Council Publ. 519, p. 6, 1957.

6. NWTC (National Waste Terminal Storage) Program Information Meeting, Proceedings, Battelle-ONWI/U.S. DOE (ONWI-62), 232 pp., 1979.

7. Shipler, D. B., Raymond, J. R., and Gnirk, P. F., "Analysis of Migration Pathways for Geologic Radioactive Waste Repositories," DOE Symp. on Environmental Control, Washington, D.C., 1978.

8. Voss, J. W., "Hazard Measure Comparison of Spent Nuclear Fuel and Uranium Ore," ONWI/Battelle Memorial Institute, Columbus, 1980.

Chapter 30

THE FUTURE: WHOSE RESPONSIBILITY FOR THE ENVIRONMENT?

Richard J. Gowen

South Dakota School of Mines and Technology
Rapid City, South Dakota

THE FUTURE

The environment in which we live appears to change so slowly that most of us are unaware that any change has occurred at all. As the years pass, we notice increasing smog in our cities, new variations in weather, the shifting landscape with changing populations, and subtle but ever present alterations in the quality of life. In considering the future of our environment, we find we must grapple with questions concerning the very future of life as we know it today. Questions of resources, technology, society, governance, and purpose. Let us focus on the future of our environment and, in particular, raise the question of who is responsible for the environment.

We know that action, or lack of action, will clearly influence our future environment. It is fashionable to catalog the many shortsightedness of our forefathers in developing the technologies which support the style of living to which we have become accustomed. However, I shall struggle to resist the temptation to list the well known ills of our present environment. Rather, let us concentrate on the future in our consideration of the priority of the multitude of options proposed to relieve our current energy crisis.

We have recently witnessed the dramatic effects of air pollution produced by Mount St. Helen's eruptions. In the state of South Dakota, we watched our blue skies turn hazy and felt our eyes sting with the presence of volcanic ash. We were appalled at the dramatic effects on the landscape, crops, transportation, and all aspects of life in communities immediately downwind of the volcanic eruption. But in a few weeks, life had returned to near normal on the earth's surface leaving the volcanic ash in the atmosphere to ultimately contribute to a cooling of our climate.

Still we on the earth are aware of the increasing problems of regional air pollution. Problems ranging from the Los Angeles smog to acid rain. While volcanic ash and manmade pollution have an effect on our environment, perhaps the greatest consequences on society may be associated with changes in the radiation balance between the earth and the atmosphere caused by atmospheric carbon dioxide (1,2,3). As the world struggles to resolve the problems of providing much needed energy, it is reasonable to anticipate that there will be an increase in the concentration of atmospheric carbon dioxide as a result of burning fossil fuels. Carbon dioxide is a relatively stable gas and approximately half of that added to the atmosphere will go into the oceans or be taken up by our forests. The other half of the carbon dioxide will remain in the atmosphere. It is estimated that the pre-industrial revolution level of carbon dioxide in the atmosphere was between 265 and 290 parts per million by volume (ppmv). The present level of carbon dioxide has 330 ppmv and is estimated to reach the level of approximately 400 ppmv by the end of this century. We could easily double the pre-industrial revolution level by the middle of the next century.

Carbon dioxide is an atmospheric trace gas which has a major effect on our world heat balance as it absorbs infrared radiation. We know that some of the infrared radiation emitted from the earth's surface passes through the atmosphere and escapes to space. But as the concentration of carbon dioxide increases, some of the earth's normal surface radiation will be absorbed and re-radiated and result in the warming of the surface of the earth with a significant potential for a change in our climate.

This process has been studied with a number of mathematical models which predict that the surface of the earth will increase at least one degree centigrade above present temperatures by the turn of the century and perhaps as much as two or three degrees centigrade by the middle of the next century (1,2,3). Scientists recognize that there are a number of gaps in our knowledge of earth climate. Therefore, there is reason to question the predicted changes in temperature. However, there seems to be general agreement that these estimates of future climate changes are correct at least within a factor or two (1). The effect of this seemingly minute change in the mean temperature of the surface of the earth is perhaps best seen in the extremes of regional temperature. This modest mean temperature change includes a regional change of ten degrees in the polar regions with resultant changes in the level of the oceans and increases in regional rainfalls throughout the world.

These predictions for dramatic changes in our future climatic conditions have a high degree of uncertainty not only because they attempt to predict the results of a highly interactive system which includes the influence of we humans but, additionally, such predictions are developed using mathematical models that are generally recognized to be still under development. Hence, there is a question concerning how much trust should be put on such predictions. We as scientists simply cannot prove that such models will produce results that are acceptable to the entire scientific community and, therefore, the predictions suffer from a loss of credibility. This inability to predict with certainty the results of policy decisions concerning the complex processes of our environment gives rise to questions of our ability to safeguard the environment.

A critical question is how much proof of an impending crisis is essential to effect a rational decision from society to forestall such a crisis. Of special interest to us is the question of whose responsibility is it to provide such proof. You will find these same questions repeated over and over again in concerns related to the development of our mineral resources--the questions of credibility, the adequacy of proof and a concern for who bears the responsibility for proving adequacy that all risks are minimized.

We know today that pollution from atmospheric carbon dioxide is increasing. We also know that this is a global problem that goes far beyond our nation's borders. Even if the more technologically advanced nations should choose to limit the use of fossil fuels in an attempt to control the buildup of carbon dioxide in the atmosphere, it is difficult to see how emerging nations could afford to limit their use of fossil fuels and still continue to develop. Thus, there is need for a global policy decision to determine just how much risk is acceptable and what are the effects of our continued need for fossil fuels. Thus, on the one hand, we have a global need for the continued use of fossil fuels and, on the other hand, we have an early alarm of potentially harmful effects to the very environment which provides us with life. What is our responsibility, yours and mine, to work toward the development of a global policy decision particularly when we know that the only way we will be able to prove the existence of the carbon dioxide "greenhouse" effect is to wait until we have actually created such a "greenhouse." This could take the next generation. The same generation must also wrestle with the issues of how we must supply energy for the very maintenance of society. If we take no conscious action to arrive at such a policy, we in fact will have made a decision. We will have chosen to let our children's children try to reverse the pollution that we will have created.

I do not speak to you today as a prophet of doom for, in fact, we really don't know how much pollution might be created, nor do I speak to you as an environmentalist. I am a concerned engineer and scientist who knows that we have a problem. I have chosen to emphasize our environmental problem because it illustrates the increasing difficulty that society faces in arriving at policy decisions, particularly policy decisions that involve technological issues for which clearcut answers are not available. Clearly, we have a responsibility in formulating such policy decision.

DEMOCRACY AT THE TEST

Over two hundred years ago, our founding fathers developed a democratic system which has served this nation well. They opted for a representative democracy rather than an Athenean democracy in which every citizen would be asked to vote on each crucial issue as it might arise. Our founding fathers recognized that such a true democracy would not function for the thirteen rapidly developing colonies soon to join together to create one nation under God. Rather they chose a representative democracy after the model of Thomas Jefferson. Our Jeffersonian democracy has served us through numerous wars, the industrial revolution, and our own societal revolutions. Yet, there are storm clouds on the horizon which urge me to encourage you to increase your vigilance, particularly since we as engineers and scientists are called upon to assume an increasingly important role in shaping our democratic processes.

Recently, John Kemeny who served as Chairman of the President's Commission on Three Mile Island commented on the Three Mile Island investigation and the subsequent implementation of the commission's recommendations. He has called for a change in the fundamental decision making process at the federal level so that we can better come to grips with the enormous and complex issues that face this nation (4). I concur in this recommendation and urge you to closely examine the very democratic process to which we all so strongly subscribe. As we consider the question of who is responsible for protecting our environment, which is in fact a very complex question involving considerations ranging from technology to economics to the very quality of life itself, we must be ever mindful that this question could possibly be so complex that even our cherished democratic system of governance may not be able to adequately provide an answer. We optimists know that we will muddle through hopefully without things getting too bad. We pessimists know we will muddle.

A NATIONAL ENERGY POLICY

An example in point is our inability to develop a national energy policy. A generation ago we became aware of our lack of superiority in space. We answered our President's call for a race to the moon, a goal which we could easily visualize night after night. The complexities of the race were left to the engineers and scientists to solve with technological breakthroughs. We became accustomed to knowing that we could develop the technology essential to placing a man on the moon. We succeeded in placing our astronauts on the moon only to find ourselves torn apart by strife over a little understood war in a far away corner of the world in which many of our relatives and friends gave their lives. A war fought with the most advanced technology available and a war that we somehow did not win.

We heard very little about an energy crisis. We saw our national pride shattered in retreat, our Presidency blackened, the credibility of our Congress, business and even society questioned as never before. We saw rising consumerism and protection against the complexities of the very industrial world which is the blood of our nation's economic wealth. We saw the growth of anti-business, anti-government, anti-science, anti-technology, anti-nukes, and anti-progress. And then along came an oil embargo that wrenched us to a halt. An embargo that heralded the outcry that this nation was in a struggle for its very existence.

I'm sure that you recognize that we did not enter an energy crisis with an oil embargo but that, in fact, we had begun the energy crisis years before. Yet it took an oil embargo to catch our attention so that we would begin to use the resources of our democratic process to identify an optimal and immediate solution. We know that such has not happened nor is it likely to happen. I shudder to think that we must await the equivalent of an oil embargo before we will take time to consider our environmental future.

The Great Debate

Instead of a cohesive energy plan, we find great debates on the approach to be followed. On almost any evening we can hear about the latest projections for energy needs or the latest threat of an increase in gasoline prices by simply turning on the evening news. The media, of course, is trying to keep us informed as part of our democratic process. But somehow, one must occasionally question just who informs the media, particularly since there seems to be so much divergence in the "facts" presented by we experts. There are those who are certain that there is an energy crisis, at least one that cannot easily be solved by a free industry. We are told that through increased conservation there will be no energy shortage and, especially, if we avail ourselves of the free energy of the sun or the wasted energy in our garbage. Yet, Dr. Arthur Bueche, Senior Vice President for Corporate Technology with the General Electric Company, in a recent article estimated that if this nation were to embark on an all out effort to develop alternate fuel sources ranging from solar to geothermal to biomass only about 15 quads of additional energy would result (5).* Even with highly optimistic estimates of the energy to be saved by increased conservation, Bueche notes that we still face a 15 quad gap in our ability to provide energy at the turn of the century. This is equivalent to the amount of energy needed to supply 15 cities the size of Boston for each year.

* One quad equals 10^{15} BTU's or 7½ billion gallons of gasoline or 48 million tons of coal.

In May 1980, a report was released from the Massachusetts Institute of Technology calling for a tripling of worldwide coal production during the next 20 years (6). The report optimistically concludes that coal can supply from half to two-thirds of all additional energy needed by the world between now and the year 2000 and that the United States should play a pivotal role in helping meet this need. Since the beginning of 1969, the basic OPEC cost of crude oil has leaped from $12.70 to $30.00 per barrel or the energy equivalent of about $144 per ton of coal. Yet, American coal costs only about $35 per ton to mine. Pollution control regulations increase the cost of using coal in the United States resulting in an added cost of about $25 per ton. Thus, even including all of the environmental and transportation costs, coal is now roughly 60% cheaper than oil. This nation sits on top of 25% of the world's known reserves of the precious black mineral, yet using coal to meet our energy needs meets with strong opposition from our environmentalists.

The Great Debate of the 70's continues as we focus on the use of nuclear energy. Perhaps you will recall when we called nuclear power the "clean option." Now we have prophets of doom who debate the pro-nukes concerning the radiation dangers of everything from chest x-rays to uranium mill tailings to the radiation exposure of flying at 36,000 feet. Do we really have a problem with the removal of and storage of nuclear waste, or do we have a problem because you can't store it in my backyard, wherever that may be.

We have the technology to produce abundant energy from nuclear power. But following the incident at Three Mile Island there is a serious question of whether we can economically afford to pursue our nuclear option. John Kemeny, in discussing the results of the Three Mile Island Presidential Commission notes, "...if recommendations like those we came up with are implemented during the next few years, I think nuclear power can be one of the energy alternatives available to humanity. I am equally convinced, however, that if recommendations like ours are not implemented--if a year from now it is business as usual--then the nuclear power industry will put itself out of business, as victim of its own attitudes." (4).

Recently, I had the honor of co-chairing a working conference jointly sponsored by the Nuclear Regulatory Commission and the Institute of Electrical and Electronics Engineers to evaluate the application of advanced electrotechnology to improving the safe operation of nuclear power plants (7). The fundamental nuclear reactor process is well understood, and our utility reactors are both well designed and well built. But the technology used to control these reactors, operating procedures and management philosophies are far out of date (4,7). I do not criticize the nuclear industry, rather, I note that for years the industry and the regula-ors have shared a "mindset" that nuclear power plants were safe. Therefore, there was really no need to question the application of new technologies to insure the safety of operations. Besides, it was a lot less expensive to build a new plant using the same technology that had been previously approved rather than battle the NRC for a new approval. Days, weeks, months, years have a special meaning when a billion dollar investment is waiting to become operational.

Unfortunately, it took a Three Mile Island to bring the industry to a halt, if only momentarily, to consider its future. The regulators and the regulated recognized the need for a change from the "mindset" about safety, but change is neither easy nor cheap.

Will We Proceed

Thus, while this nation has an ample supply of coal and the technology to produce abundant power by nuclear generation, we are unable to develop an overall policy on how to proceed. We know we need to take action today to build the 100 to 250 additional nuclear plants, the 300 to 400 coal mines, and an almost equal number of fossil fuel plants that are needed if we are to meet our projections of energy needs for the end of the century. Yet, decisions are not forthcoming.

These decisions involving the formulation of a national policy for energy development carry with them tremendous implications for our future and will require enormous resources. There is a need for billions and billions of dollars to carry out such necessary development and the implications of such a commitment to coal and nuclear technology lead to serious questions concerning the impact upon our environment. It is little wonder that we find ourselves stagnated and unable to arrive at an acceptable decision. It is a time in which we joke about a do-nothing Congress and thank God that they haven't taken the wrong action.

We find that the complexity of decisions that Congress is called to act upon has increased almost beyond comprehension. Issues that are so complex that we now have groups of experts pointing out that other groups of experts obviously are wrong. Our rapidly changing technology forces us to re-examine our roles as scientists and engineers. The classical role of the scientist can perhaps best be described in a quote from Professor Harold Urey, a Nobel laureate and University of California chemist, "To those of us who spend our lives working on scientific problems, science is a great intellectual adventure of such interest that nothing else we ever do can compare with it. We are attempting to understand the order of a physical universe vast in extent in space and time and most complicated and beautiful in its detail" (8).

While there is still a need to understand our vast, complicated universe, there is even a greater need for scientists and engineers to translate to society an understanding of the implications of the use of such scientific discoveries almost immediately after they are made. Not too long ago, if one asked a question requiring a scientific answer, it was appropriate to recommend that the questioner find a person trained as a scientist to answer the question. For at that time, the growth of our scientific knowledge base was slow enough that well trained scientists could provide answers on all the important issues that had weathered the test of time and were accepted as a consensus within the scientific community. In today's world of exploding technology, scientists and engineers find it difficult to remain current in even their own areas of specialized expertise, let alone find time to become educated in the interactions of technologies that effect multidisciplinary problems.

It is easy to see why we experts might be perceived to be a little confused when we provide advice to Congress, executive agencies and the public. It is little wonder that our representatives and government leaders find it difficult to arrive at an optimal decision if they can arrive at any decision at all.

GOVERNMENT BY REGULATION

It should come as no surprise that the response to such confusion is an increasing number of debilitating regulations, regulations that at best are attempts by our lawmakers to practice engineering and science in an effort to create order out of our apparent confusion and regulations that all too often are extensions far beyond the original intent of the Congress and have now come to represent a new kind of lawmaking. Such regulations are laws made outside of our democratic process. While many are aware of this alarming trend, fighting such regulations is difficult and expensive.

Congressman Dale Bumpers of Arkansas has proposed legislation to require federal agencies to prove why a particular regulation is essential, rather than force us to challenge obviously lopsided regulations through lengthy court battles (9). Under the Bumpers legislation, the cost of a challenge would be charged against the loser--the government or the regulated.

Such regulations can have a devastating effect on our ability to solve the nation's problems, particularly when those regulations are developed in the arena of special lobby groups. One cannot help but wonder why essentially the same level of environmental equipment is required to clean the sulfur from low-sulfur western coal as is required for eastern coal. Perhaps there is some truth in the suggestion that such a regulation creates an inequality in the market place and serves as a further example of the ability of special interest groups to function within the confusion of solving technologically-oriented problems through our present system laws developed democratically and laws in the form of regulations that are outside of the democratic process.

THE TIME IS NOW

I have shared examples of our problems with arriving at decisions concerning complex technically-oriented issues, such as an energy policy or the regulation of our nuclear power plants. We know that there are even more difficult questions to be answered concerning our environment. These are questions which have a far greater impact than even our energy crisis. Questions which are deeply shrouded in the emotions of fear and doubt. Questions which cannot be answered with anything near the 100% assurance of lack of risk that some segments of society clamor for. Yet, they are questions for which we cannot afford to let "nature take its course."

So just whose responsibility is it to answer these questions: Society? Congress? the environmentalists? the media? or our children's children? Who are we? Are we part of the problem or are we part of the solution? Are we ready to accept the fact that we have a special responsibility to lead the development of the answers to the questions of the future of our environment.

The future begins today. Although the problems are immensely complex, almost astronomic in dimensions, we must revitalize our democratic decision making process and bring the strength of free enterprise working in harmony with our engineering and scientific capabilities to find responsible solutions to our global problems such as the future of our environment.

BIBLIOGRAPHY

1. William W. Kellogg and Stephen H. Schneider, "Global Air Pollution and Climate Change," IEEE Transactions of Geoscience Electronics, Vol. GE-16, No. 1, pp. 44-49, January 1978.

2. Gerald J. Dittberner, "Climate Change: Volcanoes, Man-Made Pollution and Carbon Dioxide," IEEE Transactions on Geoscience Electronics, Vol. GE-16, No. 1, pp. 50-58, January 1978.

3. Earl W. Barrett, "Man and Climate: An Overview," IEEE Transactions on Geoscience Electronics, Vol. GE-16, No. 1, pp. 62-70, January 1978.

4. John G. Kemeny, "Saving American Democracy: The Lessons of Three Mile Island," Technology Review, pp. 65-74, June-July 1980.

5. Arthur M. Bueche, "Energy Resource Allocation for the 21st Century," Industrial Power Systems, September/December 1979.

6. "King Coal: Ready to Reign Again," Time, p. 81, May 19, 1980.

7. Report on The Working Conference on Advanced Electrotechnology Applications to Nuclear Power Plants, Vols. 1 and 2, Washington, D.C.., January 15-17, 1980.

8. Walter Sullivan, Edward Edelson, Mark Dowie, Thomas O'Toole, David Perlman, Robert Cooke, Jules Bergman, and Christine Russell, "Science, Technology, and the Press: Must the 'Age of Innocence' End?", Technology Review, pp. 46-56, March/April 1980.

9. Jill Wechsler, "Reducing Regulations," Dun's Review, p. 115, March 1980.

Environmental II

Session Chairman
Harlan E. Gellhaus
Francis-Meador-Gellhaus, Inc.
Rapid City, South Dakota

Chapter 31

AN UPDATE OF HOMESTAKE'S GRIZZLY GULCH TAILINGS DISPOSAL PROJECT

Fred D. Fox

Homestake Mining Company
Lead, South Dakota

INTRODUCTION

Approximately two years have elapsed since the first summary of the Grizzly Gulch Tailings Disposal Project was presented (1). Since that time, various physical modifications and additions to the existing treatment system have transpired. Further, Homestake and the entire mining industry have been impacted severely by additional modifications and additions concerning regulatory reform aimed at total disciplinary permitting and hazardous waste management. The following paper summarizes the past accomplishments relating to the Grizzly Gulch Tailings Disposal Project, correlating these accomplishments with future plans that are being developed to comply with regulations and to continue Homestake's commitment to environmental protection.

The famous Homestake Gold Mine, located in Lead, South Dakota, has been operating in the Black Hills region for over a century, approximately 12 years before the south half of Dakota Territory was admitted to the Union as the State of South Dakota (2).

During the 104 years of operational development, Homestake's Gold Mine, as well as the entire mining industry, have been influenced by numerous governmental regulations dating back to the Refuse Act of 1899 which regulated the deposit of debris from mines and stamp works. The enactment of the National Environmental Policy Act of 1969 may be thought of as the beginning of a long line of unplanned steps to deal with the many concerns of matters pertaining to the environment. Environmental awareness continued into the 1970's, in which the Clean Air Act, Federal Water Pollution Control Act, Safe Drinking Water Act, Toxic Substances Control Act, and the Resource Conservation and Recovery Act (RCRA), to name a few, surfaced.

In the summer of 1976, Homestake began constructing the Grizzly Gulch Tailings Dam Disposal Project. In the fall of 1976, the Resource Conservation and Recovery Act surfaced, which provided "technical and financial assistance for the development of management plans and facilities for the recovery of energy and other resources from discarded materials, and to regulate the management of hazardous waste" (3). Consequently, Homestake's tailings impoundment was designed and engineered prior to RCRA, but it has had the opportunity to grow and develop with these regulations, and others, which have done so since the problem of unsafe hazardous waste disposal came to light in 1978.

The following paper briefly describes the major highlights of the current pollution control practices at Homestake and summarizes the projects and plans that are currently being developed to satisfy the ever-changing environmental regulations governing past, present, and future mining and ore processing development.

GENERAL PROJECT DESCRIPTION (1)

The milling waste that is not utilized for stope backfill is consolidated at the slurry pump station. The pump station consists of a reinforced concrete substructure with a prefabricated steel superstructure. Tailings leave the pump station within an eight-inch (20 cm) rubber-lined steel pipeline. Starting at an elevation of 4,920 feet (1,500 m), the tailings travel 2.9 miles (4.7 km) at a minimum transport velocity of 6 fps (1.8 m/s). Maximum elevation along the pipeline route is 5,409 feet (1,650 m) before the pipeline descends to the tailings dam crest at the 5,350-foot (1,630 m) elevation. Maximum static head is 487 feet (148 m). A 230-foot (70 m) high, zoned, earth and rockfill starter dam was constructed across a steep, narrow canyon to create an impoundment area for storage of tailings.

The base width of the embankment is 1,000 feet (305 m) and the crest width is 29 feet (9 m). Crest length is 1,350 feet (411 m). See Figure 1. Slurry transport water, excess net precipitation and ground water infiltration is recycled 3.4 miles (5.5 km) to the mine-mill complex for reuse.

A 4.5 mile (7.2 km) long interceptor canal was excavated 50 feet (15 m) above the embankment crest to prevent stream flow, upper basin precipitation and storm water runoff from entering the tailings impoundment.

The Ore Mining and Dressing Effluent Guidelines require that all extraneous water which enters the 157-acre (64 ha) impoundment basin be treated prior to discharge. The ground water downstream from the tailings impoundment is collected in a gunite backed seepage collection trench excavated in bedrock. The seepage water from the impoundment basin is recycled. See Figure 2.

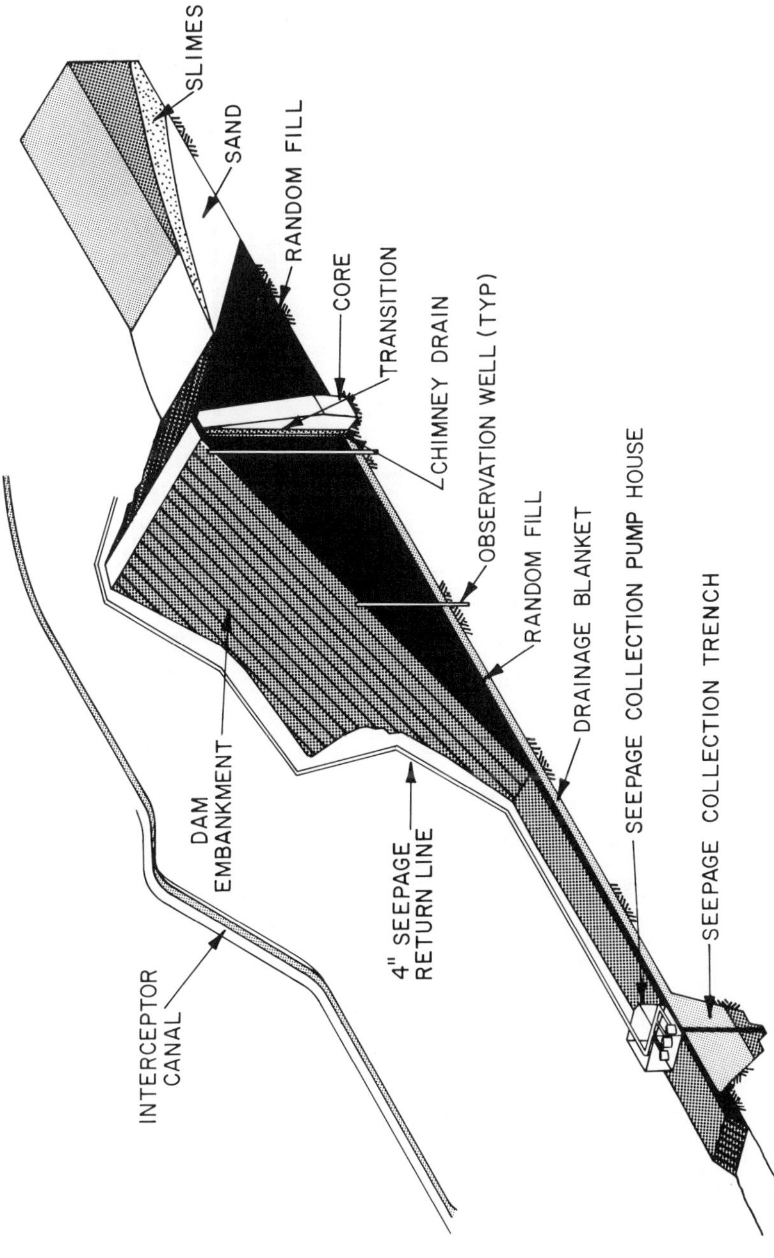

Figure 1. Homestake tailings impoundment.

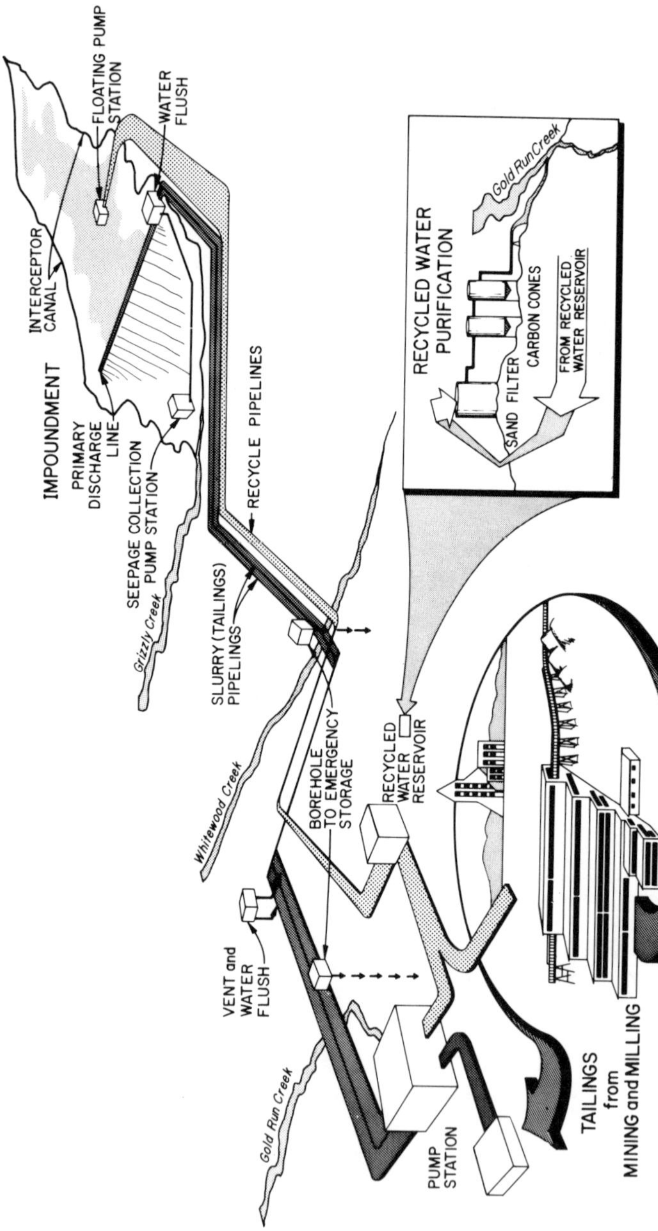

Figure 2. Homestake's Grizzly Gulch Tailings Disposal Project.

Pollution Control Requirements

Effluent Discharge

The South Dakota Board of Environmental Protection classified the receiving stream one mile downstream from the Homestake Mine end-of-pipe discharge as a "cold water permanent fish life propagation water." Water quality limitations for Homestake's discharge were developed by adjusting the 96-hour EC50 values downward by a factor of 0.01 (4) considering the 25-year, 7-day low mean flow of the receiving stream (5). The EC50 indicates that effective concentration (actual or interpolated) at which 50 percent of the 2-month-old rainbow trout survived (4).

Homestake's National Pollutant Discharge Elimination System (NPDES) permit that was issued on September 3, 1976 is described in Table 1. As noted, the discharge limitations and monitoring requirements became effective on July 1, 1977.

Hazardous Waste

On May 19, 1980, the U.S. Environmental Protection Agency promulgated hazardous waste regulations pursuant to the Resource Conservation and Recovery Act (RCRA) of 1976. These regulations impose legal obligations on facilities that generate, ship, transport, store, treat or dispose of hazardous wastes. Subpart D of RCRA specifically lists "cyanidation wastewater treatment tailings pond sediment from mineral metals recovery operations" and "spent cyanide bath solutions from mineral metals recovery operations" as hazardous wastes from nonspecific sources, both of which include tailings and decant water. Therefore, Homestake has legal obligations concerning the generation (milling process), storage, and treatment or disposal (tailings impoundment) of hazardous wastes. To fullfill these obligations, Homestake will have to satisfy numerous permit requirements that have not yet been fully developed.

STATUS ON FULLFILLMENT OF POLLUTION CONTROL REQUIREMENTS

Effluent Discharge (NPDES)

Completion of the tailings disposal project on December 2, 1977 in conjunction with the implementation of two 125-ft. (38 m) diameter thickeners in the spring of 1978 for the removal of total suspended solids (TSS), proved unsuccessful in producing the desired results of less than 10 ppm TSS. In addition, the destruction of total cyanide to the 0.02 ppm limit by utilization of ozone, followed by activated carbon cones for final clarification, failed to produce acceptable results. On July 1, 1978 Homestake was not in compliance with its

Table 1. Homestake Gold Mine - NPDES Permit.

Effluent Characteristic	Discharge Limitations (Concentration)		Monitoring Requirements (Effective July 1, 1977)	
	mg/l Daily Ave.	mg/l Daily Max.	Measurement Frequency	Sample Type
Total Suspended Solids	N/A	10.0[1]	Weekly	Composite
Total Cadmium	N/A	0.00025	Monthly	Composite[2]
Total Copper	N/A	0.009	Weekly	Composite[3]
Total Zinc	N/A	0.020	Weekly	Composite
Total Chromium	N/A	0.85	Monthly	Composite
Total Lead	0.20	0.28	Weekly	Composite
Total Mercury	N/A	0.001	Monthly	Composite
Total Nickel	N/A	1.25	Monthly	Composite
Total Silver	N/A	0.001	Weekly	Composite[4]
Total Arsenic	N/A	0.36	Monthly	Composite
Cyanide	N/A	0.02	Monthly	Composite[5]

Flow shall not exceed 825 gpm and shall be monitored continuously. Also see Part III, B. (3).

The pH shall not be less than 6.0 standard units nor greater than 9.0 standard units and shall be monitored weekly by instantaneous type of measurements.

There shall be no discharge of floating solids or visible foam in other than trace amounts.

Samples taken in compliance with the monitoring requirements specified above shall be taken at the following locations: at the discharge from the new treatment facility prior to mixture with the non-contact wastewaters.

1) In addition to the above requirements, any grab sample shall not exceed 18 mg/l for this parameter.
2) Values less than 0.01 mg/l may be reported as "less than 0.01 mg/l."
3) Values less than 0.05 mg/l may be reported as "less than 0.05 mg/l."
4) Values less than 0.02 mg/l may be reported as "less than 0.02 mg/l."
5) Frequency to become weekly upon approval of an automated analysis for this parameter.

NPDES permit limitations.

To combat the existing problems associated with the treatment of total suspended solids, Homestake initiated pilot testwork of a sand filter (50 gpm) (3.2 l/s) in the fall of 1978, which proved very successful. Construction of a full scale sand filter treatment system (3,600 gpm total capacity) (227 l/s) was well under way in the spring of 1979, and was completed and put on line August 27, 1979. Consequently, the present total suspended solids limitation of 10 ppm has been successfully met.

In March, 1979, a meeting was held with the Environmental Protection Agency (EPA) and the State Department of Water & Natural Resources (DWNR) to update them on the current status of producing an adequate wastewater treatment facility. At this meeting, it appeared to those present that Homestake should concentrate on the treatment of cyanides, which would require a major research effort. Such an effort was immediately implemented, with preliminary screening of the following potential treatment processes:

PROCESS	STATUS
1. ION EXCHANGE PROCESS	- Developed into a pilot plant study.
2. PRUSSIAN BLUE PRECIPITATION PROCESS	- Developed into a pilot plant study.
3. OZONATION PROCESS	- Developed into a pilot plant study.
4. ACIDIFICATION/VOLATILIZATION PROCESS	- Developed into a pilot plant study.
5. ALKALINE CHLORINATION PROCESS	- Developed into a pilot plant study.
6. HYDROGEN PEROXIDE PROCESS	- Under current laboratory testing.

In addition to pursuing research on developing an adequate treatment process for total cyanides, Homestake also continued research aimed at analytical methodology, with the goal of producing a method that will precisely and accurately measure total cyanide to the 0.02 ppm level in the presence of large amounts of thiocyanate and other interfering agents.

On February 12, 1980 Homestake entered into a consent decree (6) with EPA and the DWNR that committed Homestake to an ambitious program of research, development, engineering, construction and operation to solve the treatment problems associated with past copper, zinc, cyanide, and flow violations.

Homestake agreed to pay a civil penalty of forty thousand dollars ($40,000.00) to the U.S. Treasury for past permit violations. Furthermore, Homestake agreed to pay to the State of South Dakota, a sum of three hundred fifty thousand dollars ($350,000.00) for past violations. In addition, the decree committed Homestake to place into operation by September 1, 1979, a completed and functional sand filtration facility, which was fulfilled by August 27. Finally, the decree stipulated that Homestake must construct and place into continuous operation by September 1, 1981, a treatment system to treat heavy metals and total cyanide in the discharged effluent.

To date, no analytical technique or treatment system has been defined that will accurately measure and/or destroy cyanide and its heavy metal complexes to the low limits desired. However, research and development work will continue until such a technique and system are identified and proven to be acceptable to all parties involved.

To satisfy the 65° Farenheit (18.3^{o}C.) temperature requirements of the receiving stream, Homestake constructed and put on line a cooling reservoir during the month of November, 1979. The reservoir consists of a two-cell rectangular reinforced concrete structure, 80 feet (24.4 m) wide by 150 feet (45.7 m) long and 6 feet (1.8 m) in depth, with 16 vertical rise spray cooling bars, per cell. The reservoir proved to be a partial success, cooling the warm mine water at 82-85°F by approximately 10°F during winter months but only 5°F during warm summer months. Consequently, the discharge temperatures during the winter months averaged 73°F and during the summer months averaged 78°F, well above the current limitation. Therefore, additional cooling of the discharge water is required.

To further reduce the quantities of heavy metals in the discharge, a second carbon cone facility consisting of four vertical steel cones filled with 8,000 pounds (3,620 kg) of activated carbon was installed and became operational in November 1979.

In order to keep abreast of the tremendous environmental sampling load, a new analytical laboratory was constructed and became operational in May, 1980.

To comply with the June 7, 1979 Best Management Practices (BMP) requirements pursuant to the NPDES regulations, Homestake began to implement and document procedures and practices to insure that all toxic and hazardous materials were prevented from reaching pathways to the outside environment. This program, along with numerous other projects aimed at controlling the use, handling, storage, and treatment of toxic and hazardous materials, is being implemented as the new, ever-changing regulations evolve.

Hazardous Waste Management (RCRA)

The construction of a safe, environmentally sound impoundment was originally planned for the containment of the solid fraction of Homestake's wastewater, prior to the May 19 designation of such solids, and decant water, as hazardous waste. Homestake is very fortunate for having the foresight to plan and engineer an impoundment that not only serves as a tailings storage facility but also acts as a water storage impoundment. Because of these qualities, very few modifications to the existing starter dam will have to be implemented in order to comply with the legal requirements pertaining to the Resource Conservation and Recovery Act.

The tailings retention dam was originally designed to be constructed in stages (7). Three alternatives of lifting the present dam were proposed: (1) conventional centerline method, (2) downstream, and (3) upstream method using tailings. Due to an unprojected increase in the amount of tailings discharged, Homestake is presently evaluating critical factors in design of the centerline and downstream methods. See Figures 3 and 4. The upstream method of increasing the height of the dam has been ruled out due to various factors pertaining to the stability and the obvious loss of tailings storage capacity. It is projected that the second lift of the retention structure will be completed within the year 1983.

In August, 1980, field investigations were initiated to define the quantities and quality of available materials for the central core, transition zone, and subsequent drain filter. To raise the existing dam to the ultimate height of 150 feet (45.7 m) to elevation 5,500, it is estimated that there would be an increase of 2.3 million cubic yards of material if the downstream method were chosen over the conventional centerline method. In addition, the downstream method would require an extension of the underdrain and relocation of the seepage collection sump pump facilities. The decision as to which method to utilize will depend entirely on extensive pre-engineering testwork which will demonstrate the safety aspects of each method being considered.

The slurry pumping system is operating satisfactorily although various modifications have been implemented. The system was designed to operate one bank of pumps and one line at all times, keeping the other bank on standby. Due to an increase in the milling tonnage, the inability to recycle all of the sluice water, and the operational density requirements not being consistent, both slurry lines are presently being utilized. Plans are currently being developed to cope with these problems, hopefully reverting back to the use of one pipeline.

The emergency underground dump system to the 1,100-foot level has been activated numerous times during operation due to power outages. Because the capacity for these dumps at the 1,100-foot level has decreased, the system currently dumps to a new disposal area on the 1,700-foot level.

The seepage collection trench at the toe of the existing dam has proven to be very successful in controlling ground water contamination immediately downstream from the impoundment as demonstrated by monthly ground and surface water monitoring.

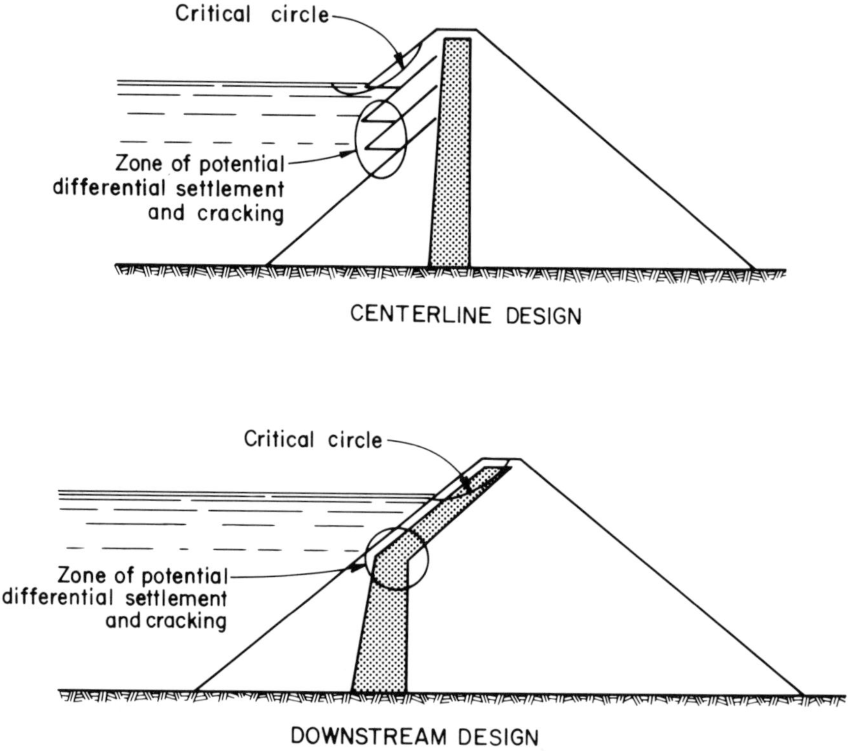

Figure 3. Critical factors in the design of the tailings impoundment.

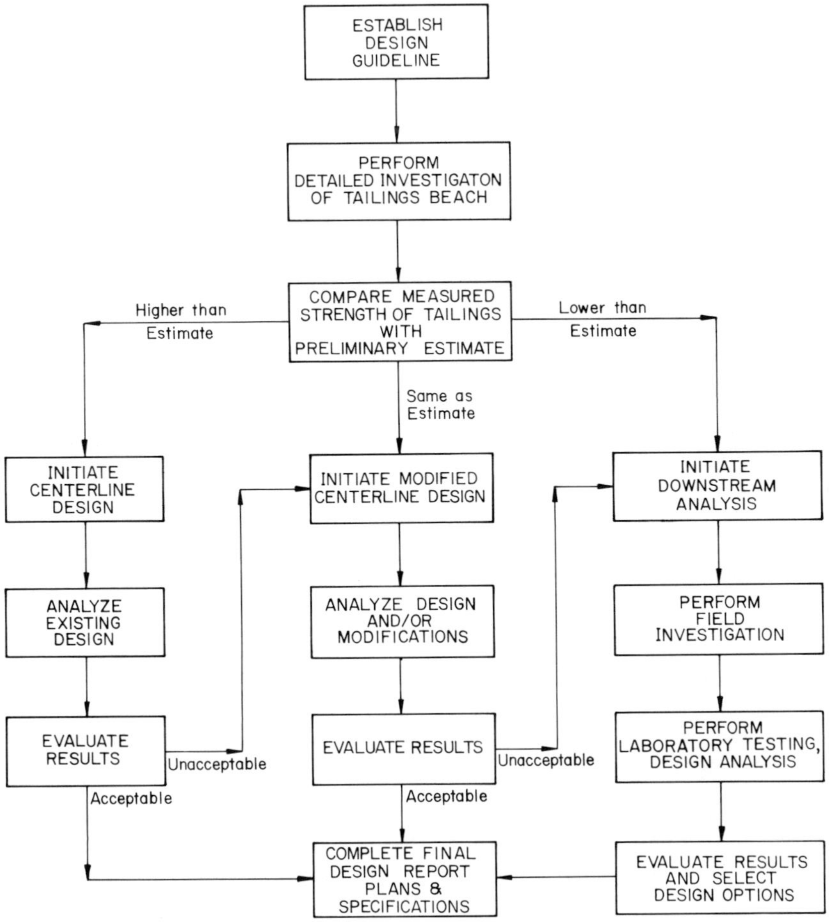

Figure 4. Flow diagram for the proposed study program.

In addition to the floating pontoon barge that decants 350 gpm (22 l/s) via a 6-inch (15 cm) steel pipeline 3.4 miles (5.5 km) to the recycle reservoir, a 12-inch (30 cm) decant line with separate pumping systems was installed to provide more flexibility. A 10-inch (25.4 cm) 1200 gpm (75.7 l/s), together with a 12-inch (30 cm) 2400 gpm (151.4 l/s) pump placed on skids, served the 12-inch (30 cm) decant line. These two pumps are presently being placed upon a floating barge, similar to the 6-inch facility, because of the fear of pump freeze-up from the encroaching slimes and sands.

Capital, Operating and Maintenance Costs

Capital costs incurred from September 1975 through December 31, 1979 are listed in Table 2 and a summary of capital costs for the year, 1979, is given in Table 3.

Operating and maintenance costs incurred from January 1, 1979 through December 31, 1979 are listed in Table 4.

Operational Experience Update

The slurry disposal system became operational December 2, 1977 incurring approximately 2.75 years of operational experience. From this experience, the following summary and conclusions can be drawn:

(1) The eight, closed-coupled, centrifugal tailings pumps in series have operated essentially trouble-free.

(2) Sand-filtered effluent will be utilized as source makeup water for gland seal cooling water, thereby eliminating a portion of city water that is presently entering into the system.

(3) The sand filter is operating satisfactorily, reducing total suspended solids below the 10 ppm requirement.

(4) The carbon cones at the final discharge are performing as filters during low flows which tends to build up solids that are dislodged during peak-flows (when two mine pumps are running), thereby creating a solids problem in the discharge.

(5) Inspection of the dual 8-inch (20 cm) rubber lined slurry lines indicate relatively little wear, thereby increasing the life expectancy of the above ground tailings transport system.

(6) Additional capacity was required for the underground emergency disposal sites due to frequent power outages (at least two per year).

(7) Research and development work continues on solving the total cyanide and heavy metal problems.

Table 2. Homestake Gold Mine Grizzly Gulch Tailings Disposal Project and Internal Plant Modifications - Summary of Capital Costs for Year 1979*.

September 1975 - December 31, 1979

	PROJECT BUDGET IN 1977	REVISED BUDGET FOR 1978
Site Preparation	$ 212,500	$ 188,056
Engineering (Outside Consultants)	838,750	817,830
Subtotal	$ 1,051,250	$ 1,005,886
Grizzly Gulch Impoundment		
Mobilization	300,000	300,082
Foundation & area preparation	678,800	603,927
Embankment	3,940,500	3,673,548
Interceptor canal	1,845,000	1,647,868
Emergency pond	2,000	1,164
Seepage collection ditch	125,000	167,061
Electrical service	15,750	15,625
Monitoring & stabilization	109,000	94,604
Demobilization	100,000	100,048
Subtotal	$ 7,116,050	$ 6,603,927
Pump House & Pipelines		
Slurry pipelines*	1,989,600	2,255,299
Slurry pump station	956,750	1,268,295
Decant pump station	53,300	100,163
Subtotal	$ 2,999,650	$ 3,623,757
TOTAL PRECONSTRUCTION, ENGINEERING DAM AND PIPELINES	$11,166,950	$11,233,570
Water Recycle & Wastewater Treatment		
Emergency disposal	$ 168,000	$ 109,974
Recycle pumps & pipelines	254,100	296,330
Carbon-in-pulp thickeners	1,100,000	1,037,900
Sand dam enlargement	381,000	401,750
Sand dam overflow	188,000	264,411
Mill & sluice reservoir	690,000	697,570
Char plant feed pumps	1,000	585
Vat overflow thickener	50,000	33,751
Washwater overflow	152,000	153,351
Lab & equipment	52,500	27,519
Carbon plant pipeline/other pipelines	22,800	98,279
Subtotal	$ 3,059,650	$ 3,121,420
TOTAL POLLUTION CONTROL PROJECT	$14,226,600	$14,354,990
Add: Decant pipeline - Approved July Board Meeting*	661,000	661,000
TOTAL REVISED POLLUTION CONTROL PROJECT	$14,887,600	$15,015,990

*Grizzly Gulch Tailings Disposal Project Closed Out December 31, 1979

Table 3. Pollution Control Expenditures

Year		Pollution Control Expense	Pollution O & M Costs	Pollution Capital Costs	TOTAL
Actual	1972	$ 3,043.00	0	0	3,043.00
"	1973	37,037.00	0	0	37,037.00
"	1974	50,789.00	0	0	50,789.00
"	1975	51,811.00	0	0	51,811.00
"	1976	214,288.00	0	0	214,288.00
"	1977	38,572.00	$ 16,336.00	$12,267,639.00	$12,322,547.00
"	1978	57,220.00	527,316.00	1,877,182.00	2,461,718.00
"	1979	67,860.00	749,274.00	1,364,100.00*	2,181,234.00
			3 YEAR GRAND TOTAL -		$17,322,467.00

* BREAKDOWN OF 1979 CAPITAL COSTS FOR POLLUTION CONTROL:

B - 150 Ozone Generator	142,328
Cooling Reservoir	84,280
Treatment Plant Trunk Pipeline .	252,126
Baker Sand Filter	385,174
Carbon Cones & Building	103,437
Power Line & Substation	11,804
Wastewater Treatment Plant Road .	29,062
Grizzly Gulch	355,862
1979 CAPITAL COSTS TOTAL	$1,364,073

Table 4. Homestake Gold Mine Pollution Control Operating & Maintenance Costs

January 1 - December 31, 1979

	Amount	Unit Cost
Pumphouse & Pump Operation	$161,562	$0.11
Sand Dam Area	18,545	0.01
Wastewater Discharge Treatment	243,517	0.17
Pipelines	78,290	0.04
Emergency Underground Disposal	23,486	0.02
Impoundment	48,881	0.04
Power System	131,923	0.10
Decant Pumping System	43,070	0.03
Total	$749,274	$ 0.52

Production 1979 - - - - - - - - - - - - - - - - 1,429,886 tons

Homestake Gold Mine
Pollution Control Operating & Maintenance Costs
By Expense Classification

January 1 - December 31, 1979

	Amount	Unit Cost
Labor	$289,071	$0.20
Supplies	174,873	0.12
Other Services	149,322	0.10
Power & Fuel	136,010	0.10
Total	$749,275	$ 0.52

(8) Work continues on establishing flow meter accuracies in terms of defining an adequate internal water balance.

(9) Monthly monitoring data indicates excellent seepage collection control recovery and recycle.

(10) The 50-foot (15.25 m) diameter overflow recycle pumping facility is planned to be replaced by more efficient flooded suction pumps, increasing the chance of operating one pipeline.

(11) The increased amount of tailings discharged has decreased the life expectancy of the initial starter dam; therefore, pre-engineering for the next lift has been implemented.

Conclusion

With very few modifications and sound engineering, the Grizzly Gulch tailings retention dam, impoundment, and slurry transport system can "grow-up" and evolve with the ever-changing environmental regulations governing the transportations, disposal and storage of hazardous wastes. Current technology is available to satisfy the majority of the requirements concerning hazardous waste management. On the other hand, current technology does not exist to treat the total cyanide found in Homestakes' solutions. Nevertheless, Homestake will continue its research efforts aimed at reducing cyanide and heavy metals to the low levels required by the present permit limitations.

REFERENCES

1. Jeffries, Lawrence F. and Tczap, Andrew, "Homestake's Grizzly Gulch Tailings Disposal Project," presented at International Mining Show, Las Vegas, October 11, 1978.

2. 1876 Homestake Centennial, Pamphlet, Homestake Mining Company, 1976.

3. Resource Conservation and Recovery Act of 1976, Public Law 94-580, 90 Stat. 2795, October 21, 1976 as amended by P.L. 95-609., November 8, 1978.

4. Evaluation of Proposed NPDES Permit Limitations for Homestake Mining Company, National Enforcement Investigations Center, Denver, Colorado, August, 1975.

5. Synopsis. Summary of Rationale. Homestake Mining Company Permit Number SD - 0000043. Environmental Protection Agency Region VIII, May 18, 1976.

6. United States of America, Plaintiff, State of South Dakota, Intervenor, v. Homestake Mining Company, Defendant. Civil No. 78 - 5094 Consent Decree., February 12, 1980.

7. Interim Report, Grizzly Gulch Tailings Disposal System, Prepared by Dames & Moore, December, 1975.

Chapter 32

CYANIDE IN RIPARIAN VEGETATION

Daniel L. Noble and Sr. Marmion Howe*

*The authors are, respectively, Systems Ecologist, USDA Forest Service, Rocky Mountain Forest and Range Experiment Station (in cooperation with South Dakota School of Mines and Technology) and Professor, Department of Biology, South Dakota School of Mines and Technology, Rapid City, South Dakota.

Riparian communities are those related to, or near a natural watercourse (or sometimes of a lake, impoundment, or tidewater). Generally, riparian communities contrast sharply with the dominant vegetation type in an area, for example, the willow-birch-aspen-dominated watercourse in the Black Hills with its associated beaver-muskrat-trout habitat as compared to the dominant monotype ponderosa pine forest on either side of the drainage.

Riparian ecosystems are valuable for the aesthetics, namely, varied landscape; photography and art; human awareness; diversity of plant and animal life, some of which may be unique to that habitat; cover, food, and water for associated wildlife and domestic animals; for recreation; for domestic water supply; and as natural outdoor laboratories.

One major negative impact on riparian communities is hazardous chemicals from effluents from agricultural, industrial, and mining-metallurgical processes. One such chemical is cyanide, long used in the processing of gold and silver, and discharged previously directly into some of our natural ecosystems. With present environmental awareness, industry and government are attempting to minimize or eliminate this contamination.

Homestake Mining Company began processing gold ore in the Lead-Deadwood area of South Dakota in 1878 and has remained in production at variable activity levels to the present time. The amount of cyanide (and heavy metals) entering the ecosystem during this time can only be crudely estimated, in part because of the variability in activity and the lack of understanding and concern about environmental pollution until the 1960's. Whitewood Creek, a tributary of the Belle Fourche River which flows into the Cheyenne River and ultimately into the Missouri River, received most of the cyanide (and heavy metals), and by the 1960's, it had become a "dead" stream biologically.

In 1977, Homestake Mining Company completed a multi-million dollar project (Grizzly Gulch), combined with improved processes, to eliminate or reduce cyanide and heavy metal pollution. The biological recovery of Whitewood Creek appears to be progressing well; however, problems remain concerning long-term effects from heavy metals and possibly cyanide along with the need to continue to control or eliminate the pollution problem.

This study developed as a result of a discussion of the need for additional research on cyanide in meetings between Homestake Mining Company officials and personnel of the Rocky Mountain Forest and Range Experiment Station. Several key points and questions were discussed at those meetings, such as:

a. What proportion of cyanide in the environment is produced naturally by plants?

b. Does the cyanide content of plants increase with increased cyanide in the environment?

c. Does excessive cyanide reduce growth or cause mortality of some plants?

d. Could plants be used to "alter or fix" cyanide in a non-toxic form, thus mitigating hazardous effects of cyanide?

OBJECTIVES

From these discussions a study to measure cyanide in the water-soil-plant matrix of a component of riparian vegetation was devised. It was hoped that the study would answer partially the following questions:

a. Is there a correlation between cyanide in the water, in the soil, and in the vegetation?

b. Do plants take up and concentrate cyanide and, if so, what types of plants?

c. What are the differences between cyanide in the natural environment and areas impacted by mining-metallurgical activities?

Referral to the literature has revealed that little is known about cyanide uptake in plants and even less is known about the mechanisms of biochemical utilization of cyanide.

Old timers knew about cyanide in plants, however; some called it prussic acid (1), and it remains a feared hazard in forage for livestock. The distribution of cyanophoric plants is widespread, at least 1,000 species have been identified (2). Cyanide in animal forage crops such as the sorghums, Johnson grass, cane, Sudan grass, flax, and lima beans has been responsible for livestock poisoning, and the root of the tapioca plant must be treated to leach out cyanogens before consumption by humans (2). Sheep ranchers have learned that chokecherry leaves and twigs can be lethal (3).

Some plants, such as the sorghums, become toxic under certain conditions. Sorghum in the mature stage does not contain appreciable amounts of cyanogens, but young plants or young shoots from roots of mature sorghum may contain considerable cyanide. Arrowgrass growing on water-covered or very wet soil is far less poisonous than arrowgrass that continues to grow on soil that has dried out (4).

The toxicity to animals depends upon a number of factors, but a lethal dose for sheep and cattle is 2.4 mg/kg body weight. In water the safety factor has been set at ten to twenty parts per billion by the Environmental Protection Agency (5).

The production of cyanide by higher plants is due to the presence of a certain glucoside that releases hydrocyanic acid when acted upon by enzymes found within the plant cell (3). Thus, hydrogen cyanide does not occur in the free state but is produced only after the plant cell has been disrupted, and a specific enzyme makes contact with the cyanogenic compound. As to the environmental conditions for cyanide production, there is some information, albeit indirect. For example, the use of certain fertilizers increases the cyanide-producing potential of sorghum. However, when $H^{14}CN$ was administered to sorghum plants, cyanide was shown to be extensively metabolized by the plants but not into cyanogenic compounds (6).

STUDY AREA AND METHODS

The general study area was in the northern Black Hills and associated plains of South Dakota with the primary area near the Homestake Gold Mine at Lead, SD. Sample zones were selected at: (1) the point of primary pollution on Whitewood Creek and at increasing distances from that point, (2) control--an area with no history of mining-milling pollution, and (3) survey samples of soil and water taken at long distances from pollution source.

The amount of funding limited the number of sample zones to four and the number of collection dates to ten (May to October) and the number of long-distance survey areas to five but sampled once only. Areas sampled were at the following sites (all distances are direct or "as the crow flies.")

Point I. Primary pollution point--on Whitewood Creek at entry of Goldrun Creek, approximately 402 meters (1,320 ft) from the ore-processing unit of Homestake Gold Mine at an elevation of 1,451 meters (4,760 ft).

Point II. Pollution point two--on Whitewood Creek about 305 meters (1,000 ft) above Point I and the entry of Goldrun Creek and immediately below the ore-processing unit at an elevation of 1,463 meters (4,800 ft).

Point III. Pollution point three--on Whitewood Creek about 14.5 kilometers (9 mi) southeast of Point I near the town of Whitewood about 402 meters (1,320 ft) upstream from Interstate 90 at an elevation of 1,052 meters (3,450 ft).

Point IV. Control point--on Yellow Creek, a tributary of Whitewood Creek about 3.2 kilometers (2 mi) southwest of Point I at an elevation of 1,509 meters (4,950 ft), about 1.6 kilometers (1 mi) southwest of the Pluma Grade School.

Long distance survey samples were collected from:

(1) Whitewood Creek, 42 kilometers (26 mi) northeast of Lead, near its entry into the Belle Fourche River.

(2) Belle Fourche River, 48.3 kilometers (30 mi) northeast of Lead and 12.9 kilometers (8 mi) downstream from entry of Whitewood Creek (near Newell, SD).

(3) Belle Fourche River, 55.6 kilometers (34.5 mi) northeast of Lead and 27.4 kilometers (17 mi) downstream from entry of Whitewood Creek (east of Sturgis, SD on Highway 34).

(4) Cheyenne River, 108.7 kilometers (67.5 mi) east of Lead and 27.4 kilometers (17 mi) downstream from entry of Belle Fourche River (near Wasta, SD).

(5) Cheyenne River, 154.6 kilometers (96 mi) east of Lead and 43.5 kilometers (27 mi) downstream from entry of Belle Fourche River (near Howe's Corner).

The riparian vegetation was sampled from three major types--rooted herbaceous aquatics, herbaceous shoreline forbs, and deciduous shoreline shrubs or trees.

In order to obtain an adequate number of samples of each type of vegetation and still keep the analyses within the budget, it was necessary to limit the number of species to four--an aquatic grass (arrowgrass) growing directly in the stream; two common forbs, tansy

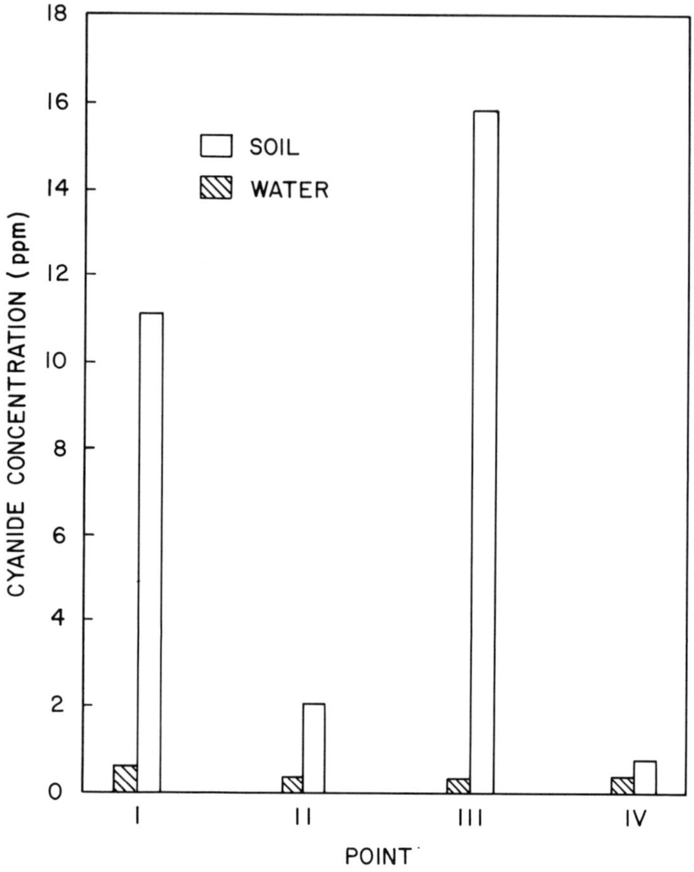

Figure 1. Relative concentration of cyanide in water and soil at the following points:

I. Pollution source
II. Pollution 300 m from source
III. Pollution 14.5 km from source
IV. Control site

and dandelion; and the tree, chokecherry. At first, willows were tested, but they did not contain cyanide, so tests were discontinued. Soil samples were collected from the A_0 and A_1 horizon (5 to 10 cm depth), and water samples were from the surface area of flowing water.

Specimens were collected in plastic bags and glass jars, sealed and refrigerated; tests were run within 24 hours of collection. To date, 7 of the 10 sampling runs have been completed.

LABORATORY TECHNIQUES FOR CYANIDE DETERMINATIONS

Total cyanide determinations were run at the EPA-approved Travis Laboratories in Rapid City, SD using Standard Methods procedure. Samples are boiled in an acid medium, treated with several reagents ($MgCl_2$, NaH_2PO_4, chloramine-T), and Pyridine-Barbituric Acid Reagent is added for color development, and transmittance is read. Colorimetric determinations are more sensitive than the titration method.

PRELIMINARY RESULTS

The study is incomplete, including the statistical analyses to indicate significant and nonsignificant results. However, we might look at a few trends which may prove to be interesting in the final analysis (Table 1).

Table 1. Average cyanide concentration, μg/g (ppm), in water, soil, and four plant species taken at four major sampling points.*

Sampling point	Water	Soil	Dandelion	Tansy	Aquatic grass	Chokecherry
I	0.42	11.30	10.25	7.6	4.53	21.40
II	0.03	2.04	0.50	2.0	(--)	81.08
III	0.17	16.20	11.30	3.8	1.20	137.60
IV	0.02	0.70	0.50	0.5	0.50	(--)

*Point I, primary pollution point; Point II, pollution point 300 meters from primary; Point III, pollution point 14.5 kilometers from primary; and Point IV, control.

a. Cyanide in water never exceeded one part per million, and at Points II and IV there was less than 0.03 μg/gram. Cyanide in the water does not appear to be higher when cyanide in the soil is higher (Figure 1). (This should not be interpreted as saying there is not exchange mechanisms from water to soil or from soil to water.)

b. Soil concentrations averaged 11.5 and 16.5 μg/gram at Points I and III, respectively. Point II averaged 2.04 μg/gram and was near zero at the control site.

c. As expected, chokecherry trees were high in cyanide and potentially dangerous whether they grew in cyanogenic soil or elsewhere. Our tests show that most of the cyanide is in the stem and leaves, with lesser amounts in the fruit seeds and only trace amounts in the fruit pulp.

d. Aquatic arrowgrasses showed small increases in cyanide in areas where it was present in the soil.

e. Tansies were variable in their cyanide concentrations but may have a tendency to concentrate cyanide where soil concentrations are higher.

f. Dandelions seem to be positively correlated with the presence of cyanide in the soil, more so than grass or tansy. They may turn out to be an index plant.

g. The long distance survey samples from Whitewood Creek and Belle Fourche River near Newell had a concentration of 15.4 and 3.1 ppm of cyanide in the soil and < 0.02 ppm in the water. The other samples from Belle Fourche and Cheyenne River were all < 0.05 ppm of cyanide.

DISCUSSION

Although the Homestake mining processes have put little cyanide in Whitewood Creek since 1977, residual cyanide is still found in the soils, probably in the form of trace-metal complexes. However, cyanide concentrations in water overlying these soils is minimal. It would seem that increases in cyanide in riparian vegetation, other than plants such as chokecherry which are naturally high in cyanide, pose little risk to animals. Additional studies will be necessary to determine effects of cyanide pollution on animal life. Wildlife and livestock drink the water, and some forms of aquatic invertebrates were observed at all sampling points. However, there were observable decreases in number and kinds near Points I and II.

The analytical test for cyanide used in this study was accurate to a hundredth of a part per million (0.01 ppm). Since this is greater than the 10 to 20 ppb limits set by the Environmental Protection Agency, any conclusions or discussions are difficult concerning the safety for animal use or possible limnological effects.

In summary, this study, presently in progress, will provide guidelines and methods for the development of future studies on cyanide in natural ecosystems as well as provide additional information about cyanide in riparian habitats and will establish plant species that may concentrate cyanide from increased cyanide in their environment when completed.

ACKNOWLEDGMENTS

Funds provided by the USDA Forest Service and the Homestake Mining Company are gratefully acknowledged.

BIBLIOGRAPHY

1. Lange, N. D., and Farker, G. M., Lange's Handbook of Chemistry, Tenth Edition, McGraw-Hill, New York, 2001 pp., 1967.

2. Conn, E. C., "Cyanogenic Glycosides," Journal Agricultural Food Chemistry, 17, 519-526, 1969.

3. Couch, J. F., "Poisoning of Livestock by Plants that Produce Hydrocyanic Acid," USDA Leaflet No. 88, 4 p., Washington, D.C., 1934.

4. Huffman, W. T., Moran, E. A., and Binns, W., "Poisonous Plants," p. 118-130. In Animal Diseases: The Yearbook of Agriculture 1956. The United States Government Printing Office, The United States Department of Agriculture, Washington, D.C.

5. Ecological Analysts, Inc., "Cyanide: An Overview and Analysis of the Literature on Chemistry, Fate, Toxicity, and Detection in Surface Waters," Inter-Industry Cyanide Group Report, 109 pp., Baltimore, Maryland, 1979.

6. Blumenthal-Goldschmidt, S., Butler, G. W., and Conn, E. E., "Incorporation of Hydrocyanic Acid Labelled with Carbon-14 into Asparagine in Seedlings," Nature, 197, 718-719, 1963.

Chapter 33

RECLAMATION AT THE DAVE JOHNSTON MINE

D. G. Deveraux

Northern Energy Resources Company
Portland, Oregon

A BRIEF SKETCH OF THE DAVE JOHNSTON MINE

The Dave Johnston coal mine is located near the southwestern extremity of the Powder River Basin in central Wyoming. This mine is operated entirely for the purpose of furnishing fuel to the 750,000 Kw Dave Johnston power generation station located five miles east of Glenrock, Wyoming on the North Platte River about 24 miles east of Casper, Wyoming. Coal is delivered to the power plant by a company-owned railroad moving the coal 15 miles downgrade from the mine to the power plant. The mine first belonging to Pacific Power & Light Company has now been transferred to Northern Energy Resources Company (NERCO), a mining subsidiary of Pacific Power & Light Company.

The coal beds which make up the field at the Dave Johnston Mine are locally referred to as the School and Badger seams and occur beneath gently rolling terrain in the Fort Union formation. The School seam is found approximately 180 feet beneath the Badger in the southern portion of the field and about 110 feet beneath the Badger towards the north. The coal beds dip down toward the northeast on about a four percent average inclination. The School seam has an average thickness of about 35 feet and the Badger seam about 16 feet.

The climate at the Dave Johnston Mine is semi-arid, with annual precipitation varying between 10 and 12 inches with 70 percent occurring during the months of April to September, with the maximum occurring in May. A few storms occur in July and August which are usually torrential in character, often causing erosion damage to unprotected soils.

Snowfall averages 50 inches a year at the mine with 12 to 15 inches of snow yielding one inch of moisture. However, snow seldom covers the ground for more than a few days after each storm due to the windiness of the area during this season. The frost-free period varies from 100-125 days, and temperature extremes vary from -40 to 105°F on the site.

Historical usage of the land surface at the Dave Johnston Mine prior to mining was those portions of the coal deposits at Dave Johnston not actively being mined are still committed to this land use. During mining, the active mine areas are fenced off, preventing domestic livestock foraging. Currently, a small herd of cows are grazing on reclaimed land in the north part of the Dave Johnston Mine. This is part of a grazing research study being conducted at the mine by our company research staff. It should be noted that wildlife, such as deer and antelope, have found the reclaimed areas to contain grasses and forbs to their liking.

The mine was opened in 1958, utilizing scrapers and dozers as the primary tools for overburden removal. This method of stripping continued until the spring of 1972. At this time, a medium-sized walking dragline (39 cu. yd.) was erected on site and placed into operation stripping overburden.

MINE RECLAMATION

Pacific Power's land reclamation program at the Dave Johnston Mine began in 1964 nearly ten years before the Wyoming Mine Reclamation Law was passed into law. As of March 31, 1978, approximately 1,700 acres of strip mined land have been reclaimed. Of this 1,700 acres, about 200 acres of the early efforts were reseeded a second time because initial results were not satisfactory to the company.

The reclamation methods now practiced at the Dave Johnston Mine have provided excellent reclamation results. That portion of the mine spoil bank that will not be disturbed with subsequent mining operations is leveled and contoured with scrapers and dozers to rolling contour to match the surrounding area. Reclaimed slopes of not more than 3H:1V are constructed with most slopes 5H:1V or less. After leveling and contouring, the spoil is deep-ripped 3 to 4 feet on the contour with a D-9 equipped with double rippers. This technique is used to loosen the compaction caused by the scrapers and dozers during the leveling and shaping operations. Topsoil is then spread on the ripped spoils approximately 8 inches deep with motor scrapers. After topsoiling, the topsoiled area is again ripped on the contour with a dryland chisel plow; this is done both to prepare a seedbed and to loosen compaction caused by the topsoiling operations. To provide protection against wind and water erosion, a grass hay or small grain straw mulch is blown on the topsoil at the rate of 2 to 3 tons per acre, depending on inclination of slope. The straw is disced lightly to anchor the straw. In late fall, October through November, a seed mixture consisting of grasses and forbs selected from species native to the Dave Johnston site and from improved introduced varieties are planted with a standard farm grain drill equipped with depth bands and spring packers. All of the work on the reclaimed spoil area, from the leveling to final seeding, is done with the equipment traveling the same direction as the land contour. This technique aids to control wind and water erosion, perpetuates runoff water to enter and be absorbed into the soil and subsoil, and to re-establish contour configuration to the land.

Topsoil

One of the most important factors in successful reclamation at Dave Johnston is the use of topsoil applied over the spoil. It appears that even an inch of the topsoil material is a significant aid in establishing new vegetation on reclaimed sites. At the mine topsoil is applied by motor scrapers to a depth of approximately 8 inches over the prepared spoil. Long-range mine planning is required to allow for the maximum recovery and utilization of all available topsoil and to handle it in the most economical and efficient fashion.

In the Dave Johnston Mine site, topsoil depth varies from a few inches (1"-2") on the ridge tops to several feet in valley areas. All useable soil horizons are recovered and utilized for topsoiling mine spoil. If the soil horizon contains any organic material and is not toxic, it is worth saving for topsoil use.

Topsoil often contains a good source of native seeds and rhizome material. When topsoil is spread on the spoil directly upon its removal in the spring of the year, a large number of the native seeds germinate, and much of the rhizome material will sprout and grow. This is particularly valuable in restoring the land to its former character and life support.

Some researchers say that storing topsoil appears to decrease the germinating qualities of the entrapped native seeds. However, there are numerous studies that indicate "stored" topsoil retains most of its seeds throughout storage if topsoil piles are compacted to exclude O_2 and, thus, prevent composting. In many cases, the topsoil storage enhances germination by breaking down the seeds natural dormance.

Borrowed Topsoil

Topsoil borrow areas were established on lands adjoining the active mine to provide topsoil for areas mined prior to the time when the topsoil was saved. A self-loading motor scraper was used to pick up topsoil and vegetation to a depth of 4 to 6 inches and then spread this material over the old reworked spoil. The scrapers took 10-11-foot swaths following the land contour, leaving a three foot or wider strip of native vegetation. This technique was done to provide a source for natural reseeding and for a buffer protection for the stripped area. The borrow strips were later prepared and seeded to perennial grasses. The regrowth of both native and planted species have been highly successful in the borrow areas.

Ripping

Soil compaction of the overburden spoil during the leveling and contouring operations is a problem. Equipment weighing anywhere from 40,000 to 160,000 lbs makes many passes over the spoil that is being leveled and contoured for reclamation, and packing of the spoil occurs resulting in the compaction problem.

The principal problem with compaction is that it reduces the infiltration of moisture. Studies conducted by the Central States Forest Experiment Stations show that the moisture infiltration rate on ungraded spoil banks was more than four times greater than on spoils that had been sloped and graded with equipment. Since moisture cannot penetrate the compacted surface readily, it accumulates on the surface and forms heavy runoff, resulting in serious erosion often reaching gully proportions.

At Dave Johnston the compaction problem is eliminated by ripping the spoil with a D-9 equipped with a double ripper after the shaping, leveling, and contouring has been completed and before topsoil is applied. Ripping, like all of the other operations, is done by working on the contour of the land to enhance the contour structure of the reclaimed spoil. Ripping furrows running on the contour also aid in trapping moisture that would otherwise run off.

Ripping has been done both before topsoiling and after. We prefer ripping prior to topsoiling for the following reasons:

1) After ripping, the spoil surface is rough, providing an excellent physical characteristic to "marry" the topsoil and the spoil. This technique reduces both wind and water erosion considerably.

2) Ripping after topsoiling pulled up considerable spoil through the topsoil, actually leaving spots void of topsoil.

Ripping is also used as a repair procedure on slopes that have had erosion problems and also to re-establish contour lines on slopes that were not worked on the contour in the original reclamation. The small amount of compaction occurring during the topsoiling process is taken care of in the preparation of the seed bed.

Mulch

Pacific Power and Light Company has applied grass hay or small grain straw as a mulch since 1964. Two to three tons of straw mulch are applied by blower to each acre and then lightly disced to anchor and prevent the straw from blowing. Attempts are made to apply greater amounts of mulch on steeper slopes. The primary purpose of the mulch is to prevent erosion, both wind and water, and is also an aid in moisture retention.

Mulch is not always necessary when the topsoil material contains large quantities of native vegetation; the uprooted vegetation will act as a mulch.

Cover Crops

We are currently evaluating a two step planting operation whereby an annual grain, such as wheat, barley, or millet is planted early in the spring and grown as a cover crop on the prepared area. Permanent grasses are directly seeded into the cover crop stubble in the fall. This procedure helps to reduce the repair needed for eroded areas and replacement of topsoil lost to wind and water erosion before permanent seeding can occur. Standing stubble left by the cover crop catches more snow through the winter. Thus, more moisture is available to germinating reclamation plants in the spring. The use of cover crops to replace straw mulch reduces mulching costs by at least one half.

Seeding

The seeding operation is one of the most critical steps of the entire reclamation procedure. Seed species, seed quality, timing with respect to weather, depth of seed placement, seeding rate per acre, direction of drilling on slopes, width of drill rows, equipment weight, all, are factors that will affect the success of the reclamation planting. Fairway crested wheatgrass and Critana Thickspike wheatgrass appear to be the best site adapted grasses for the Dave Johnston Mine spoil revegetation. I am sure that Fairway and Critana have proven to be superior grass selections because of their excellent germination and seeding vigor which are characteristics that cannot be overlooked for successful plant establishment on mine spoils.

At Dave Johnston we have used planting rates as high as 18 pounds per acre or higher. One might think that with our conditions, i.e. low rainfall and poor soil conditions, that a high seeding rate would increase the chance of successful plant establishment. However, it works just the opposite. The standard dryland seeding rate recommended for the Dave Johnston area is 10 to 12 pounds/acre. Increasing the seeding rate over the recommended dryland standard will result in too many plants competing for the moisture and nutrients that are available. As a result of over seeding, many or all of the plants could die due to a weakened condition caused by interplant competition.

Pure live seed count is another important factor that the reclamationist should consider. Pure live seed is the actual number of live seeds of the species wanted in a pound of seed. It is calculated by multiplying percent purity times percent of germination for each species being planted. For example, Western wheatgrass at 86 percent purity and 85 percent germination would be 73 percent pure live seed. Our last order of seed averaged 83.3 percent pure live seed. We paid $3.74 per pound, bulk rate, seeded. On a pure live seed basis, we actually paid $4.33 per pound. A contractor could furnish a seed mixture with a pure live seed percentage that was high or one that was low. If it is the latter, and you pay for the seed on a bulk rate, you could be grossly overpaying for seed, not to mention the poor established stand you would obtain. To aid the reclamation seed buyer, many of the seed companies are now quoting and selling their seed on the pure live seed basis.

In brief, seed selection and rate of seeding are the most important factors in obtaining good reclamation. Remember the following:

1) Reclamation seed mixtures should be selected upon the specific adaptability of each species to the ecosystem being seeded.

2) If there are limited resources of soil moisture and nutrients to establish the plant species, it should be borne in mind that,

   (a) a given site can raise only a given number of healthy plants successfully, depending upon the amount and availability of the soil moisture and nutrients, and

   (b) overseeding can result in too much competition among seeded plants. Too much competition between these plants can actually result in a very poor established grass or shrub stand or even in the death of all plants.

3) Planting too deep can kill the germinating grass seed. Planting depth is critical, especially to small seeded grass species. Usually, over 3/4" is too deep.

Time of Seeding

Successful seeding has been done in both the fall and spring at the Dave Johnston Mine. The preferable planting time for permanent seeding is late October through November. Seeding should be done late enough so that the seed will not germinate until the next spring, thus taking advantage of both the winter and early spring moisture received on the site. Dave Johnston's spring weather conditions are very unpredictable; final ground preparation and seeding must usually be accomplished in a narrow time span, requiring close coordination of equipment, manpower and materials. Considerable soil moisture is lost in spring plantings because the seedbed and planting cannot be done until the soil has dried sufficiently to allow equipment in the area. Another precaution should be mentioned, namely, if plants are seeded too early in the fall, seedings are particularly susceptible to premature sprouting caused by early fall moisture followed by warm weather. Should this occur and should extreme cold and freezing weather follow, a large portion if not all of the sprouted seedlings can be killed.

Land Use

Historically, the land mined at the Dave Johnston Mine was used for livestock range and wildlife before it was mined. Observations have indicated that the reclaimed land has as much or more carrying capacity than the previous native vegetation. This has been confirmed by plant and animal scientists at the Bureau of Land Management and Soil Conservation Service of the University of Wyoming. The reclaimed land has also returned to a suitable habitat for antelope, deer and other wildlife species.

Chapter 34

DUST GENERATION CONTROL AT SURFACE COAL MINES

Charles A. Kliche

Department of Mining Engineering
South Dakota School of Mines and Technology
Rapid City, South Dakota

INTRODUCTION

The Northern Great Plains Coal Province which occupies approximately 36.6 ha contains about one-half of the nation's total coal resources. About one million ha are underlain by coal that is considered to be mineable from the surface.

The topography of the province is predominately plains, rolling hills, and some badland areas characterized by breaks or sharply eroded hills. Elevations range from about 1500 m above sea level on the west to about 600 m above sea level on the eastern boundary. The average slope of the ground is approximately 2 m per km. Streams in the province drain generally easterly and southeasterly (1).

In the United States, the Northern Great Plains Province lies within the states of North Dakota, South Dakota, Montana, Wyoming, Nebraska, and Colorado. The coal within the province is located almost entirely in the states of North Dakota, Montana, and Wyoming.

Although it is relatively low rank, the coal within the province encompasses a large portion of the total remaining U. S. reserves.

The U. S. Geological Survey estimates the total remaining coal resources of the United States to be 2.6 trillion mt within 1000 m of the surface, and 2.9 trillion tons within 1830 m of the surface. Of this total, only about 177 billion mt are classified as reserves meaning that they are (1) known in location, quantity and quality from geologic evidence supported by engineering measurements, and (2) economically recoverable based on currently available technologies and current market value (2).

As shown in Figure 1, the Northern Great Plains Coal Province is made up of six regions. Almost 50 percent of the remaining U. S. reserves is located in the province (Table 1), and most of this tonnage is contained in the Fort Union and Powder River regions.

Table 1. Coal Resources in U. S. Geological Survey Provinces (1)[a] (billions of metric tons)

	Identified	Undiscovered	Total
Eastern	250	41	291
Interior	251	235	486
Northern Great Plains	631	692	1,323
Rocky Mountains	170	358	528
Other	132	164	297

[a]Because available estimates are by state and USGS Provinces cross state boundaries, the figures for these provinces are only approximate.

Most of the coal within the Northern Great Plains Province is relatively low rank lignite or subbituminous, although near the edge of the Rocky Mountains the coal is somewhat higher in rank. The coal tends to be relatively high in moisture and volatile matter content and low in energy value. However, about 44 percent of the coal in the province is low sulfur (3) (coal with a sulfur content of one percent or less).

The Fort Union and Wasatch Formations are the two most important coal-bearing formations in the Northern Great Plains Province. Lignite from the Fort Union Formation occurs mostly in North Dakota and Montana. It is a low-sulfur, high moisture, non-agglomerating coal with a heating value averaging about 15.8×10^6 joule per kg. Ash content averages about 6.2%, and sulfur content averages about 0.6%. Lignite coal beds of commercial thickness range from 2 to 9 m thick (4).

Most of the subbituminous coal in the province occurs in the Powder River Basin of Wyoming and Montana. The most persistent and thickest coals in the Fort Union Formation occur in the Tongue River Member. The Anderson and Canyon seams merge west of Gillette, Wyoming to form the Wyodak-Anderson coal beds. These coal beds average about 21 m in thickness with an average heating value of 19.2×10^6 joules per kg. Sulfur content averages 0.5%, moisture content averages 29.8%, and ash content averages 6.0% (4).

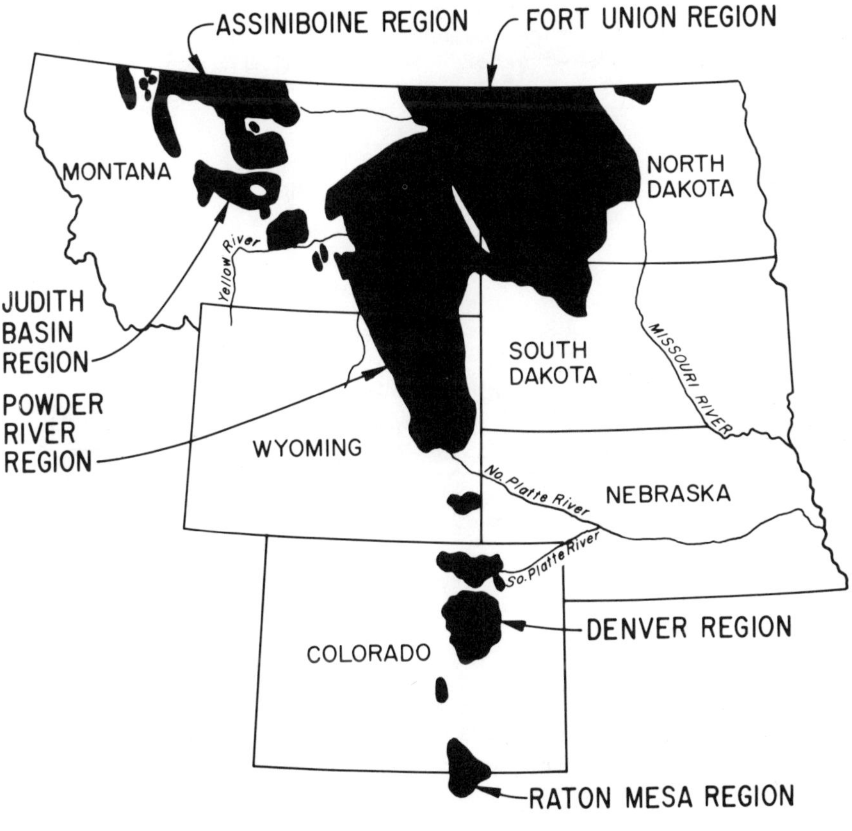

Figure 1. Northern Great Plains Coal Province.

DUST GENERATION

Particle Transportation By Wind

Transportation of particles by wind consists of three types, namely, saltation, surface creep and suspension. See Figure 2. They are not independent of each other and nearly always occur simultaneously. Of the three means of soil movement, saltation accounts for up to three-quarters of the total movement.

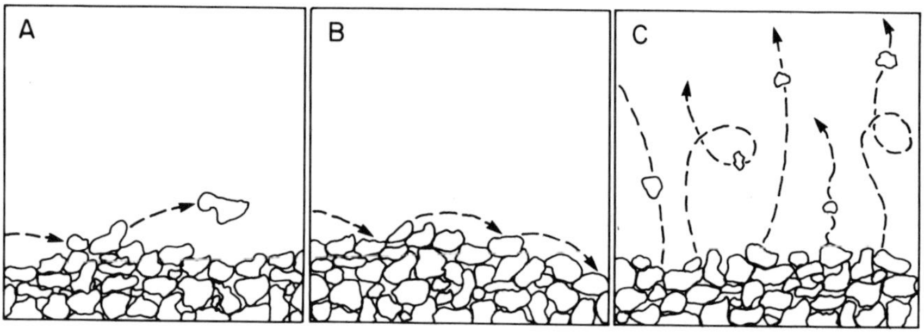

Figure 2. Modes of soil movement: (a) saltation, (b) surface creep, and (c) suspension (5).

Saltation is the bouncing, jumping motion of sand grains driven by wind close to the ground surface. When the wind blows over an exposed surface of soil, the velocity at the very surface is zero because of friction. Particles of between 0.1 mm and 0.5 mm in diameter extend up into the higher velocity wind layers. This creates a spin where the top of the particle moves at a greater speed than the wind, resulting in spinning of the particle and a horizontal force. The spinning of the particle causes a partial vacuum above the particle and a zone of compressed air below it. The particle then rises at an angle of 75 to 90° for 15 to 30 cm, gains forward momentum, and descends by gravity. The height reached and the distance traveled depends on the wind velocity (6).

Surface creep is the slow movement of the soil surface, caused primarily by the impact of saltating particles on grains that are too heavy to be dislodged and bounced into the air. Suspension is caused mainly when saltating particles dislodge small grains of less than 0.1 mm in diameter. Particles of this size are light enough to be carried by eddies in all directions. These small particles can be carried long distances by the air currents.

Dust clouds are the result of small particles moving in suspension. Dust is raised mainly by the impacts of grains moving in suspension, but it can also be raised by vehicles, animals, blasting and other forces.

Once raised up into the air, dust is raised high in the atmosphere by the upward velocity of turbulent wind. The upward velocity of erosive wind is about 4 km per hour which is sufficient to lift clay, silt, and very fine sand particles. The degree of turbulence varies directly with the roughness of the ground and inversely with height. The wind near the ground is characterized by eddies of extremely variable velocity moving in all directions. This eddying, or turbulence, is what makes the soil erode (3).

Saltation provides most of the driving force in wind erosion after soil movement has started. This is because the velocity of the wind some distance above the ground is much greater than the velocity at the ground. The amount of soil erosion by wind therefore depends primarily on the velocity of wind within one foot of the ground.

Dust Emission Equations

Western residual soils are typically sandy or clayey, relatively infertile, often highly erodible, and sometimes high in salts. The sandy soils have a low moisture-holding capability, whereas the clay soils tend to be very hard when dry (7). Once the protective covering of native flora is removed from the soil, or the protective covering of overburden is removed from the coal, the soil and coal become susceptible to wind erosion. This can be shown using the wind erosion equation (1):

$$E = I\ K\ C\ L\ V$$

where E is the gross erosion (tons per acre per year); I is the soil erodibility index; K is the field roughness; C is the climate factor; L is the field length; and V is the vegetative cover.

Soil erodibility index, I, is a function of the ability of the soil to form clods. It depends largely on the amount of coarse materials and clays present. Sandy soils are especially subject to wind erosion because they often contain too little silt and clay to form clods resistant to wind action. Soils high in clay are subject to deterioration due to repeated freezing and thawing, and wetting and drying.

The roughness factor, K, takes into account the fact that rough surfaces absorb and deflect wind energy and also trap some material.

The climate factor, C, is related to wind speed and soil moisture content decreases. This factor is very important in areas of high average wind speed and low annual precipitation, such as the Northern Great Plains Province.

The field length index, L, and the vegetative factor, V, are related. As the length of an area void in vegetation or low in vegetation increases, wind erosion also increases. Wind erosion decreases as the protective effects of good ground cover increases.

Equations have also been developed for determining dust emissions from unpaved roads and aggregate storage piles (1).

$$E(\text{unpaved roads}) = 0.81\ S(V)$$

where E is the emission factor (pounds per vehicle-mile); S is the silt content of road surface material in percent; and V is the average vehicle speed in miles per hour.

$$E(\text{storage piles}) = 0.33/(PE/100)2$$

where E is the emission factor (pounds per ton placed in storage) and PE is Thornthwaite's precipitation-evaporation index.

Source and Nature of Airborne Dust

Results of a study prepared for the EPA of a 5,700 mt per day coal strip mine model is shown in Table 2. Assumptions used in the model such as haul distances from the mine to the production site, average overburden thickness, and average coal seam thickness were based on national averages.

In the study, four possible sources of air emissions associated with the strip mining model were considered. They are:

1. Emissions from diesel-powered vehicles.
2. Particulate emissions from the preparation plant and loading facility.
3. Fugitive dust emissions from the operations within the mine.
4. Emissions from thermal drying (if used).

Table 2. Strip Coal Mining Module Estimated Air Emissions (8)
(Basis: 5,700 mt/day R.O.M. Coal)

Source	Particulates (kg/day)
Vehicle emissions	8.3
Overburden removal*	526.2
Primary crushing*	57.2
Loading and unloading at the preparation plant	17.7
Loading in the pit*	229.5
Vehicular travel*	18.1
Thermal drying**	229.1
Vehicle emissions from refuse hauling operations	0.6
Rounded Total	1087.0

* 80% particulate control by water spraying and dust control techniques

** 90% particulate control by wet scrubbers

Contact with coal mining companies in the Northern Great Plains Coal Province indicates several dust sources. The two main sources of dust are the stripping operation, including areas stripped of topsoil and other barren areas, and haul roads.

Areas stripped of vegetative cover and backfilled areas are subject to wind erosion. When heavy equipment passes over surface material void in vegetation, they crush the soil clods. Therefore, the removal of vegetative cover, the crushing of the clods, and the action of the wind, which tends to dry out the soil structure, all contribute to an increase in erodibility.

Dust is raised on haul roads by the passage of vehicles. The motion of wheels or treads on a vehicle lifts the fine particles into the air where they are kept aloft by eddies or air currents generated by the movement of vehicles. When there is no wind, the particles will eventually settle back to the ground, but with even a slight breeze, fine particles can be carried long distances.

Truck loading and dumping operations are also major sources of fugitive dust. These operations generate both coal dust and dust from spoil materials. Other important sources of dust include blasts, train loading, stockpiles, and dust generated at the working face.

DUST MITIGATION

Mitigation Methods

Little can be done to reduce turbulent flow of air across a field, but surface velocity can be reduced by various means. The best way to reduce the surface velocity is by establishing plant growth. Other methods include roughening the ground surface, covering the exposed areas with mulches, wetting the area, or using a chemical stabilizer.

Control methods for mineland stabilization are generally grouped into the following categories: physical, chemical, vegètative, and combinations of these methods. A U.S. Bureau of Mines report defines the stabilization procedures as follows (9).

1. Physical -- the covering of the tailings with soil or other restraining materials.
2. Chemical -- the use of a material to interact with fine-sized minerals to form a crust.
3. Vegetative -- the growth of plants in the tailings.

Although this system of classification was originally developed by the Bureau of Mines to describe methods of stabilizing tailings basins, it can be expanded to include all mineland stabilization.

Physical Stabilization. Water is the most used substance for the mitigation of wind-blown dust. Other materials used include straw and hay mulches, rock, wood residues, and fabric or mats. Water is used mainly because of its availability and its low application cost. The use of water has two main disadvantages. First, it has a limited duration after application. Secondly, when water evaporates, the soil becomes more fragile and thus dustier (5).

Mulches can be used for temporary dust control on stockpiles, dumps, backfilled areas, and other areas awaiting permanent reclamation. Straw or hay usually needs to be held in place to prevent the wind from blowing it away. Common methods of holding straw or hay in place are crimping, discing or rolling it into the soil; covering it with a net or wire; or spraying it with a chemical takifier (a natural, organic chemical binder).

Fabric or mats, including jute, excelsior, and woven paper or plastic fibers, are provided in rolls to be fastened to the soil with wire or staples. They require large amounts of labor for installation, cost at least four times as much as tacked straw, and are not adapted to fitting to rough surfaces or rocky areas (1).

Chemical Stabilization. Chemical binders form a matted surface which prevents saltation and inhibits water erosion. Limiting factors (1) in the use of chemical binders are: (1) the cost of the binder, (2) the breakdown of the treated surface due to weathering, (3) inability to withstand traveling and handling in some instances, and (4) chemical contamination of the material being protected in some instances. Chemical stabilizers, in general cannot, withstand the heavy loads and traffic of a haul road but can be very effective on storage piles, material being transported, tailings and refuse piles, bare soils, and inactive haul roads (1). The chemical binder selected for temporary wind erosion control will depend on the rate of soil movement that is damaging and the degree of control needed.

Chemical binders for temporary wind erosion control have been tested by the U. S. Bureau of Mines (9), the Bureau of Reclamation (10), and the U. S. Department of Agriculture (11). Results of the tests are listed in Appendix A.

Vegetative Stabilization. Vegetation is one of the best methods of controlling wind erosion. The development of good vegetative cover in adequate moisture areas where fugitive dust is most prevalent is a major problem. Analyses of spoil materials from surface coal mines in the Northern Great Plains Province have consistently shown nutrient deficiencies, particularly in nitrogen and phosphorous (1). Fertilization of nutrient deficient spoils can often produce more forage than the amount yielded by native range.

A rough ground surface is more resistant to wind erosion than a smooth one. Vegetation and vegetative residues act both as a rough surface and as a protective cover to limit soil movement. The amount of vegetative material required to protect the soil from wind depends primarily on surface roughness, degree of cloddiness, size of field, and strength of the wind. Often 560 kg per ha of wheat stubble is sufficient on slightly erodable soil, 2242 kg on moderately erodable soil, and 8967 kg per ha on highly erodable dune material (3).

A problem to overcome when establishing vegetation is that of sandblasting the plants by windblown dust and sand. One method of preventing sandblasting of the growing plants is the use of a chemical binder with the seed mix that does not inhibit plant growth. Hydromulching is another commonly used method. Hydromulching is the application of a mulch in a water slurry. The slurry may also contain fertilizer, erosion-control compounds, growth regulators, soil amendments, and herbacides. The most important quality of a hydromulching material is that it must adhere to the soil even on steep slopes and hold the seed in place during heavy rainfall impacts and wind.

Procedures for Dust Control

Mineland reclamation is the most common procedure employed to suppress dust. All companies in the Northern Great Plains Province are required to reclaim disturbed lands.

All of the coal mining companies who responded to a questionnaire concerning dust problems at their mine sites reported the use of water to suppress dust. Water is most commonly used on haul roads. It is also used at the working face and, in some operations, at the train loading facility.

Chemical binders are being used or considered for use at most of the operations. The chemicals are generally used to reduce dust on haul roads. One coal mining company reported that it had tried several types of binding agents, but the agents had proved ineffective after a short time due to road maintenance and slickness when wetted.

APPENDIX A

Bureau of Mines (9)

Seventy chemicals have been tested in the laboratory by the Bureau of Mines. The most effective chemicals of those tested are listed below in order of effectiveness based on the cost in dollars required to stabilize 1 ha of tailings. The costs are in 1973 dollars.

1. Coherex, a resinous adhesive, furnished good wind-resistant tailings surface when applied in quantities equivalent to 2,245 l per ha, costing approximately $161 per ha, but good resistance to water jet testing was not achieved until reagent quantities costing $1,600 per ha were applied.

2. Lignosulfonates -- calcium, sodium, and ammonium lignosulfonates -- were effective stabilizers in 2,690 kg per ha quantities costing about $321 to $420 per ha dependent on the specific lignosulfonate used.

3. Compound SP-400, Soil Gard, and DCA-70, elastomeric polymers, produced wind- and water- resistant surfaces using 514, 842 and 468 l per ha, respectively, costing about $321 per ha. All were exceptionally effective on sandy tailings and produced satisfactory results on both acidic and alkaline tailings.

4. Cement and milk-of-lime additives were effective in stabilizing the surface when applied at costs of about $470 per ha.

5. Paracol TC 1842, a resin emulsion, was an effective stabilizer at costs of about $618 per ha.

6. Pamak WTP, a wax, tar, and pitch product, was an effective stabilizing agent at $618 per ha.

7. Petroset SB-1, an elastomeric polymer, was an effective stabilizer at $618 per ha.

8. Potassium silicate, having a SiO_2 to K_2O ratio of 2.5 was an effective stabilizer where applied at costs of $1,112 to $2,347 per ha.

9. PB-4601, a polymeric stabilizing agent, was an effective stabilizing agent at costs of $1,235 per ha.

10. A cationic neoprene emulsion and Rezosol, an elastomeric polymer, were effective stabilizers at a cost of about $1,235 per ha.

11. Dresinol-TC 1843, an ammonium casein of tall oil pitch, was an effective stabilizer at $1235 per ha.

12. Sodium silicates, having ratios of 2.4 to 2.9 of SiO_2 to Na_2O were effective stabilizers when applied in quantities costing about $494 per ha. Calcium chloride was an effective additive to the sodium silicate and permitted a reduction in the quantities of sodium silicate needed for effective stabilization.

Department of Agriculture (11)

Thirty-four commercially available materials for erosion control were tested in 1971 by the Agriculture Research Service, U. S. Department of Agriculture. The materials were evaluated in terms of four criteria:

1. Costs less than $124 per ha.
2. No adverse effect on plant growth or emergence.
3. Prevents erosion initially and provides reduced erosion for at least two months.
4. Easy to apply without special equipment.

Table 3. Materials that Meet Criteria for Temporary Wind Erosion Control

Product	Chemical Composition	No-Erosion Rate	Cost ($)
COHEREX	Liquid, petroleum resin-in water emulsion	1,590	85.50
DCA-70	Liquid polymer of some type	159	90.40
Petroset SB	Liquid rubber emulsion	206	122.30
Polyco 2460	Liquid, stryene-butadiene copolymer	159	66.50
Polyco 2605	Liquid, vinylchloride-vinylidene chloride co-polymer	159	100.80
SBR Latex S02105	Liquid, styrene-butadiene latex	159	67.70

Bureau of Reclamation (10)

Thirty liquid soil stabilizing products were tested by the U. S. Bureau of Reclamation to determine the effectiveness of the products for erosion control and soil stabilization. The following tests were conducted:

1. Compressive strength.
2. Water erosion.
3. Penetration of the stabilizer.
4. Wind erosion.
5. Outdoor exposure (weathering).

Table 4. Materials Recommended by the Bureau of Reclamation for Various Soil Stabilization Application

Item	Application	Materials[a]
1	Dust control	Asphalt cutback B-5494, and petroleum resin emulsion B-6638
2	Temporary erosion control	PVA emulsions B-6471, B-6524, B-6602, and B-6654; vinyl acrylic copolymer emulsions B-6738, and B-6745; and copolymer emulsion of methacrylates and acrylates B-5778
3	Takifiers for organic mulches	Asphalt emulsion, type SS-1; and plastic emulsions listed under Item 1
4	Stabilization of secondary roads	Asphalt cutback B-5494, and asphalt emulsion B-6656
5	Waterproofing of problem soils	PVC plastic sheeting, asphalt membrane materials
6	Stabilization of clay soils	Hydrated lime and cement

[a]All materials in the report are referred to by their laboratory sample numbers.

REFERENCES

1. Schaller, F. W., and Sulton, Paul, eds., Reclamation of Drastically Disturbed Lands, Madison, WI, The American Society of Agronomy, Inc., 1978.

2. White, Irvin L., et. al., "Energy From the West: Energy Resource Development Systems Report," Vols. I & II, EPA Interagency Energy/Environment R & D Program Report No. EPA-600/7-79-06 a & b, March 1979.

3. Chepil, W. S., "Erosion of Soil by Wind," Soil--The Yearbook of Agriculture 1957, The United States Department of Agriculture, Washington, D.C.

4. 1978 Keystone Coal Industry Manual, Mining Information Services of the McGraw-Hill Publications, New York, 1978.

5. Canessa, William, "Chemical Retardants Control Fugitive Dust Problems," Pollution Engineering, Vol. 9, No. 7, pp. 24-26, 1977.

6. Nelson, C. R., Yardley, D. H., and Lacabanne, W. D., "Mitigation of Dust From Taconite Tailings," Proc. 50th Annual Mtg. Minn. Sect. AIME and 39th Annual Mining Symp., Duluth, MN, January 1978.

7. Cook, Frank, "Evaluation of the Environmental Effects of Western Coal Mining," Vol. 1, EPA Interagency Energy/Environmental R & D Program Report No. EPA-600/7-79-110, May 1979.

8. Jones, D. C., et. al., "Monitoring Environmental Impacts of the Coal and Oil Shale Industries: Research and Development Needs," EPA Interagency Energy/Environment R & D Program Report No. EPA-600/7-77-015, Feb. 1977.

9. Dean, K. C., Havens, Richard, and Glantz, M. W., "Methods and Costs for Stabilizing Fine-Sized Mineral Wastes," BuMines RI 7896, 1974.

10. Morrison, W. R. and Simmons, L. R., "Chemical and Vegetative Stabilization of Soils," Bureau of Reclamation Report No. REC-ERC-73-13, Jan. 1977.

11. Armburst, D. V., and Dickerson, J. V., "Temporary Wind Erosion Control: Cost and Effectiveness of 34 Commercial Materials," Journal of Soil and Water Conservation, Vol. 25, No. 4, pp. 154-157, 1971.